T0136973

Smart Innovation, Systems and Technologies

Volume 90

Series editors

The Smart Innovation, Systems and Technologies book series encompasses the topics of knowledge, intelligence, innovation and sustainability. The aim of the series is to make available a platform for the publication of books on all aspects of single and multi-disciplinary research on these themes in order to make the latest results available in a readily-accessible form. Volumes on interdisciplinary research combining two or more of these areas is particularly sought.

The series covers systems and paradigms that employ knowledge and intelligence in a broad sense. Its scope is systems having embedded knowledge and intelligence, which may be applied to the solution of world problems in industry, the environment and the community. It also focusses on the knowledge-transfer methodologies and innovation strategies employed to make this happen effectively. The combination of intelligent systems tools and a broad range of applications introduces a need for a synergy of disciplines from science, technology, business and the humanities. The series will include conference proceedings, edited collections, monographs, handbooks, reference books, and other relevant types of book in areas of science and technology where smart systems and technologies can offer innovative solutions.

High quality content is an essential feature for all book proposals accepted for the series. It is expected that editors of all accepted volumes will ensure that contributions are subjected to an appropriate level of reviewing process and adhere to KES quality principles.

More information about this series at http://www.springer.com/series/8767

Alena V. Favorskaya · Igor B. Petrov
Editors

Innovations in Wave Processes Modelling and Decision Making

Grid-Characteristic Method and Applications

Editors
Alena V. Favorskaya
Department of Computer Science and Numerical Mathematics
Moscow Institute of Physics and Technology
Dolgoprudny, Moscow Region
Russia

Igor B. Petrov
Department of Computer Science and Numerical Mathematics
Moscow Institute of Physics and Technology
Dolgoprudny, Moscow Region
Russia

ISSN 2190-3018 ISSN 2190-3026 (electronic)
Smart Innovation, Systems and Technologies
ISBN 978-3-319-89289-4 ISBN 978-3-319-76201-2 (eBook)
https://doi.org/10.1007/978-3-319-76201-2

Printed on acid-free paper

Check for updates

This Springer imprint is published by the registered company Springer International Publishing AG part of Springer Nature
The registered company address is: Gewerbestrasse 11, 6330 Cham, Switzerland

Preface

Recently, the applications of wave processes modelling have created tremendous interest among researchers and practitioners in many areas of science and technology for intelligent decision-making. Examples of computer simulation using intelligent grid-characteristic method and decision-making in geophysics, seismic prospecting, global seismic, medicine, aircraft and railway industry are included in this book. Grid-characteristic method is a numerical method for solving hyperbolic systems of equations (for example, the elastic and acoustic wave equations). This method allows to calculate the wave processes in heterogeneous media accurately and physically correctly. Grid-characteristic method permits to use the correct boundary and interface conditions in integral regions. Problems of seismic prospecting, earthquake stability, global seismic on Earth and Mars, medicine, railway ultrasonic non-destructive testing, aircraft composites modelling, and other applications had been reported using the developed grid-characteristic method in the last 10 years.

We are grateful to the researchers for inventing numerical methods, parallel algorithms, physical correct problems statements and performing computer simulation for research and development reported in this book. Our thanks are due to Springer-Verlag for the opportunity to publish our book.

The book is directed to the students, researchers, practitioners and professors interested in numerical mathematics, computer science, computer simulation, high-performance computer systems, unstructured meshes, interpolation, seismic prospecting, geophysics, medicine, non-destructive testing and composite materials.

Moscow, Russia
January 2018

Alena V. Favorskaya
Igor B. Petrov

Contents

Chapter 1
Theory and Practice of Wave Processes Modelling

Alena V. Favorskaya and Igor B. Petrov

Abstract This chapter presents a brief description of chapters devoted to the innovations in wave processes modelling and decision making (grid-characteristic method and applications). Grid-characteristic method is a direct finite-difference numerical method for obtaining full-wave solution of hyperbolic systems of equations. One can use different types of grids such as the regular, triangular, tetrahedral, and nested grids. This method is often used for modelling of the acoustic and isotropic/anisotropic elastic waves in heterogeneous media. Also, the original analytical algorithms for interpolation on the unstructured triangular and tetrahedral meshes are developed. The study of wave processes might be used in different applied areas, e.g. geophysics, non-destructive testing of different objects and materials, ultrasonic testing, seismic stability investigation, and ultrasonic operations modelling. Original investigation of composite materials delamination and non-destructive testing is done. The geological faults zones study is performed. Also, the migration techniques of acoustic and elastic fields are developed.

Keywords Interpolation · Computer simulation · Full-wave modelling
Wave processes · Acoustic waves · Elastic waves · Anisotropy
Finite-difference method · Numerical mathematics · Seismic prospecting
Migration · Composite material

A. V. Favorskaya (✉)
Non-State Educational Institution "Educational Scientific and Experimental Center of Moscow Institute of Physics and Technology", 9, Institutsky Pereulok St., Dolgoprudny, Moscow Region 141700, Russian Federation
e-mail: aleanera@yandex.ru

A. V. Favorskaya · I. B. Petrov
Department of Computer Science and Numerical Mathematics, Moscow Institute of Physics and Technology, 9, Institutsky Pereulok St., Dolgoprudny, Moscow Region 141700, Russian Federation
e-mail: petrov@mipt.ru

A. V. Favorskaya · I. B. Petrov
Scientific Research Institute for System Studies of the Russian Academy of Sciences, 36(1), Nahimovskij Ave., Moscow 117218, Russian Federation

A. V. Favorskaya and I. B. Petrov (eds.), *Innovations in Wave Processes Modelling and Decision Making*, Smart Innovation, Systems and Technologies 90,
https://doi.org/10.1007/978-3-319-76201-2_1

1.1 Introduction

As high-performance computer systems are developed, the physical laboratory experiments are replaced by numerical modelling. A feature of numerical simulation is not only its cheapness but also the possibility of a more detailed study of physical processes and phenomena. However, the results of computer modelling always contain some inconsistencies. Therefore, in order to make it possible to conduct the numerical experiments, it is necessary to develop the high-precision numerical methods that will allow to model the physical processes under consideration. Investigation of wave processes in complex heterogeneous media is applicable for solving a large number of applied problems in such areas as medicine, construction industry, oil and gas industry, aircraft industry, among others.

1.2 Chapters Included in the Book

The main purpose of this book is to describe the crucial issues of the proposed innovative approach for applied study of wave processes.

Chapters 2–4 are closely related to each other. They disclose the process of interpolation on unstructured triangular (Chap. 2) and tetrahedral (Chaps. 3 and 4) grids. Chapters can be valuable to all researchers who use these types of meshes. Since these Chapters include analytical algorithms, the software implementation of these analytical formulae and algorithms allows to reduce the amount of computing operations spent on the application of these types of grids significantly. The process of interpolation on unstructured grids has a significant number of applications. It can be used independently in the process of rebuilding grids. For example, the process of restructuring unstructured mesh into an unstructured one and the process of restructuring the unstructured grid into a structured one are solved by applying the interpolation discussed in Chaps. 2–4. The interpolation can also be used as an algorithmic stage in the numerical method using transformable or adaptive unstructured grids. In addition, the interpolation is closely related to the family of grid-characteristic methods discussed in the Book. Because the adaptation of the grid-characteristic method to unstructured grids is reduced to the interpolation problem [1–3]. Chapters 2 and 3 give ready to use analytical expressions for polynomial interpolation from 1 to 5 orders inclusive. Also, the algorithms for piecewise linear interpolation into arbitrary shaped triangular and tetrahedron are considered in Chaps. 2 and 4, respectively. In addition, Chaps. 2 and 3 describe methods of hybrid interpolation and interpolation with different limiters. Also nested unstructured triangular and tetrahedral meshes are discussed in Chaps. 2 and 3, respectively. There are a lot of ready to use analytical formulae in Chaps. 2–4. These analytical formulae can significantly reduce the amount of high-performance computer system power and time needed for simulations.

Chapter 5 outlines the crucial points of a family of grid-characteristic methods [4–13]. This family of direct finite-difference method is well suited for full-wave simulation of acoustic and isotropic/anisotropic elastic wave propagation in complex heterogeneous environments with apparent selection of boundaries and interfaces [14]. It is possible to use different types of grids: structured, hexahedral, unstructured, nested, and mixed. It is also possible to combine the grid-characteristic method and, for example, a Smoothed Particles Hydrodynamics (SPH) method during the same calculation [15]. These features of a family of grid-characteristic methods allow to solve a large class of applied problems. Systems of equations describing wave processed in anisotropic elastic, isotropic elastic, and acoustic media are considered. Also, the mathematical statements and implementation of various boundary and interface conditions for 2D and 3D cases are presented in Chap. 5. The condition of complete continuity of velocity and traction, free sliding, non-reflecting boundary conditions, boundary conditions with a given traction, with a given velocity of the interface, mixed boundary conditions, interface conditions between the elastic and acoustic media are discussed in this chapter.

Chapter 6 presents the results of numerical modeling of delamination of composite materials [16] with a complex heterogeneous structure, as well as some approaches to the development of methods for non-destructive testing of their state. The wave processes of a complex nature are also played a key role in both studies. Difficulties arise even at the stage of creating a mathematical model for the delamination of composite material. Nowadays, none of these delamination mathematical models is consistent with the laboratory experiments. Composite materials are being actively developed and used in a wide range of applied fields. The complex structure of composite materials ensures their high strength at low weight. This feature allows to consider them well applicable in the aircraft industry. However, to do this, it is necessary to develop fast and effective methods for non-destructive testing of composite materials for the presence of destructions in them. The testing methods suitable for homogeneous media are not suitable for composite materials in principle. Because the methods of nondestructive testing are closely related to the wave processes occurring within the material under investigation. The wave processes in a composite material are arranged essentially differently in comparison with homogeneous materials. Therefore, the development of methods for testing composite materials is closely related to the study of wave processes. To study the wave processes today, a full-wave simulation is the best suited approach discussed in Chap. 6.

Chapter 7 demonstrates that the grid-characteristic method is applicable for direct modelling of the seismic prospecting process on typical seismogeological models, such as Marmousi and SEG/EAGE Salt Model. The actual problems of the oil industry, as well as the problems of detection and investigation of geological faults, are considered. In Chap. 7, the application of full wave modelling of seismic waves arising in the process of seismic exploration of these fractured zones is discussed. The possibility to predict effects on seismograms by analyzing spatial wave patterns for a specially developed series of numerical experiments is demonstrated. Since other techniques of observing spatial dynamic wave processes arising in geological environments are not possible, these spatial wave fields can be obtained only by full-wave numerical simulation. However, the specificity of the problem imposes the special demands on both the numerical methods used and development of a series of problem statements for numerical experiments. Features of this innovative approach are set out in Chap. 7. It was also shown in this Chapter that typical analytical tests cannot guarantee that software gives an opportunity for the geologist to develop right conclusions. This problem can be solved only by understanding the physical basis of the phenomena under consideration and the peculiarities of the operation of the difference methods used in the software, simultaneously. This suggests the method called Wave Logica, fragments of which are also given in the Chap. 7.

Chapter 8 discusses the methods of seismic migration [17]. Seismic migration is actively used in seismic surveys to detect contrast interfaces between different geological rocks based on field seismograms. In Chap. 8, the difference between the acoustic wave and elastic wave approaches is demonstrated using the example of adjoint operator [17, 18] and Born approximation [19–24]. The detailed description of the relevant analytical expressions is given in this Chapter. The types of interfaces that are equally displayed by the migration of acoustic fields and the migration of elastic fields are considered in detail. Also, types of interfaces have been identified, which reveal better a migration of elastic fields. The application of migration of elastic fields for seismograms obtained during the use of equipment for single-component seismic exploration of the Earth's interior is considered. Even for this case, the advantage of the migration of elastic fields is shown in comparison with the migration of acoustic fields.

Chapter 9 deals with seismic migration in a case of an elastic wave field. Formulae for seismic migration are obtained on the basis of the corresponding integrals of Rayleigh and Kirchhoff [17]. In Chap. 9, one can find a comparison of the results obtained with the help of Rayleigh integrals and grid-characteristic method. The meaning of such comparison is that the grid-characteristic method is finite-difference, while the Rayleigh-based method is predominantly analytical. That means that all calculations are related to the calculation of analytic expressions for integrals and derivatives, respectively. Due to the fact that these methods are two qualitatively different approaches, their comparison is of interest. Also, in this Chapter the multiple interfaces that are necessary for applying migration based on the Rayleigh and Kirchhoff formulae for the elastic case are described in detail. The possibility of analytically predicting the position of these interfaces on the obtained migration images is considered, what further can give a chance of their elimination using, for example, image recognition methods [25–29].

1.3 Conclusions

The chapter has provided a briefly description of eight chapters with original investigations of wave processes. All included Chapters provide the latest achievements in numerical methods and their applications.

A key feature of the numerical methods outlined in the Chapters is a focus on solving applied problems of various fields, such as seismic, seismic prospecting, investigation of composite materials, study of the human body, ultrasound and laser operations on the human body, etc. Also, the investigation of associated wave processes allows to develop new engineering solutions with following testing. Thus, the approaches for computer modelling proposed in this book are developed as a more perfect, convenient, efficient and cheaper alternative to laboratory and field research and testing. Also, a method of purposeful development of higher-quality engineering solutions based on the analysis of wave processes is proposed in this book.

References

1. Favorskaya AV, Petrov IB (2016) A study of high-order grid-characteristic methods on unstructured grids. Numer Anal Appl 9(2):171–178
2. Petrov IB, Favorskaya AV, Muratov MV, Biryukov VA, Sannikov AV (2014) Grid-characteristic method on unstructured tetrahedral grids. Dokl Math 90(3):781–783
3. Petrov IB, Favorskaya AV, Sannikov AV, Kvasov IE (2013) Grid-characteristic method using high-order interpolation on tetrahedral hierarchical meshes with a multiple time step. Math Models Comput Simul 5(5):409–415
4. Favorskaya AV, Petrov IB (2016) Wave responses from oil reservoirs in the Arctic shelf zone. Doklady Earth Sciences 466 (2):214–217
5. Favorskaya AV, Petrov IB (2017) Numerical modeling of dynamic wave effects in rock masses. Dokl Math 95(3):287–290
6. Favorskaya A, Petrov I, Golubev V, Khokhlov N (2017) Numerical simulation of earthquakes impact on facilities by grid-characteristic method. Proc Comp Sci 112:1206–1215
7. Petrov IB, Favorskaya AV, Khokhlov NI, Miryakha VA, Sannikov AV, Golubev VI (2015) Monitoring the state of the moving train by use of high performance systems and modern computation methods. Math Models Comput Simul 7(1):51–61
8. Magomedov KM, Kholodov AS (1988) Grid characteristic methods. Nauka, Moscow
9. Kholodov AS (1978) Construction of difference schemes with positive approximation for hyperbolic equations. USSR Comput Math Math Phys 18(6):116–132
10. Kholodov AS (1980) The construction of difference schemes of increased order of accuracy for equations of hyperbolic type. USSR Comput Math Math Phys 20(6):234–253
11. Kholodov AS, Kholodov YaA (2006) Monotonicity criteria for difference schemes designed for hyperbolic equations. Comput Math Math Phys 46(9):1560–1588
12. Alekseenko AE, Kholodov AS, Kholodov YA (2016) Boundary control problems for quasilinear systems of hyperbolic equations. Comput Math Math Phys 56(6):916–931
13. Belotserkovskii OM, Popov FD, Tolstykh AI, Fomin VN, Kholodov AS (1970) Numerical solution of some problems in gas dynamics. USSR Comput Math and Math Phys 10(2):158–177
14. Beklemysheva KA, Petrov IB, Favorskaya AV (2014) Numerical simulation of processes in solid deformable media in the presence of dynamic contacts using the grid-characteristic method. Math Models Comput Simul 6(3):294–304
15. Vasyukov AV, Ermakov AS, Potapov AP, Petrov IB, Favorskaya AV, Shevtsov AV (2014) Combined grid-characteristic method for the numerical solution of three-dimensional dynamical elastoplastic problems. Comput Math Math Phys 54(7):1176–1189
16. Petrov I, Vasyukov A, Beklemysheva K, Ermakov A, Favorskaya A (2016) Numerical modeling of non-destructive testing of composites. Proc Comput Sci 96:930–938
17. Zhdanov MS (2002) Geophysical inverse theory and regularization problems. Methods in Geochemistry and Geophysics, vol 36, Elsevier, Netherlands
18. Luo Y, Tromp J, Denel B, Calandra H (2013) 3D coupled acoustic-elastic migration with topography and bathymetry based on spectral-element and adjoint methods. Geophysics 78(4):193–202

19. Morse PM, Feshbach H (1953) Methods of theoretical physics. McGraw-Hill Book Co. Inc, New York, Toronto, London
20. Moser TJ (2012) Review of ray-Born forward modeling for migration and diffraction analysis Stud Geophys Geod 56(2):411–432
21. Cervený V, Klimeš L, Pšenčík I (2007) Seismic ray method: recent developments. Adv Geophys 48:1–126
22. Thierry P, Operto S, and Lambar´e G (1999) Fast 2-D ray—born migration/inversion in complex media. Geophysics 64(1):162–181
23. Beydoun WB, Mendes M (1989) Elastic ray-Born L2-migration/inversion. Geophys J Int 97(1):151–160
24. Beylkin G, Burridge R (1990) Linearized inverse scattering problems in acoustics and elasticity. Wave Motion 12(1):15–52
25. Favorskaya MN, Buryachenko VV (2018) Warping techniques in video stabilization. Intell Syst Ref Librar 135:177–215
26. Favorskaya MN, Jain LC (2017) Large scene rendering. Intell Syst Ref Librar 122:281–320
27. Favorskaya M, Buryachenko V, Tomilina A (2017) Structure-based improvement of scene warped locally in digital video stabilization. Proc Comput Sci 112:1062–1071
28. Favorskaya M, Jain LC (2016) Recognition of pedestrian active events by robust to noises boost algorithm. Adv Intell Syst Comput 357:863–873
29. Favorskaya M, Jain LC, Proskurin A (2016) Unsupervised clustering of natural images in automatic image annotation systems. Intell Syst Ref Lib 108:23–155

Chapter 2
Interpolation on Unstructured Triangular Grids

Alena V. Favorskaya

Abstract The chapter develops the analytical formulae for high-order interpolation on the unstructured triangular grids, such as the polynomial interpolation, piecewise linear interpolation, and hybrid interpolation. The interpolation might be used during the creation of new unstructured triangular or regular gird instead of previous ones as an element of numerical method for finding 2D solutions on the unstructured triangular grids and visualization of some 2D field, as well as during the images' creation or converting. Also the hierarchical nested unstructured triangular grids are discussed in this chapter. This type of grids can be used as an element of numerical method on the unstructured triangular grids for the visualization, creation, and transformation of 2D images. The more the numerical mathematicians work, the faster the software executes and the lesser hardware resources are needed for obtaining the same solution. Analytical formulae reduce the recourses needed, for example the software operation time and amount of dynamic computer memory. In this chapter, one can find the analytical expressions ready for use. The deduced analytical expressions and formed tables help to achieve the huge numerical modelling results in a case of the hardware resources' deficiency.

Keywords Interpolation · High-order interpolation
Unstructured grids · Triangular grids · Polynomial interpolation
Piecewise interpolation · Hybrid interpolation · Nested meshes

---

A. V. Favorskaya (✉)
Non-State Educational Institution "Educational Scientific and Experimental Center of Moscow Institute of Physics and Technology", 9, Institutsky Pereulok St, Dolgoprudny, Moscow Region 141700, Russian Federation
e-mail: aleanera@yandex.ru

A. V. Favorskaya
Department of Computer Science and Numerical Mathematics, Moscow Institute of Physics and Technology, 9, Institutsky Pereulok St, Dolgoprudny, Moscow Region 141700, Russian Federation

A. V. Favorskaya
Scientific Research Institute for System Studies of the Russian Academy of Sciences, 36(1), Nahimovskij ave, Moscow 117218, Russian Federation

A. V. Favorskaya and I. B. Petrov (eds.), *Innovations in Wave Processes Modelling and Decision Making*, Smart Innovation, Systems and Technologies 90,
https://doi.org/10.1007/978-3-319-76201-2_2

2.1 Introduction

Interpolation on the unstructured triangular and tetrahedral grids [1] makes it possible to perform a full-wave modelling in order to minimize the risks of different nature and allows to describe the complex geometric shapes. The interpolation plays a fundamental role in studying the causes of seismogram pattern [2–4], destruction in buildings under the intensive dynamic loads from the earthquakes or man-made nature actions [5], causes of the integrity of the tissues of the human body [6–8] in the conduct of ultrasonic operations, ultrasonic diagnostics, and injuries of the human body parts including the influence of implants, and modelling of non-destructive control [9] of various designs and materials including the composite materials [10], and other examples of studying of dynamical wave processes, called Wave Logica. Very often, 2D analysis is sufficient for the research of dynamic wave processes and identification of the reasons for formation of the certain regularities. Thus, interpolation on the triangular grids is not less relevant than interpolation on the tetrahedral grids.

The problem for obtaining the solutions of systems of equations is in that the computers do not deal with the continuity, continuous values and fields; the computers deal only with the discrete values. Thus, computers provide the solutions only in finite points. However, one needs to know a solution in any point of space. This problem is solved using interpolation. In this chapter, the interpolation on the unstructured triangular meshes is discussed. There are two reasons for such discussion. The first reason is due to the fact that a lot of problems contain only some 2D physical effects and do not contain any 3D effect. Also the interpolation technique on the unstructured triangular and tetrahedral grids can be used for the images' construction and transformation. The second reason to discuss the interpolation on the unstructured triangular grids is a complicity of the interpolation technique on the unstructured tetrahedral meshes. Therefore, it is simpler to understand these techniques using two-dimensional triangular grids with the following investigation on the unstructured tetrahedral grids.

The interpolation on the unstructured triangular and tetrahedral grids has several applications. For example, the interpolation can be used for creation of a new unstructured triangular, tetrahedral, or regular grid instead of previous one. The interpolation is applied as an element of numerical method for finding 2D solutions and 3D solutions on the unstructured triangular and tetrahedral meshes [1], for example a grid-characteristic method [1–10]. Also, the interpolation is useful during visualization of some space fields.

The problem of diminishing the value of mistake is secondary. The purpose of development of the software should not be forgotten. The problem of increasing the order of approximation and convergence is tertiary. The purpose of development of the numerical method should not be forgotten. The primary problem is the problem of obtaining the mathematically and physically correct and detailed solutions. The high-order approximation schemes and methods are tested qualitatively and detailed in a one-dimensional case on the structured uniform grids, with known solutions, and without vast amount of complex interface and boundary conditions. However, these problems have not interest of practice because the analytical solutions of them are well known. Also, the criteria of approximation, stability, and convergence are not perfect. The real order of approximation and convergence could be much smaller than the mathematically investigated one. And nobody knows could these high-order schemes and methods provide the correct results and solutions of real problems and how less the real order of approximation and convergence will be regarding the theoretical order. The high-order schemes are usually tested only for inner points, but the order of approximation and convergence or non-physical oscillations are decreased on the boundaries and interfaces. Usually the practical problems contain the numerous complex boundaries and interfaces and differ from the models by this vast amount of boundaries and interfaces.

Remember that the problem of obtaining the mathematically and physically correct and detailed solutions is primary. There are two ways to solve this primary problem. The first way is to increase the order of approximation and convergence. The second way is the use of more detailed grids. The real alternative to increasing of the order of approximation and convergence is to use more detailed meshes. The advantages of this alternative solution are the guarantee of absence of the non-physical oscillations not only for the model problems but also for practical ones.

In a case of the unstructured grids, it is hard to use the second way due to two reasons. The first reason is a long time to create more detailed unstructured grid. The second reason is a long time to find the nearest point for the point under consideration. The material in Chaps. 2–4 eliminates both of these two reasons.

There is an initial polynomial interpolation function. In a case of high-order interpolation, this function can have the erroneous oscillations. The hybridization is used to eliminate these oscillations. There are two stages of hybridization. The first stage is to select the situations, where the hybridization is needed. The second stage is to apply the hybridization. Also there are two ways of hybridization. The first way is to increase the order of interpolation and the second one is to use more detailed unstructured grid and smaller order of interpolation. The both ways and their mixing are discussed in Chaps. 2–4.

The problem of interpolation arises because the object and result of numerical modelling of physical processes are continuous dynamic spatial physical quantities although it is necessary to operate with their discrete counterparts during computing. Thus, it is necessary to restore the value of function at any point of space-time from its values in a given discrete coordinate range and in the discrete set of reference points. This process is called interpolation. The importance of this issue is so great that the regular researches and developments in this field are conducted. A series of crucial recent researches are reviewed below.

Interpolation of rational matrix functions was discussed by Ball et al. [11]. Lama and Kwon proposed new interpolation method based on the combination of discrete cosine transform and wavelet transform [12]. A review of spatial interpolation methods applied in the environmental sciences was done in [13]. Three dimensional image reconstruction using interpolation of distance and image registration was discussed in [14]. Image interpolation based on sparse representation with nonlocal autoregressive modelling was presented in [15]. Multi patch B-spline interpolation was applied for Kirchhoff–Love space rod in [16].

Efficient algorithm for numerical approximation of metrics used for the anisotropic mesh adaptation on the triangular meshes with finite element computations was suggested in [17]. Wen et al. [18] suggested fast Fourier transformation based on the triangular self-convolution window interpolation as an efficient algorithm for power system harmonic estimation, which can eliminate the errors caused by spectral leakage and picket fence effect. Triangular Shepard interpolation was discussed in [19]. Hydrostatic atmospheric dynamical core based on the triangular C-grids using relatively small discretization stencils for interpolation was proposed in [20].

Note that interpolation on the unstructured grids meets the special difficulties. A significant version of the unstructured grids is the triangular grids. The triangular grids are regularly used due to Delaunay triangulation algorithm [21], which is very comfortable for programming and using in the hardware implementation. The triangular and tetrahedral grids are often used in solving of some applied problems, especially medical ones [6–8] and seismic prospecting and exploration of hydrocarbons, oils, and gas [2–4]. Unfortunately, the computational mathematicians and programmers have to abandon the use of the unstructured triangular and tetrahedral grids because such type of interpolation is mathematically complex. At the same time, the mathematically simplified versions do not allow to achieve the accuracy of calculations sufficient for carrying out the research saving the requirements of high computer costs. The analytical expressions and tables represented in this chapter successfully solve this problem. Also, a software library was developed on this basis. The designed library implements the presented algorithms and is actively used for the practical solution of the above-mentioned applied problems [1–10].

More the numerical mathematician work, faster the software will execute and lesser of hardware resources will be needed for obtaining the same solution. Analytical formulae reduce the recourses needed, i.e. the software operation time and amount of dynamic computer memory. Thus, in this chapter one can find a vast amount of equations and tables specially developed to create the analytical algorithms.

Note that in this chapter only the case of scalar field $u\left(\vec{r}\right)$ is considered but one can use the same expressions for vector field $\vec{u}\left(\vec{r}\right)$ if these expressions are applied to the components of vector field $u_1\left(\vec{r}\right)$, $u_2\left(\vec{r}\right)$, $u_3\left(\vec{r}\right)$, $u_4\left(\vec{r}\right)$, … one by one.

Also note that all formulae in this chapter are true for triangles of arbitrary shape despite on the rectangular triangles are drew in all figures of this chapter. The similar formulae for tetrahedrons of arbitrary shapes are discussed in Chaps. 3 and 4.

The chapter's structure is as follows. In Sect. 2.2, the analytical expressions for polynomial interpolation on the unstructured triangular grids for orders from 1 to 5 inclusive are discussed. In Sect. 2.3, the analytical formulae for piecewise interpolation on the unstructured triangular grids divided into small triangles are presented. Using this type of interpolation one can obtain the continuous piecewise linear interpolation and construct different types of hybrid interpolation on the unstructured

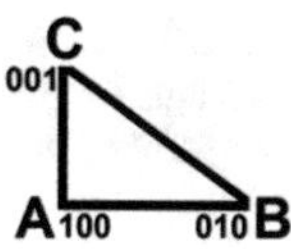

Fig. 2.1 Reference points in a triangle for polynomial interpolation with order 1

triangular grids or use the hierarchical nested unstructured triangular grids discussed in Sect. 2.4. Section 2.5 with the conclusions finalizes the chapter.

2.2 Polynomial Interpolation

In this section, the analytical formulae for polynomial interpolation on the unstructured triangular grids are considered. Using this type of interpolation, one can obtain the continuous piecewise polynomial field and continuous differentiable in each triangle [1]. The proposed method for obtaining these analytical formulae for any given polynomial degree is discussed in Sect. 2.2.1, while the lists with the analytical formulae for degrees N from 1 to 5 inclusive are given in Sects. 2.2.2–2.2.6, respectively.

2.2.1 Obtaining the Analytical Formulae

In order to determine a polynomial field with degree N, which depends on x and y, the values at several points called reference points should be known. The amount of these reference points is equal to $\frac{(N+1)(N+2)}{2}$.

The following method of arranging the reference points is suggested. The lines parallel to the sides of a triangle ABC, which divide each of its sides into N equal parts, are drawn within a triangle. The reference points are numbered in the way shown in Figs. 2.1, 2.2, 2.3, 2.4 and 2.5. These lines divide a triangle into similar smaller triangles.

The interpolation polynomials for finding some scalar field $u\left(\vec{r}\right)$ in the triangle can be written as Eq. 2.1.

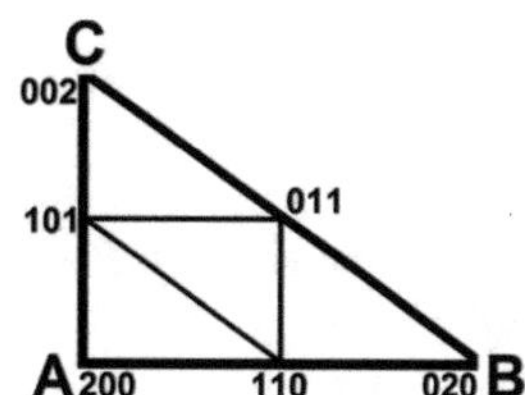

Fig. 2.2 Reference points in a triangle for polynomial interpolation with order 2

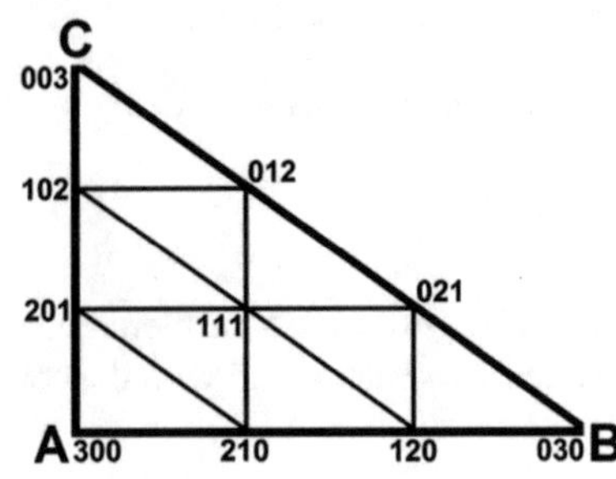

Fig. 2.3 Reference points in a triangle for polynomial interpolation with order 3

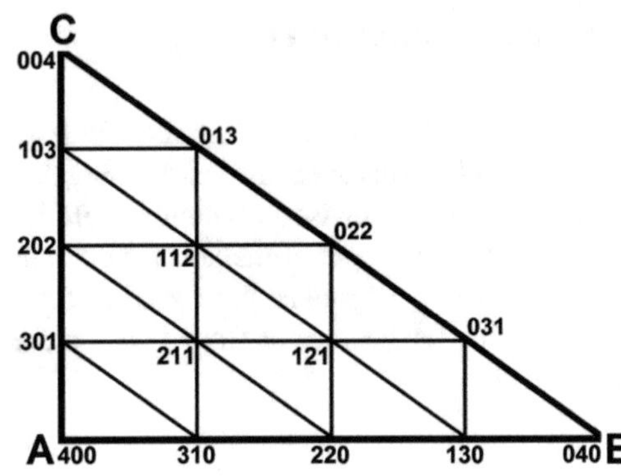

Fig. 2.4 Reference points in a triangle for polynomial interpolation with order 4

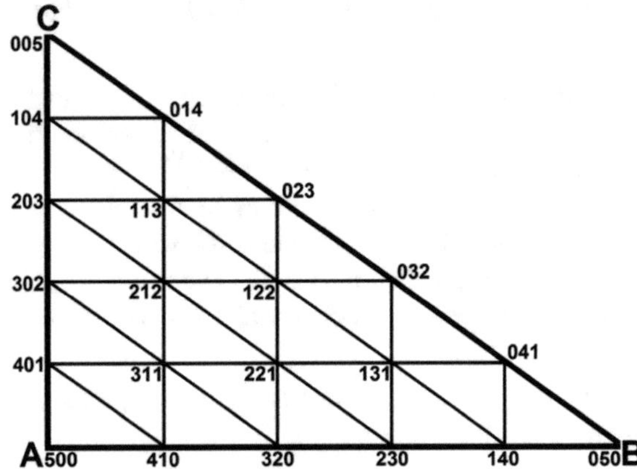

Fig. 2.5 Reference points in a triangle for polynomial interpolation with order 5

$$u\left(\vec{r}\right) = \sum_{i,j} u_{ij} x^i y^j \tag{2.1}$$

Assume that values of the field in the reference points are known and defined by Eq. 2.2.

$$u_{abc} = u\left(\vec{r}_{abc}\right) = \sum_{i,j} u_{ij} x^i_{abc} y^j_{abc} \tag{2.2}$$

One can write the solution of the system of linear Eq. 2.2 as follows:

$$u_{ij} = \sum_{a,b,c} a^{ij}_{abc} u_{abc}. \tag{2.3}$$

Thus, one can determine the weights of the reference points $w_{abc}\left(\vec{r}\right)$ using Eq. 2.4

$$w_{abc}\left(\vec{r}\right) = \sum_{i,j} a^{ij}_{abc} x^i y^j \tag{2.4}$$

and rewrite Eq. 2.1 in more applicable form:

$$u\left(\vec{r}\right) = \sum_{a,b,c} w_{abc}\left(\vec{r}\right) u_{abc}. \tag{2.5}$$

The field in the point R with radius-vector $\vec{r}$ into the triangle ABC should be found. Also 3 triangles BCR, CAR, and ABR are considered. One can find areas of these triangles using Eqs. 2.6–2.9.

$$S_A\left(\vec{r}\right) = S_{BCR} = \frac{1}{2}\left[\vec{r}_C - \vec{r}_B, \vec{r} - \vec{r}_B\right] \tag{2.6}$$

$$S_B\left(\vec{r}\right) = S_{CAR} = \frac{1}{2}\left[\vec{r}_A - \vec{r}_C, \vec{r} - \vec{r}_C\right] \tag{2.7}$$

$$S_C\left(\vec{r}\right) = S_{ABR} = \frac{1}{2}\left[\vec{r}_B - \vec{r}_A, \vec{r} - \vec{r}_A\right] \tag{2.8}$$

$$S = S_{ABC} = \frac{1}{2}\left[\vec{r}_B - \vec{r}_A, \vec{r}_C - \vec{r}_A\right] \tag{2.9}$$

Also one can find the relative areas using Eq. 2.10, where $T = A,\ B,\ C$.

$$s_T\left(\vec{r}\right) = \frac{S_T\left(\vec{r}\right)}{S} \tag{2.10}$$

The relative areas have several properties. Their sum is equal to 1 (Eq. 2.11).

$$s_A\left(\vec{r}\right) + s_B\left(\vec{r}\right) + s_C\left(\vec{r}\right) = 1 \tag{2.11}$$

If a point R is one of the reference points with indices abc, then for any b and any c Eq. 2.12 is true:

$$s_A\left(\vec{r}_{abc}\right) = \frac{a}{N}, \tag{2.12}$$

for any a and any c Eq. 2.13 is true:

$$s_B\left(\vec{r}_{abc}\right) = \frac{b}{N}, \quad (2.13)$$

and for any a and any b Eq. 2.14 is true:

$$s_C\left(\vec{r}_{abc}\right) = \frac{c}{N}. \quad (2.14)$$

The weights $w_{abc}\left(\vec{r}\right)$ of the reference points are written in a view of Eq. 2.15.

$$w_{abc}\left(\vec{r}\right) = \frac{\prod_{i=1}^{N}\left(s_{T_i}\left(\vec{r}\right) - \frac{n_i}{N}\right)}{\prod_{i=1}^{N}\left(s_{T_i}\left(\vec{r}_{abc}\right) - \frac{n_i}{N}\right)} \quad (2.15)$$

In Eq. 2.15, the letter-indices $T_i = A,\ B,\ C$ and natural numbers n_i should be founded from Eq. 2.16.

$$w_{abc}\left(\vec{r}_{a'b'c'}\right) = \delta_{aa'}\delta_{bb'}\delta_{cc'} \quad (2.16)$$

2.2.2 Weights of Reference Points for N = 1

Consider some example of finding the letter-indices and natural numbers in Eq. 2.15 for the case $N = 1$. In this case, 3 reference points 100, 010, and 001 exist. These points are represented in Fig. 2.1.

Firstly, the weight $w_{100}\left(\vec{r}\right)$ should be found using Eq. 2.17.

$$w_{100}\left(\vec{r}\right) = \frac{s_{T_1}\left(\vec{r}\right) - n_1}{s_{T_1}\left(\vec{r}_{100}\right) - n_1} \quad (2.17)$$

Suppose that $\vec{r} = \vec{r}_{001}$. Then Eq. 2.18 is obtained.

$$0 = \frac{s_{T_1}\left(\vec{r}_{001}\right) - n_1}{s_{T_1}\left(\vec{r}_{100}\right) - n_1} = \frac{s_A\left(\vec{r}_{001}\right) - n_1}{s_A\left(\vec{r}_{100}\right) - n_1} = \frac{0 - n_1}{1 - n_1} = \frac{0 - 0}{1 - 0} = s_A\left(\vec{r}_{001}\right) \quad (2.18)$$

Thus, the weight in reference point 100 is found using Eq. 2.19.

$$w_{100}\left(\vec{r}\right) = s_A\left(\vec{r}\right) \quad (2.19)$$

Similarly, one can find the weights $w_{010}\left(\vec{r}\right)$ and $w_{001}\left(\vec{r}\right)$ provided by Eqs. 2.20–2.21.

$$w_{010}\left(\vec{r}\right) = s_B\left(\vec{r}\right) \tag{2.20}$$

$$w_{001}\left(\vec{r}\right) = s_C\left(\vec{r}\right) \tag{2.21}$$

Therefore, the formula for linear interpolation in a triangle can be written in the form of Eq. 2.22.

$$u\left(\vec{r}\right) = s_A\left(\vec{r}\right)u_{100} + s_B\left(\vec{r}\right)u_{010} + s_C\left(\vec{r}\right)u_{001} \tag{2.22}$$

2.2.3 *Weights of Reference Points for N = 2*

In the case of $N = 2$, the weights of reference points represented in Fig. 2.2 are defined using Eqs. 2.23–2.28, respectively.

$$w_{200}\left(\vec{r}\right) = s_A\left(\vec{r}\right)\left(2s_A\left(\vec{r}\right) - 1\right) \tag{2.23}$$

$$w_{020}\left(\vec{r}\right) = s_B\left(\vec{r}\right)\left(2s_B\left(\vec{r}\right) - 1\right) \tag{2.24}$$

$$w_{002}\left(\vec{r}\right) = s_C\left(\vec{r}\right)\left(2s_C\left(\vec{r}\right) - 1\right) \tag{2.25}$$

$$w_{110}\left(\vec{r}\right) = 4s_A\left(\vec{r}\right)s_B\left(\vec{r}\right) \tag{2.26}$$

$$w_{011}\left(\vec{r}\right) = 4s_B\left(\vec{r}\right)s_C\left(\vec{r}\right) \tag{2.27}$$

$$w_{101}\left(\vec{r}\right) = 4s_C\left(\vec{r}\right)s_A\left(\vec{r}\right) \tag{2.28}$$

2.2.4 *Weights of Reference Points for N = 3*

In the case of $N = 3$, the weights of reference points represented in Fig. 2.3 are computed using Eqs. 2.29–2.38, respectively.

$$w_{300}\left(\vec{r}\right) = \frac{1}{2}s_A\left(\vec{r}\right)\left(3s_A\left(\vec{r}\right) - 1\right)\left(3s_A\left(\vec{r}\right) - 2\right) \tag{2.29}$$

$$w_{030}\left(\vec{r}\right) = \frac{1}{2}s_B\left(\vec{r}\right)\left(3s_B\left(\vec{r}\right) - 1\right)\left(3s_B\left(\vec{r}\right) - 2\right) \tag{2.30}$$

$$w_{003}\left(\vec{r}\right) = \frac{1}{2}s_C\left(\vec{r}\right)\left(3s_C\left(\vec{r}\right) - 1\right)\left(3s_C\left(\vec{r}\right) - 2\right) \tag{2.31}$$

$$w_{210}\left(\vec{r}\right) = \frac{9}{2}s_A\left(\vec{r}\right)s_B\left(\vec{r}\right)\left(3s_A\left(\vec{r}\right) - 1\right) \quad (2.32)$$

$$w_{021}\left(\vec{r}\right) = \frac{9}{2}s_B\left(\vec{r}\right)s_C\left(\vec{r}\right)\left(3s_B\left(\vec{r}\right) - 1\right) \quad (2.33)$$

$$w_{102}\left(\vec{r}\right) = \frac{9}{2}s_C\left(\vec{r}\right)s_A\left(\vec{r}\right)\left(3s_C\left(\vec{r}\right) - 1\right) \quad (2.34)$$

$$w_{120}\left(\vec{r}\right) = \frac{9}{2}s_B\left(\vec{r}\right)s_A\left(\vec{r}\right)\left(3s_B\left(\vec{r}\right) - 1\right) \quad (2.35)$$

$$w_{012}\left(\vec{r}\right) = \frac{9}{2}s_C\left(\vec{r}\right)s_B\left(\vec{r}\right)\left(3s_C\left(\vec{r}\right) - 1\right) \quad (2.36)$$

$$w_{201}\left(\vec{r}\right) = \frac{9}{2}s_A\left(\vec{r}\right)s_C\left(\vec{r}\right)\left(3s_A\left(\vec{r}\right) - 1\right) \quad (2.37)$$

$$w_{111}\left(\vec{r}\right) = 27s_A\left(\vec{r}\right)s_B\left(\vec{r}\right)s_C\left(\vec{r}\right) \quad (2.38)$$

2.2.5 *Weights of Reference Points for N = 4*

In the case of $N = 4$, the weights of reference points represented in Fig. 2.4 are determined by Eqs. 2.39–2.53, respectively.

$$w_{400}\left(\vec{r}\right) = \frac{1}{3}s_A\left(\vec{r}\right)\left(4s_A\left(\vec{r}\right) - 1\right)\left(2s_A\left(\vec{r}\right) - 1\right)\left(4s_A\left(\vec{r}\right) - 3\right) \quad (2.39)$$

$$w_{040}\left(\vec{r}\right) = \frac{1}{3}s_B\left(\vec{r}\right)\left(4s_B\left(\vec{r}\right) - 1\right)\left(2s_B\left(\vec{r}\right) - 1\right)\left(4s_B\left(\vec{r}\right) - 3\right) \quad (2.40)$$

$$w_{400}\left(\vec{r}\right) = \frac{1}{3}s_C\left(\vec{r}\right)\left(4s_C\left(\vec{r}\right) - 1\right)\left(2s_C\left(\vec{r}\right) - 1\right)\left(4s_C\left(\vec{r}\right) - 3\right) \quad (2.41)$$

$$w_{310}\left(\vec{r}\right) = \frac{16}{3}s_A\left(\vec{r}\right)s_B\left(\vec{r}\right)\left(4s_A\left(\vec{r}\right) - 1\right)\left(2s_A\left(\vec{r}\right) - 1\right) \quad (2.42)$$

$$w_{031}\left(\vec{r}\right) = \frac{16}{3}s_B\left(\vec{r}\right)s_C\left(\vec{r}\right)\left(4s_B\left(\vec{r}\right) - 1\right)\left(2s_B\left(\vec{r}\right) - 1\right) \quad (2.43)$$

$$w_{103}\left(\vec{r}\right) = \frac{16}{3}s_C\left(\vec{r}\right)s_A\left(\vec{r}\right)\left(4s_C\left(\vec{r}\right) - 1\right)\left(2s_C\left(\vec{r}\right) - 1\right) \quad (2.44)$$

$$w_{130}\left(\vec{r}\right) = \frac{16}{3}s_B\left(\vec{r}\right)s_A\left(\vec{r}\right)\left(4s_B\left(\vec{r}\right) - 1\right)\left(2s_B\left(\vec{r}\right) - 1\right) \quad (2.45)$$

$$w_{013}\left(\vec{r}\right) = \frac{16}{3}s_C\left(\vec{r}\right)s_B\left(\vec{r}\right)\left(4s_C\left(\vec{r}\right) - 1\right)\left(2s_C\left(\vec{r}\right) - 1\right) \quad (2.46)$$

$$w_{301}\left(\vec{r}\right) = \frac{16}{3}s_A\left(\vec{r}\right)s_C\left(\vec{r}\right)\left(4s_A\left(\vec{r}\right) - 1\right)\left(2s_A\left(\vec{r}\right) - 1\right) \quad (2.47)$$

$$w_{220}\left(\vec{r}\right) = 4s_A\left(\vec{r}\right)s_B\left(\vec{r}\right)\left(4s_A\left(\vec{r}\right) - 1\right)\left(4s_B\left(\vec{r}\right) - 1\right) \quad (2.48)$$

$$w_{022}\left(\vec{r}\right) = 4s_B\left(\vec{r}\right)s_C\left(\vec{r}\right)\left(4s_B\left(\vec{r}\right) - 1\right)\left(4s_C\left(\vec{r}\right) - 1\right) \quad (2.49)$$

$$w_{202}\left(\vec{r}\right) = 4s_C\left(\vec{r}\right)s_A\left(\vec{r}\right)\left(4s_C\left(\vec{r}\right)-1\right)\left(4s_A\left(\vec{r}\right)-1\right) \tag{2.50}$$

$$w_{211}\left(\vec{r}\right) = 32s_A\left(\vec{r}\right)s_B\left(\vec{r}\right)s_C\left(\vec{r}\right)\left(4s_A\left(\vec{r}\right)-1\right) \tag{2.51}$$

$$w_{121}\left(\vec{r}\right) = 32s_B\left(\vec{r}\right)s_C\left(\vec{r}\right)s_A\left(\vec{r}\right)\left(4s_B\left(\vec{r}\right)-1\right) \tag{2.52}$$

$$w_{112}\left(\vec{r}\right) = 32s_C\left(\vec{r}\right)s_A\left(\vec{r}\right)s_B\left(\vec{r}\right)\left(4s_C\left(\vec{r}\right)-1\right) \tag{2.53}$$

2.2.6 *Weights of Reference Points for N = 5*

In the case of $N = 5$, the weights of reference points represented in Fig. 2.5 are identified by Eqs. 2.54–2.74, respectively.

$$w_{500}\left(\vec{r}\right) = \frac{1}{24}s_A\left(\vec{r}\right)\left(5s_A\left(\vec{r}\right)-1\right)\left(5s_A\left(\vec{r}\right)-2\right)\left(5s_A\left(\vec{r}\right)-3\right)\left(5s_A\left(\vec{r}\right)-4\right) \tag{2.54}$$

$$w_{050}\left(\vec{r}\right) = \frac{1}{24}s_B\left(\vec{r}\right)\left(5s_B\left(\vec{r}\right)-1\right)\left(5s_B\left(\vec{r}\right)-2\right)\left(5s_B\left(\vec{r}\right)-3\right)\left(5s_B\left(\vec{r}\right)-4\right) \tag{2.55}$$

$$w_{005}\left(\vec{r}\right) = \frac{1}{24}s_C\left(\vec{r}\right)\left(5s_C\left(\vec{r}\right)-1\right)\left(5s_C\left(\vec{r}\right)-2\right)\left(5s_C\left(\vec{r}\right)-3\right)\left(5s_C\left(\vec{r}\right)-4\right) \tag{2.56}$$

$$w_{410}\left(\vec{r}\right) = \frac{25}{24}s_A\left(\vec{r}\right)s_B\left(\vec{r}\right)\left(5s_A\left(\vec{r}\right)-1\right)\left(5s_A\left(\vec{r}\right)-2\right)\left(5s_A\left(\vec{r}\right)-3\right) \tag{2.57}$$

$$w_{041}\left(\vec{r}\right) = \frac{25}{24}s_B\left(\vec{r}\right)s_C\left(\vec{r}\right)\left(5s_B\left(\vec{r}\right)-1\right)\left(5s_B\left(\vec{r}\right)-2\right)\left(5s_B\left(\vec{r}\right)-3\right) \tag{2.58}$$

$$w_{104}\left(\vec{r}\right) = \frac{25}{24}s_C\left(\vec{r}\right)s_A\left(\vec{r}\right)\left(5s_C\left(\vec{r}\right)-1\right)\left(5s_C\left(\vec{r}\right)-2\right)\left(5s_C\left(\vec{r}\right)-3\right) \tag{2.59}$$

$$w_{140}\left(\vec{r}\right) = \frac{25}{24}s_B\left(\vec{r}\right)s_A\left(\vec{r}\right)\left(5s_B\left(\vec{r}\right)-1\right)\left(5s_B\left(\vec{r}\right)-2\right)\left(5s_B\left(\vec{r}\right)-3\right) \tag{2.60}$$

$$w_{014}\left(\vec{r}\right) = \frac{25}{24}s_C\left(\vec{r}\right)s_B\left(\vec{r}\right)\left(5s_C\left(\vec{r}\right)-1\right)\left(5s_C\left(\vec{r}\right)-2\right)\left(5s_C\left(\vec{r}\right)-3\right) \tag{2.61}$$

$$w_{401}\left(\vec{r}\right) = \frac{25}{24}s_A\left(\vec{r}\right)s_C\left(\vec{r}\right)\left(5s_A\left(\vec{r}\right)-1\right)\left(5s_A\left(\vec{r}\right)-2\right)\left(5s_A\left(\vec{r}\right)-3\right) \tag{2.62}$$

$$w_{320}\left(\vec{r}\right) = \frac{25}{12}s_A\left(\vec{r}\right)s_B\left(\vec{r}\right)\left(5s_A\left(\vec{r}\right)-1\right)\left(5s_B\left(\vec{r}\right)-1\right)\left(5s_A\left(\vec{r}\right)-2\right) \tag{2.63}$$

$$w_{032}\left(\vec{r}\right) = \frac{25}{12}s_B\left(\vec{r}\right)s_C\left(\vec{r}\right)\left(5s_B\left(\vec{r}\right)-1\right)\left(5s_C\left(\vec{r}\right)-1\right)\left(5s_B\left(\vec{r}\right)-2\right) \tag{2.64}$$

$$w_{203}\left(\vec{r}\right) = \frac{25}{12}s_C\left(\vec{r}\right)s_A\left(\vec{r}\right)\left(5s_C\left(\vec{r}\right)-1\right)\left(5s_A\left(\vec{r}\right)-1\right)\left(5s_C\left(\vec{r}\right)-2\right) \tag{2.65}$$

$$w_{230}\left(\vec{r}\right) = \frac{25}{12}s_B\left(\vec{r}\right)s_A\left(\vec{r}\right)\left(5s_B\left(\vec{r}\right)-1\right)\left(5s_A\left(\vec{r}\right)-1\right)\left(5s_B\left(\vec{r}\right)-2\right) \tag{2.66}$$

$$w_{023}\left(\vec{r}\right) = \frac{25}{12}s_C\left(\vec{r}\right)s_B\left(\vec{r}\right)\left(5s_C\left(\vec{r}\right)-1\right)\left(5s_B\left(\vec{r}\right)-1\right)\left(5s_C\left(\vec{r}\right)-2\right) \tag{2.67}$$

$$w_{302}\left(\vec{r}\right) = \frac{25}{12}s_A\left(\vec{r}\right)s_C\left(\vec{r}\right)\left(5s_A\left(\vec{r}\right)-1\right)\left(5s_C\left(\vec{r}\right)-1\right)\left(5s_A\left(\vec{r}\right)-2\right) \tag{2.68}$$

$$w_{311}\left(\vec{r}\right) = \frac{125}{6}s_A\left(\vec{r}\right)s_B\left(\vec{r}\right)s_C\left(\vec{r}\right)\left(5s_A\left(\vec{r}\right)-1\right)\left(5s_A\left(\vec{r}\right)-2\right) \tag{2.69}$$

$$w_{131}\left(\vec{r}\right) = \frac{125}{6}s_B\left(\vec{r}\right)s_C\left(\vec{r}\right)s_A\left(\vec{r}\right)\left(5s_B\left(\vec{r}\right)-1\right)\left(5s_B\left(\vec{r}\right)-2\right) \tag{2.70}$$

$$w_{113}\left(\vec{r}\right) = \frac{125}{6} s_C\left(\vec{r}\right) s_A\left(\vec{r}\right) s_B\left(\vec{r}\right)\left(5s_C\left(\vec{r}\right) - 1\right)\left(5s_C\left(\vec{r}\right) - 2\right) \tag{2.71}$$

$$w_{221}\left(\vec{r}\right) = \frac{125}{4} s_A\left(\vec{r}\right) s_B\left(\vec{r}\right) s_C\left(\vec{r}\right)\left(5s_A\left(\vec{r}\right) - 1\right)\left(5s_B\left(\vec{r}\right) - 1\right) \tag{2.72}$$

$$w_{122}\left(\vec{r}\right) = \frac{125}{4} s_B\left(\vec{r}\right) s_C\left(\vec{r}\right) s_A\left(\vec{r}\right)\left(5s_B\left(\vec{r}\right) - 1\right)\left(5s_C\left(\vec{r}\right) - 1\right) \tag{2.73}$$

$$w_{212}\left(\vec{r}\right) = \frac{125}{4} s_C\left(\vec{r}\right) s_A\left(\vec{r}\right) s_B\left(\vec{r}\right)\left(5s_C\left(\vec{r}\right) - 1\right)\left(5s_A\left(\vec{r}\right) - 1\right) \tag{2.74}$$

2.3 Piecewise Linear Interpolation

Let us consider the analytical formulae for the piecewise interpolation on the unstructured triangular grids divided into small triangles [1]. Using this type of interpolation, one can obtain the continuous piecewise linear interpolation and construct different types of the hybrid interpolation on the unstructured triangular grids. Consider the piecewise interpolation on N^2 small triangles discussed in Sect. 2.2 and formed by reference points for polynomial interpolation with degree N varied from 2 to 5. One can find these small triangles in Figs. 2.6, 2.7, 2.8 and 2.9. They are numbered using grey color.

Note that in this Sect. 2.3 only the case is considered, when a point under consideration lays in the big triangle ABC and relative areas discussed in Sect. 2.2. Therefore, the following inequalities for these points are always satisfied:

$$(s_A \in [0, 1]) \wedge (s_B \in [0, 1]) \wedge (s_C \in [0, 1])\, beforethevaluesinreference. \tag{2.75}$$

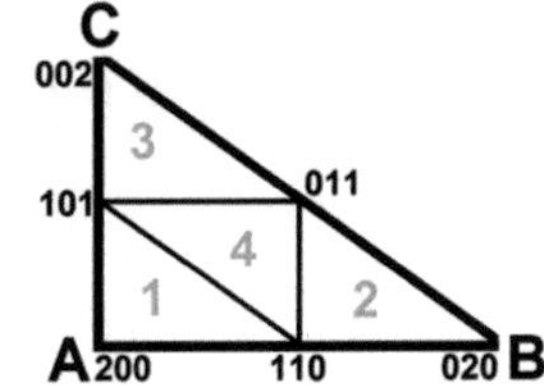

Fig. 2.6 Small triangles based on reference points for polynomial interpolation with order 2

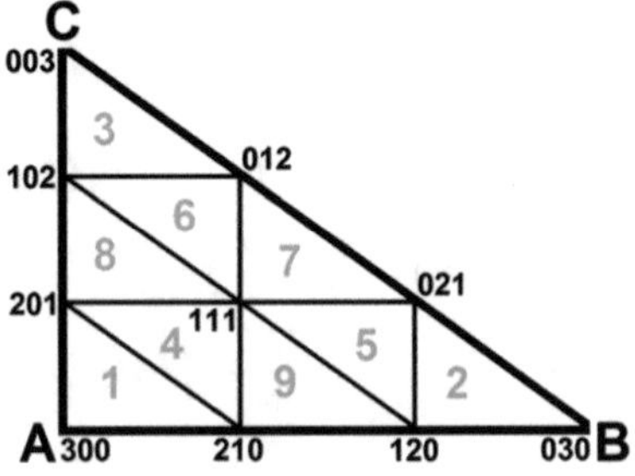

Fig. 2.7 Small triangles based on reference points for polynomial interpolation with order 3

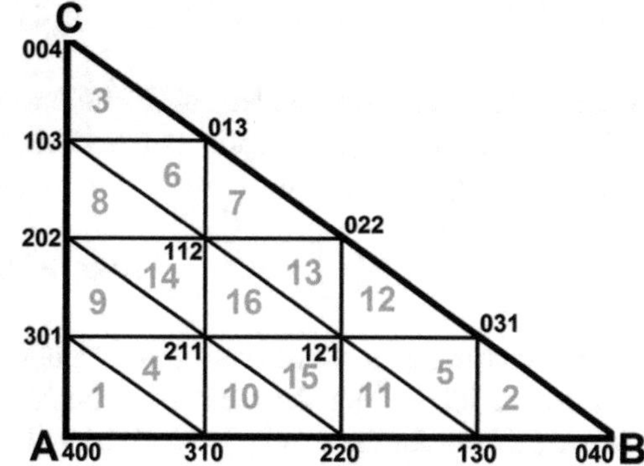

Fig. 2.8 Small triangles based on reference points for polynomial interpolation with order 4

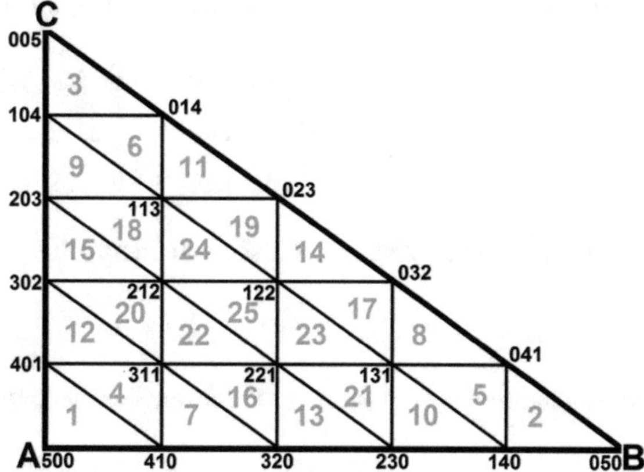

Fig. 2.9 Small triangles based on reference points for polynomial interpolation with order 5

All mathematical expressions for $N = 2$ and $N = 3$ are adduced in Sects. 2.3.1 and 2.3.4, respectively. In order to diminish the amount of mathematical expressions, four types of tables are determined in Sect. 2.3.2. These tables for N varied from 2 to 5 are adduced in Sects. 2.3.3, 2.3.5–2.3.7, respectively.

2.3.1 Algorithms and Analytical Formulae for N = 2

Let us consider degree $N = 2$. 4 small triangles are numbered in Fig. 2.6.

The algorithm includes the following steps:

Step 1. If the relative area satisfies the inequality:

$$2s_A > 1, \tag{2.76}$$

then all the relative areas satisfy the following inequalities:

$$(2s_A \in (1,2]) \wedge (2s_B \in [0,1]) \wedge (2s_C \in [0,1])\,, \tag{2.77}$$

and, therefore, the point under consideration lays in the small triangle 200, 110, 101 numbered 1 in Fig. 2.6 and one can determine a value of the piecewise linear interpolation using Eq. 2.78.

$$u\left(\vec{r}\right) = (2s_A - 1)\,u_{200} + 2s_B u_{110} + 2s_C u_{101} \tag{2.78}$$

Note that the coefficients before the values in reference points u_{abc} (Eq. 2.78) and, hereinafter, in Sect. 2.3 mean the relative areas (Eq. 2.10) that were discussed in Sect. 2.2, but for the relatively small triangle. For example, the relative areas for small triangle 200, 110, 101 numbered 1 in Fig. 2.6 can be found on the basis of the relative areas for the initial big triangle 200, 020, 002 using Eqs. 2.79–2.81.

$$(s_A)_{200,110,101} = 2\,(s_A)_{200,020,002} - 1 \tag{2.79}$$

$$(s_B)_{200,110,101} = 2\,(s_B)_{200,020,002} \tag{2.80}$$

$$(s_C)_{200,110,101} = 2\,(s_C)_{200,020,002} \tag{2.81}$$

Step 2. If the relative area satisfies the inequality:

$$2s_B > 1, \tag{2.82}$$

then all the relative areas satisfy the following inequalities

$$(2s_A \in [0,1]) \wedge (2s_B \in (1,2]) \wedge (2s_C \in [0,1])\,, \tag{2.83}$$

and, therefore, the point under consideration lays in the small triangle 110, 020, 011 numbered 2 in Fig. 2.6 and one can determine a value of the piecewise linear interpolation using Eq. 2.84.

$$u\left(\vec{r}\right) = 2s_A u_{110} + (2s_B - 1)\,u_{020} + 2s_C u_{011} \tag{2.84}$$

Step 3. If the relative area satisfies the inequality:

$$2s_C > 1, \tag{2.85}$$

then all the relative areas satisfy the following inequalities:

$$(2s_A \in [0,1]) \wedge (2s_B \in [0,1]) \wedge (2s_C \in (1,2])\,, \tag{2.86}$$

and, therefore, the point under consideration lays in the small triangle 101, 011, 002 numbered 3 in Fig. 2.6 and one can determine a value of the piecewise linear interpolation using Eq. 2.87.

Table 2.1 Inequalities for finding the small triangle, in which the point under consideration lies for $N = 2$, *table of the first type*

No	$2s_A$	$2s_B$	$2s_C$
1	>1	–	–
2	–	>1	–
3	–	–	>1
4	–	–	–

$$u\left(\vec{r}\right) = 2s_A u_{101} + 2s_B u_{011} + (2s_C - 1)\, u_{002} \tag{2.87}$$

Step 4. If all the inequalities 2.76, 2.82, and 2.85 are false, then all the relative areas satisfy the following inequalities:

$$(2s_A \in [0, 1]) \wedge (2s_B \in [0, 1]) \wedge (2s_C \in [0, 1])\,, \tag{2.88}$$

and, therefore, the point under consideration lays in the small triangle 011, 101, 110 numbered 4 in Fig. 2.6 and one can determine a value of the piecewise linear interpolation using Eq. 2.89.

$$u\left(\vec{r}\right) = (1 - 2s_A)\, u_{011} + (1 - 2s_B)\, u_{101} + (1 - 2s_C)\, u_{110} \tag{2.89}$$

2.3.2 *Types of Tables and Their Description*

In order to summarize the information in Sects. 2.3.1 and 2.3.4 and to diminish the amount of mathematical expressions in the cases of $N = 4$ and $N = 5$, four types of tables are defined. The lines in the tables help to catch the topological similarity of the formulae and due to this it is suitable to search for errors and misprints in the program code developed based on these tables and appropriate analytical formulae.

Tables of the first type contain the inequalities for finding the small triangle. Table 2.1 in Sect. 2.3.3, Table 2.5 in Sect. 2.3.4, Table 2.9 in Sect. 2.3.6, and Table 2.13 in Sect. 2.3.7 are the tables of the first type.

In order to obtain an inequality based on the filled cell in row I and in column J from the table of the first type, one can take the relative area in the zero row and column J and put it before the condition in the cell IJ. For example, the inequality based on the cell (3, 3) in Table 2.1 could be written as follows:

$$2s_C > 1. \tag{2.90}$$

In order to obtain a logical expression based on the row I in the table of first type, one can write the conjunction of all inequalities based on the cells in this row. For example, the logical expression based on the row 9 from Table 2.13 could be written as follows:

Table 2.2 Logical expressions being true if the point under consideration lies into the small triangle for $N = 2$, *table of the second type*

No	$2s_A \in$	$2s_B \in$	$2s_C \in$
1	$(1, 2]$	$[0, 1]$	$[0, 1]$
2	$[0, 1]$	$(1, 2]$	$[0, 1]$
3	$[0, 1]$	$[0, 1]$	$(1, 2]$
4	$[0, 1]$	$[0, 1]$	$[0, 1]$

Table 2.3 The vertices of the small triangles for $N = 2$, *table of the third type*

No	A'	B'	C'
1	200	110	101
2	110	020	011
3	101	011	002
4	011	101	110

$$(5s_B < 1) \wedge (5v_C > 3). \tag{2.91}$$

Tables of the second type contain the logical expressions that are being true if the point under consideration is into the small triangle with number in the zero column. Table 2.2 in Sect. 2.3.3, Table 2.6 in Sect. 2.3.4, Table 2.10 in Sect. 2.3.6, and Table 2.14 in Sect. 2.3.7 are the tables of the second type.

In order to obtain a logical expression based on the cell in row I and in column J from the table of the second type, one can take the relative area in the zero row and column J and put it before the segment in the cell IJ. For example, the logical expression based on the cell (2, 3) in Table 2.2 could be written as follows:

$$2s_C \in [0, 1]. \tag{2.92}$$

In order to obtain a logical expression based on the row I in the table of the second type, one can write the conjunction of all logical expression based on the cells in this row I. For example, the logical expression based on the row 3 in Table 2.2 could be written as follows:

$$(2s_A \in [0, 1]) \wedge (2s_B \in [0, 1]) \wedge (2s_C \in (1, 2]). \tag{2.93}$$

Tables of the third type contain the reference points corresponding to the vertices of the small triangles. Table 2.3 in Sect. 2.3.3, Table 2.7 in Sect. 2.3.4, Table 2.11 in Sect. 2.3.6, and Table 2.15 in Sect. 2.3.7 are the tables of the third type.

Tables of the fourth type contain the formulae for relative areas corresponding to the appropriate vertices of the small triangles. Table 2.4 in Sect. 2.3.3, Table 2.8 in Sect. 2.3.4, Table 2.12 in Sect. 2.3.6, and Table 2.16 in Sect. 2.3.7 are the tables of the fourth type.

Note that only the values Ns_A, Ns_B, Ns_C are used in all types of tables.

Table 2.4 Formulae for relative areas corresponding to the appropriate vertices of the small triangles for $N = 2$, *table of the fourth type*

No	$s_{A'}$	$s_{B'}$	$s_{C'}$
1	$2s_A - 1$	$2s_B$	$2s_C$
2	$2s_A$	$2s_B - 1$	$2s_C$
3	$2s_A$	$2s_B$	$2s_C - 1$
4	$1 - 2s_A$	$1 - 2s_B$	$1 - 2s_C$

2.3.3 Tables and Algorithms for N = 2

If all of the logical expressions corresponding to the rows 1, 2, …, $I - 1$ in Table 2.1 are false and the logical expression corresponding to the row I in Table 2.1 is true, then the logical expression corresponding to the row I in Table 2.2 is true and, therefore, the point under consideration lies into the small triangle I with vertices given in Table 2.3. One can find the relative areas for its vertices in Table 2.4.

2.3.4 Algorithms and Analytical Formulae for N = 3

Let us consider degree $N = 3$. 9 small triangles are numbered in Fig. 2.7.

The algorithm includes the following steps:

Step 1. If the relative area satisfies the inequality:

$$3s_A > 2, \tag{2.94}$$

then all the relative areas satisfy the following inequalities:

$$(3s_A \in (2, 3]) \wedge (3s_B \in [0, 1]) \wedge (3s_C \in [0, 1]), \tag{2.95}$$

and, therefore, the point under consideration lays in the small triangle 300, 210, 201 numbered 1 in Fig. 2.7 and one can determine a value of the piecewise linear interpolation using Eq. 2.96.

$$u\left(\vec{r}\right) = (3s_A - 2)\, u_{300} + 3s_B u_{210} + 3s_C u_{201} \tag{2.96}$$

Step 2. If the relative area satisfies the inequality:

$$3s_B > 2, \tag{2.97}$$

then all the relative areas satisfy the following inequalities:

$$(3s_A \in [0, 1]) \wedge (3s_B \in (2, 3]) \wedge (3s_C \in [0, 1])\,, \tag{2.98}$$

and, therefore, the point under consideration lays in the small triangle 120, 030, 021 numbered 2 in Fig. 2.7 and one can determine a value of the piecewise linear interpolation using Eq. 2.99.

$$u\left(\vec{r}\right) = 3s_A u_{120} + (3s_B - 2)\, u_{030} + 3s_C u_{021} \tag{2.99}$$

Step 3. If the relative area satisfies the inequality:

$$3s_C > 2, \tag{2.100}$$

then all the relative areas satisfy the following inequalities:

$$(3s_A \in [0, 1]) \wedge (3s_B \in [0, 1]) \wedge (3s_C \in (2, 3]), \tag{2.101}$$

and, therefore, the point under consideration lays in the small triangle 102, 012, 003 numbered 3 in Fig. 2.7 and one can determine a value of the piecewise linear interpolation using Eq. 2.102.

$$u\left(\vec{r}\right) = 3s_A u_{102} + 3s_B u_{012} + (3s_C - 2)\, u_{003} \tag{2.102}$$

Step 4. If the inequality 2.94 is false and the following inequalities are true:

$$(3s_B < 1) \wedge (3s_C < 1), \tag{2.103}$$

then all the relative areas satisfy the following inequalities:

$$(3s_A \in [1, 2]) \wedge (3s_B \in [0, 1)) \wedge (3s_C \in [0, 1)), \tag{2.104}$$

and, therefore, the point under consideration lays in the small triangle 111, 201, 210 numbered 4 in Fig. 2.7 and one can determine a value of the piecewise linear interpolation using Eq. 2.105.

$$u\left(\vec{r}\right) = (2 - 3s_A)\, u_{111} + (1 - 3s_B)\, u_{201} + (1 - 3s_C)\, u_{210} \tag{2.105}$$

Step 5. If the inequality 2.97 is false and the following inequalities are true:

$$(3s_A < 1) \wedge (3s_C < 1), \tag{2.106}$$

then all the relative areas satisfy the following inequalities:

$$(3s_A \in [0, 1)) \wedge (3s_B \in [1, 2]) \wedge (3s_C \in [0, 1))\,, \tag{2.107}$$

and, therefore, the point under consideration lays in the small triangle 021, 111, 120 numbered 5 in Fig. 2.7 and one can determine a value of the piecewise linear interpolation using Eq. 2.108.

$$u\left(\vec{r}\right) = (1 - 3s_A)\,u_{021} + (2 - 3s_B)\,u_{111} + (1 - 3s_C)\,u_{120} \tag{2.108}$$

Step 6. If the inequality 2.100 is false and the following inequalities are true:

$$(3s_A < 1) \wedge (3s_B < 1), \tag{2.109}$$

then all the relative areas satisfy the following inequalities:

$$(3s_A \in [0, 1)) \wedge (3s_B \in [0, 1)) \wedge (3s_C \in [1, 2]), \tag{2.110}$$

and, therefore, the point under consideration lays in the small triangle 012, 102, 111 numbered 6 in Fig. 2.7 and one can determine a value of the piecewise linear interpolation using Eq. 2.111.

$$u\left(\vec{r}\right) = (1 - 3s_A)\,u_{012} + (1 - 3s_B)\,u_{102} + (2 - 3s_C)\,u_{111} \tag{2.111}$$

Step 7. If the inequalities 2.97, 2.100, 2.106, and 2.109 are false and

$$3s_A < 1, \tag{2.112}$$

then all the relative areas satisfy the following inequalities:

$$(3s_A \in [0, 1)) \wedge (3s_B \in [1, 2]) \wedge (3s_C \in [1, 2]), \tag{2.113}$$

and, therefore, the point under consideration lays in the small triangle 111, 021, 012 numbered 7 in Fig. 2.7 and one can determine a value of the piecewise linear interpolation using Eq. 2.114.

$$u\left(\vec{r}\right) = 3s_A u_{111} + (3s_B - 1)\,u_{021} + (3s_C - 1)\,u_{012} \tag{2.114}$$

Step 8. If the inequalities 2.94, 2.100, 2.103, and 2.109 are false and

$$3s_B < 1, \tag{2.115}$$

then all the relative areas satisfy the following inequalities:

$$(3s_A \in [1, 2]) \wedge (3s_B \in [0, 1)) \wedge (3s_C \in [1, 2]), \tag{2.116}$$

Table 2.5 Inequalities for finding the small triangle, in which the point under consideration lies for $N = 3$, *table of the first type*

No	$3s_A$	$3s_B$	$3s_C$	No	$3s_A$	$3s_B$	$3s_C$	No	$3s_A$	$3s_B$	$3s_C$
1	>2	–	–	4	–	<1	<1	7	<1	–	–
2	–	>2	–	5	<1	–	<1	8	–	<1	–
3	–	–	>2	6	<1	<1	–	9	–	–	–

and, therefore, the point under consideration lays in the small triangle 201, 111, 102 numbered 8 in Fig. 2.7 and one can determine a value of the piecewise linear interpolation using Eq. 2.117.

$$u\left(\vec{r}\right) = (3s_A - 1)\,u_{201} + 3s_B u_{111} + (3s_C - 1)\,u_{102} \tag{2.117}$$

Step 9. If the inequalities 2.94, 2.97, 2.103, and 2.106 are false and

$$3s_C \le 1, \tag{2.118}$$

then all the relative areas satisfy the following inequalities:

$$(3s_A \in [1, 2]) \wedge (3s_B \in [1, 2]) \wedge (3s_C \in [0, 1]), \tag{2.119}$$

and, therefore, the point under consideration lays in the small triangle 210, 120, 111 numbered 9 in Fig. 2.7 and one can determine a value the of piecewise linear interpolation using Eq. 2.120.

$$u\left(\vec{r}\right) = (3s_A - 1)\,u_{210} + (3s_B - 1)\,u_{120} + 3s_C u_{111} \tag{2.120}$$

Note that, when all the inequalities 2.94, 2.97, 2.100, 2.103, 2.106, 2.109, 2.112, and 2.115 are false, the inequalities 2.119 are true and one can also use Eq. 2.120.

2.3.5 *Tables and Algorithms for N = 3*

If all of the logical expressions corresponding to the rows 1, 2, …, $I - 1$ in Table 2.5 are false and the logical expression corresponding to the row I in Table 2.5 is true, then the logical expression corresponding to the row I in Table 2.6 is true and, therefore, the point under consideration lies into the small triangle I with vertices given in Table 2.7. One can find the relative areas for its vertices in Table 2.8.

Table 2.6 Logical expressions being true if the point under consideration lies into the small triangle for $N = 3$, *table of the second type*

No	$3s_A \in$	$3s_B \in$	$3s_C \in$	No	$3s_A \in$	$3s_B \in$	$3s_C \in$	No	$3s_A \in$	$3s_B \in$	$3s_C \in$
1	(2, 3]	[0, 1]	[0, 1]	4	[1, 2]	[0, 1)	[0, 1)	7	[0, 1)	[1, 2]	[1, 2]
2	[0, 1]	(2, 3]	[0, 1]	5	[0, 1)	[1, 2]	[0, 1)	8	[1, 2]	[0, 1)	[1, 2]
3	[0, 1]	[0, 1]	(2, 3]	6	[0, 1)	[0, 1)	[1, 2]	9	[1, 2]	[1, 2]	[0, 1]

Table 2.7 The vertices of the small triangles for $N = 3$, *table of the third type*

No	A'	B'	C'	No	A'	B'	C'	No	A'	B'	C'
1	300	210	201	4	111	201	210	7	111	021	012
2	120	030	021	5	021	111	120	8	201	111	102
3	102	012	003	6	012	102	111	9	210	120	111

Table 2.8 Formulae for relative areas corresponding to the appropriate vertices of the small triangles for $N = 3$, *table of the fourth type*

No	$s_{A'}$	$s_{B'}$	$s_{C'}$	No	$s_{A'}$	$s_{B'}$	$s_{C'}$	No	$s_{A'}$	$s_{B'}$	$s_{C'}$
1	$3s_A - 2$	$3s_B$	$3s_C$	4	$2 - 3s_A$	$1 - 3s_B$	$1 - 3s_C$	7	$3s_A$	$3s_B - 1$	$3s_C - 1$
2	$3s_A$	$3s_B - 2$	$3s_C$	5	$1 - 3s_A$	$2 - 3s_B$	$1 - 3s_C$	8	$3s_A - 1$	$3s_B$	$3s_C - 1$
3	$3s_A$	$3s_B$	$3s_C - 2$	6	$1 - 3s_A$	$1 - 3s_B$	$2 - 3s_C$	9	$3s_A - 1$	$3s_B - 1$	$3s_C$

Table 2.9 Inequalities for finding the small triangle, in which the point under consideration lies for $N = 4$, *table of the first type*

No	$4s_A$	$4s_B$	$4s_C$	No	$4s_A$	$4s_B$	$4s_C$	No	$4s_A$	$4s_B$	$4s_C$
1	>3	–	–	7	<1	–	>2	13	<1	–	–
2	–	>3	–	8	–	<1	>2	14	–	<1	–
3	–	–	>3	9	>2	<1	–	15	–	–	<1
4	–	<1	<1	10	>2	–	<1	16	–	–	–
5	<1	–	<1	11	–	>2	<1				
6	<1	<1	–	12	<1	>2	–				

2.3.6 *Tables and Algorithms for N = 4*

Let us consider degree $N = 4$. 16 small triangles are numbered in Fig. 2.8.

If all of the logical expressions corresponding to the rows 1, 2, ..., $I - 1$ in Table 2.9 are false and the logical expression corresponding to the row I in Table 2.9 is true, then the logical expression corresponding to the row I in Table 2.10 is true and therefore the point under consideration lies into the small triangle I with vertices given in Table 2.11. One can find the relative areas for its vertices in Table 2.12.

For example, $I = 2$. If

$$(4s_A \leq 1) \wedge (4s_B > 3), \tag{2.121}$$

Table 2.10 Logical expressions being true if the point under consideration lies into the small triangle for $N = 4$, *table of the second type*

No	$4s_A \in$	$4s_B \in$	$4s_C \in$	No	$4s_A \in$	$4s_B \in$	$4s_C \in$	No	$4s_A \in$	$4s_B \in$	$4s_C \in$
1	(3, 4]	[0, 1]	[0, 1]	7	[0, 1)	[1, 2]	(2, 3]	13	[0, 1)	[1, 2]	[1, 2]
2	[0, 1]	(3, 4]	[0, 1]	8	[1, 2]	[0, 1)	(2, 3]	14	[1, 2]	[0, 1)	[1, 2]
3	[0, 1]	[0, 1]	(3, 4]	9	(2, 3]	[0, 1)	[1, 2]	15	[1, 2]	[1, 2]	[0, 1)
4	[2, 3]	[0, 1)	[0, 1)	10	(2, 3]	[1, 2]	[0, 1)	16	[1, 2]	[1, 2]	[1, 2]
5	[0, 1)	[2, 3]	[0, 1)	11	[1, 2]	(2, 3]	[0, 1)				
6	[0, 1)	[0, 1)	[2, 3]	12	[0, 1)	(2, 3]	[1, 2]				

Table 2.11 The vertices of the small triangles for $N = 4$, *table of the third type*

No	A'	B'	C'	No	A'	B'	C'	No	A'	B'	C'
1	400	310	301	7	112	022	013	13	022	112	121
2	130	040	031	8	202	112	103	14	112	202	211
3	103	013	004	9	301	211	202	15	121	211	220
4	211	301	310	10	310	220	211	16	211	121	112
5	031	121	130	11	220	130	121				
6	013	103	112	12	121	031	022				

then

$$(4s_A \in [0, 1]) \wedge (4s_B \in (3, 4]) \wedge (4s_C \in [0, 1]) \quad (2.122)$$

and, therefore, the point under consideration lies into the small triangle with the number 2 and vertices 110, 020, 011 and one can find a value of the piecewise linear interpolation using Eq. 2.123.

$$u\left(\vec{r}\right) = 4s_A\left(\vec{r}\right)u_{110} + \left(4s_B\left(\vec{r}\right) - 3\right)u_{020} + 4s_C\left(\vec{r}\right)u_{011} \quad (2.123)$$

2.3.7 Tables and Algorithms for N = 5

Let us consider degree $N = 5$. 25 small triangles are numbered in Fig. 2.9.

If all of the logical expressions corresponding to the rows 1, 2, …, $I - 1$ in Table 2.13 are false and the logical expression corresponding to the row I in Table 2.13 is true, then the logical expression corresponding to the row I in Table 2.14 is true and, therefore, the point under consideration lies into the small triangle I with vertices given in Table 2.15. One can find the relative areas for its vertices in Table 2.16.

For example, $I = 12$. If all of the logical expressions corresponding to the rows 1, 2, 3, 4, 5, 6, 7, 8, 9, 10, and 11 in Table 2.13 are false, and

Table 2.12 Formulae for relative areas corresponding to the appropriate vertices of the small triangles for $N = 4$, *table of the fourth type*

No	$s_{A'}$	$s_{B'}$	$s_{C'}$	No	$s_{A'}$	$s_{B'}$	$s_{C'}$	No	$s_{A'}$	$s_{B'}$	$s_{C'}$
1	$4s_A - 3$	$4s_B$	$4s_C$	7	$4s_A$	$4s_B - 1$	$4s_C - 2$	13	$1 - 4s_A$	$2 - 4s_B$	$2 - 4s_C$
2	$4s_A$	$4s_B - 3$	$4s_C$	8	$4s_A - 1$	$4s_B$	$4s_C - 2$	14	$2 - 4s_A$	$1 - 4s_B$	$2 - 4s_C$
3	$4s_A$	$4s_B$	$4s_C - 3$	9	$4s_A - 2$	$4s_B$	$4s_C - 1$	15	$2 - 4s_A$	$2 - 4s_B$	$1 - 4s_C$
4	$3 - 4s_A$	$1 - 4s_B$	$1 - 4s_C$	10	$4s_A - 2$	$4s_B - 1$	$4s_C$	16	$4s_A - 1$	$4s_B - 1$	$4s_C - 1$
5	$1 - 4s_A$	$3 - 4s_B$	$1 - 4s_C$	11	$4s_A - 1$	$4s_B - 2$	$4s_C$				
6	$1 - 4s_A$	$1 - 4s_B$	$3 - 4s_C$	12	$4s_A$	$4s_B - 2$	$4s_C - 1$				

Table 2.13 Inequalities for finding the small triangle, in which the point under consideration lies for $N = 5$, *table of the first type*

No	$5s_A$	$5s_B$	$5s_C$	No	$5s_A$	$5s_B$	$5s_C$	No	$5s_A$	$5s_B$	$5s_C$
1	>4	–	–	10	–	>3	<1	19	<1	–	>2
2	–	>4	–	11	<1	–	>3	20	>2	<1	–
3	–	–	>4	12	>3	<1	–	21	–	>2	<1
4	–	<1	<1	13	>2	>2	–	22	>2	–	–
5	<1	–	<1	14	–	>2	>2	23	–	>2	–
6	<1	<1	–	15	>2	–	>2	24	–	–	>2
7	>3	–	<1	16	>2	–	<1	25	–	–	–
8	<1	>3	–	17	<1	>2	–				
9	–	<1	>3	18	–	<1	>2				

Table 2.14 Logical expressions being true if the point under consideration lies into the small triangle for $N = 5$, *table of the second type*

No	$5s_A \in$	$5s_B \in$	$5s_C \in$	No	$5s_A \in$	$5s_B \in$	$5s_C \in$	No	$5s_A \in$	$5s_B \in$	$5s_C \in$
1	(4, 5]	[0, 1]	[0, 1]	10	[1, 2]	(3, 4]	[0, 1)	19	[0, 1)	[1, 2]	(2, 3]
2	[0, 1]	(4, 5]	[0, 1]	11	[0, 1)	[1, 2]	(3, 4]	20	(2, 3]	[0, 1)	[1, 2]
3	[0, 1]	[0, 1]	(4, 5]	12	(3, 4]	[0, 1)	[1, 2]	21	[1, 2]	(2, 3]	[0, 1)
4	[3, 4]	[0, 1)	[0, 1)	13	(2, 3]	(2, 3]	[0, 1]	22	(2, 3]	[1, 2]	[1, 2]
5	[0, 1)	[3, 4]	[0, 1)	14	[0, 1]	(2, 3]	(2, 3]	23	[1, 2]	(2, 3]	[1, 2]
6	[0, 1)	[0, 1)	[3, 4]	15	(2, 3]	[0, 1]	(2, 3]	24	[1, 2]	[1, 2]	(2, 3]
7	(3, 4]	[1, 2]	[0, 1)	16	(2, 3]	[1, 2]	[0, 1)	25	[1, 2]	[1, 2]	[1, 2]
8	[0, 1)	(3, 4]	[1, 2]	17	[0, 1)	(2, 3]	[1, 2]				
9	[1, 2]	[0, 1)	(3, 4]	18	[1, 2]	[0, 1)	(2, 3]				

Table 2.15 The vertices of the small triangles for $N = 5$, *table of the third type*

No	*A'*	*B'*	*C'*	No	*A'*	*B'*	*C'*	No	*A'*	*B'*	*C'*
1	500	410	401	10	230	140	131	19	023	113	122
2	140	050	041	11	113	023	014	20	212	302	311
3	104	014	005	12	401	311	302	21	131	221	230
4	311	401	410	13	320	230	221	22	311	221	212
5	041	131	140	14	122	032	023	23	221	131	122
6	014	104	113	15	302	212	203	24	212	122	113
7	410	320	311	16	221	311	320	25	122	212	221
8	131	041	032	17	032	122	131				
9	203	113	104	18	113	203	212				

Table 2.16 Formulae for relative areas corresponding to the appropriate vertices of the small triangles for $N = 5$, *table of the fourth type*

No	$s_{A'}$	$s_{B'}$	$s_{C'}$	No	$s_{A'}$	$s_{B'}$	$s_{C'}$	No	$s_{A'}$	$s_{B'}$	$s_{C'}$
1	$5s_A - 4$	$5s_B$	$5s_C$	10	$5s_A - 1$	$5s_B - 3$	$5s_C$	19	$1 - 5s_A$	$2 - 5s_B$	$3 - 5s_C$
2	$5s_A$	$5s_B - 4$	$5s_C$	11	$5s_A$	$5s_B - 1$	$5s_C - 3$	20	$3 - 5s_A$	$1 - 5s_B$	$2 - 5s_C$
3	$5s_A$	$5s_B$	$5s_C - 4$	12	$5s_A - 3$	$5s_B$	$5s_C - 1$	21	$2 - 5s_A$	$3 - 5s_B$	$1 - 5s_C$
4	$4 - 5s_A$	$1 - 5s_B$	$1 - 5s_C$	13	$5s_A - 2$	$5s_B - 2$	$5s_C$	22	$5s_A - 2$	$5s_B - 1$	$5s_C - 1$
5	$1 - 5s_A$	$4 - 5s_B$	$1 - 5s_C$	14	$5s_A$	$5s_B - 2$	$5s_C - 2$	23	$5s_A - 1$	$5s_B - 2$	$5s_C - 1$
6	$1 - 5s_A$	$1 - 5s_B$	$4 - 5s_C$	15	$5s_A - 2$	$5s_B$	$5s_C - 2$	24	$5s_A - 1$	$5s_B - 1$	$5s_C - 2$
7	$5s_A - 3$	$5s_B - 1$	$5s_C$	16	$3 - 5s_A$	$2 - 5s_B$	$1 - 5s_C$	25	$2 - 5s_A$	$2 - 5s_B$	$2 - 5s_C$
8	$5s_A$	$5s_B - 3$	$5s_C - 1$	17	$1 - 5s_A$	$3 - 5s_B$	$2 - 5s_C$				
9	$5s_A - 1$	$5s_B$	$5s_C - 3$	18	$2 - 5s_A$	$1 - 5s_B$	$3 - 5s_C$				

$$(5s_A > 3) \wedge (5s_B < 1)\,, \tag{2.124}$$

then

$$(5s_A \in (3,4]) \wedge (5s_B \in [0,1)) \wedge (5s_C \in [1,2]) \tag{2.125}$$

and, therefore, the point under consideration lies into the small triangle with the number 12 and vertices 401, 311, 302 and one can find a value of the piecewise linear interpolation using Eq. 2.126.

$$u\left(\vec{r}\right) = \left(5s_A\left(\vec{r}\right) - 3\right)u_{401} + 5s_B\left(\vec{r}\right)u_{311} + \left(5s_C\left(\vec{r}\right) - 1\right)u_{302} \tag{2.126}$$

2.4 Several Approaches for Hybrid Interpolation

In this section, several approaches for the hybrid interpolation on the unstructured triangular grids [1] are considered. Using algorithms for the piecewise linear interpolation discussed in Sect. 2.3, one can obtain the hierarchical nested unstructured triangular grids described in Sect. 2.4.1. In Sect. 2.4.2, the recurrent formulae for recalculation of the local reference points' indices to the global reference points' indices are discussed. In Sect. 2.4.3, an example of hybridization called the hybrid parabolic–linear interpolation is considered. In Sect. 2.4.4, another example of hybridization based on a limiter is discussed. In Sect. 2.4.5, an example of approach based on both hierarchical nested unstructured grids and hybridization called parabolic interpolation on the reference points for interpolation of fourth order is offered.

2.4.1 Hierarchical Nested Unstructured Triangular Grids

Using algorithms for the piecewise linear interpolation discussed in Sect. 2.3, one can obtain the hybrid hierarchical unstructured grids. The application of these types of grids allows to avoid the resources' spending like time of software execution for the detailed unstructured triangular grid. This result is achieved due to the analytical expressions given in Sect. 2.3 and allows to understand, in which triangle the point under consideration lies without using the search algorithms. One can use the hierarchical nested unstructured triangular grids applying the following algorithm for the point under consideration $\vec{r}$:

Step 0. Firstly, the relative areas $s_A\left(\vec{r}\right)$, $s_B\left(\vec{r}\right)$, and $s_C\left(\vec{r}\right)$ corresponding to the big triangle ABC should be calculated using Eqs. 2.6–2.10, respectively.

Step 1. For the first level of hybridization with N_1 using the algorithms discussed in Sect. 2.3, one can find the reference points associated with the big triangle ABC

forming the small triangle $A_1B_1C_1$ and calculate the corresponding relative areas $s_{A,1}(s_A)$, $s_{B,1}(s_B)$, and $s_{C,1}(s_C)$.

Step 2. For the second level of hybridization with N_2 using the algorithms discussed in Sect. 2.3, one can find the reference points associated with the triangle $A_1B_1C_1$ forming the small triangle $A_2B_2C_2$ and calculate the corresponding relative areas $s_{A,2}\left(s_{A,1}\right)$, $s_{B,2}\left(s_{B,1}\right)$, and $s_{C,2}\left(s_{C,1}\right)$.

Step 3. For the third level of hybridization with N_3 using the algorithms discussed in Sect. 2.3, one can find the reference points associated with the triangle $A_2B_2C_2$ forming the small triangle $A_3B_3C_3$ and calculate the corresponding relative areas $s_{A,3}\left(s_{A,2}\right)$, $s_{B,3}\left(s_{B,2}\right)$, and $s_{C,3}\left(s_{C,2}\right)$.

Step k. For the k level of hybridization with N_k using the algorithms discussed in Sect. 2.3, one can find the reference points associated with the triangle $A_{k-1}B_{k-1}C_{k-1}$ forming the small triangle $A_kB_kC_k$ and calculate the corresponding relative areas $s_{A,k}\left(s_{A,k-1}\right)$, $s_{B,k}\left(s_{B,k-1}\right)$, and $s_{C,k}\left(s_{C,k-1}\right)$.

Note that one can find the coordinates of the reference point *abc* associated with the triangle *ABC* using Eq. 2.127.

$$\vec{r}_{abc} = \frac{a}{N}\left(\vec{r}_A - \vec{r}_C\right) + \frac{b}{N}\left(\vec{r}_B - \vec{r}_C\right) + \vec{r}_C \tag{2.127}$$

Figure 2.10 shows a visualization of this algorithm.

2.4.2 *Formulae for Recalculation from Local to Global Indices*

In order to diminish the amount of calculations, the values of radius-vectors corresponding to the reference point into the big triangle *ABC* of the hierarchical nested unstructured triangular grid are defined using the recurrent formulae for recalculation from local reference points' indices to global reference points' indices. One can use this recurrent formulae before calculations and save the non-recurrent formulae for recalculation from local to global reference points' indices as a table and then use Eq. 2.127 with global three-digital index *abc* and $N = N_1N_2 \ldots N_K$.

There are $\frac{(N+1)(N+2)}{2}$ reference points into the triangle if the appropriate order is equal to N. One can determine the rule for recalculation from serial reference points' index $I \in \left[1, \frac{(N+1)(N+2)}{2}\right]$ into three-digital index *abc* introduced in Sect. 2.2 (Eq. 2.128) and vice versa (Eq. 2.129).

$$I \in \left[1, \frac{(N+1)(N+2)}{2}\right] \mapsto \{a(I), b(I), c(I)\} \tag{2.128}$$

$$\{a \in [0, N], b \in [0, N], c \in [0, N], a + b + c = N\} \mapsto I(a, b, c) \tag{2.129}$$

The steps of this algorithm are mentioned below.

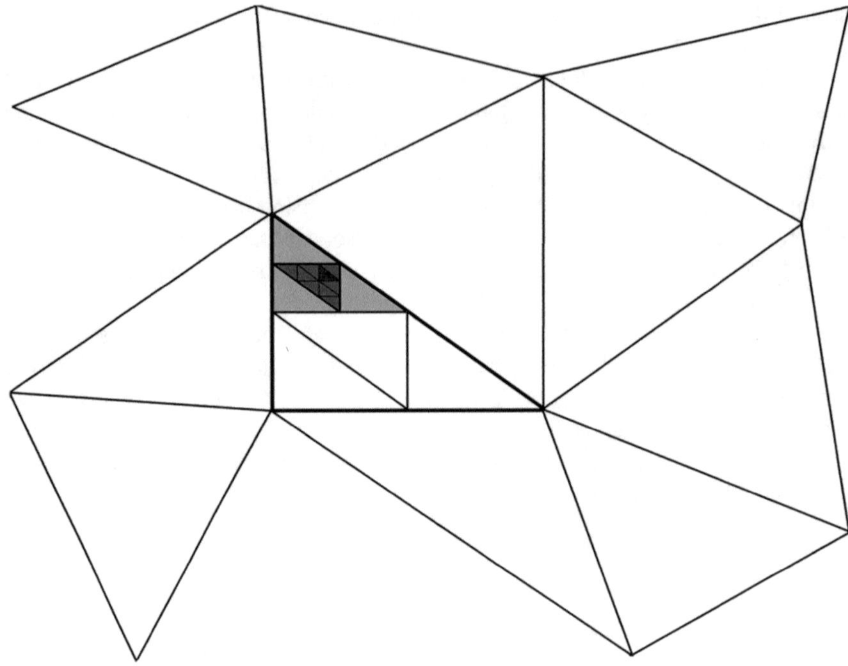

Fig. 2.10 Use of the hierarchical nested unstructured triangular grid

Step 1. N_1^2 small triangles $A_1B_1C_1$ with the serial indices $J_1 \in \left[1, N_1^2\right]$ are known into the initial triangle ABC. The set of the three-digital indices is defined as:

$$a\,(a,\,J_1) \quad b\,(a,\,J_1) \quad c\,(a,\,J_1) \tag{2.130}$$

$$a\,(b,\,J_1) \quad b\,(b,\,J_1) \quad c\,(b,\,J_1) \tag{2.131}$$

$$a\,(c,\,J_1) \quad b\,(c,\,J_1) \quad c\,(c,\,J_1) \tag{2.132}$$

relatively the initial triangle ABC for all of the three vertices A_1 (Eq. 2.130), B_1 (Eq. 2.131), and C_1 (Eq. 2.132) of each of the triangles $A_1B_1C_1$ with the serial indices $J_1 \in \left[1, N_1^2\right]$ are known and represented into the tables of the third type in Sect. 2.3.

Step 2.0. All steps starting from Step 2 are carried out for each of the small triangles $A_1B_1C_1$ with the serial indices $J_1 \in \left[1, N_1^2\right]$.

Step 2.1. For each of the small triangles $A_1B_1C_1$ with the serial indices $J_1 \in \left[1, N_1^2\right]$, the local indices of $\frac{(N_2+1)(N_2+2)}{2}$ reference points into the small triangle $A_1B_1C_1$ are known. For each of these reference points with the serial indices $I_2 \in \left[1, \frac{(N_2+1)(N_2+2)}{2}\right]$, one can calculate the three-digital indices $a_1\,(I_2)$, $b_1\,(I_2)$, and $c_1\,(I_2)$ relatively the small triangle $A_1B_1C_1$ with the serial index $J_1 \in \left[1, N_1^2\right]$ using Eq. 2.128 for N_2.

Thus, for each of these $\frac{(N_2+1)(N_2+2)}{2}$ reference points with the serial indices I_2 one can calculate new three-digital indices relatively the initial triangle ABC using Eqs. 2.133–2.135.

$$a(I_2, J_1) = a_1(I_2)\, a(a, J_1) + b_1(I_2)\, a(b, J_1) + c_1(I_2)\, a(c, J_1) \tag{2.133}$$

$$b(I_2, J_1) = a_1(I_2)\, b(a, J_1) + b_1(I_2)\, b(b, J_1) + c_1(I_2)\, b(c, J_1) \tag{2.134}$$

$$c(I_2, J_1) = a_1(I_2)\, c(a, J_1) + b_1(I_2)\, c(b, J_1) + c_1(I_2)\, c(c, J_1) \tag{2.135}$$

Step 2.2 Into each of the small triangles $A_1B_1C_1$ with the serial indices $J_1 \in [1, N_1^2]$, N_2^2 small triangles $A_2B_2C_2$ with the serial indices $J_2 \in [1, N_2^2]$ are known. The set of the three-digital indices:

$$a_1(a, J_2) \quad b_1(a, J_2) \quad c_1(a, J_2) \tag{2.136}$$

$$a_1(b, J_2) \quad b_1(b, J_2) \quad c_1(b, J_2) \tag{2.137}$$

$$a_1(c, J_2) \quad b_1(c, J_2) \quad c_1(c, J_2) \tag{2.138}$$

relatively the small triangle $A_1B_1C_1$ with the serial index $J_1 \in [1, N_1^2]$ for all of the three vertices A_2 (Eq. 2.136), B_2 (Eq. 2.137), and C_2 (Eq. 2.138) of each of the triangles $A_2B_2C_2$ with the serial indices $J_2 \in [1, N_2^2]$ independent on J_1 are known and represented into the tables of the third type in Sect. 2.3.

Thus, one can find the set of the serial indices $I_2 \in \left[1, \frac{(N_2+1)(N_2+2)}{2}\right]$ for all of the three vertices A_2 (Eq. 2.139), B_2 (Eq. 2.140), and C_2 (Eq. 2.141) of each of the triangles $A_2B_2C_2$ with the serial indices $J_2 \in [1, N_2^2]$ using Eq. 2.129 for N_2.

$$I_2(a, J_2) = I_2(a_1(a, J_2), b_1(a, J_2), c_1(a, J_2)) \tag{2.139}$$

$$I_2(b, J_2) = I_2(a_1(b, J_2), b_1(b, J_2), c_1(b, J_2)) \tag{2.140}$$

$$I_2(c, J_2) = I_2(a_1(c, J_2), b_1(c, J_2), c_1(c, J_2)) \tag{2.141}$$

Therefore, using calculations carries out during Step 2.1 the set of the three-digital indices:

$$a(a, J_1, J_2) \quad b(a, J_1, J_2) \quad c(a, J_1, J_2) \tag{2.142}$$

$$a(b, J_1, J_2) \quad b(b, J_1, J_2) \quad c(b, J_1, J_2) \tag{2.143}$$

$$a(c, J_1, J_2) \quad b(c, J_1, J_2) \quad c(c, J_1, J_2) \tag{2.144}$$

relatively the initial triangle ABC for all of the three vertices A_2 (Eq. 2.142), B_2 (Eq. 2.143), and C_2 (Eq. 2.144) of each of the triangles $A_2B_2C_2$ with the serial indices $J_2 \in [1, N_2^2]$ are known. For example, using Eq. 2.133 one can find:

$$b(a, J_1, J_2) = b(I_2(a, J_2), J_1). \tag{2.145}$$

Step 3.0. All steps starting from Step 3 are carried out for each of the small triangles $A_2B_2C_2$ with the serial indices $J_2 \in \left[1, N_2^2\right]$.

Step 3.1. For each of the small triangles $A_2B_2C_2$ with the serial indices $J_2 \in \left[1, N_2^2\right]$, the local indices of $\frac{(N_3+1)(N_3+2)}{2}$ reference points into the small triangle $A_2B_2C_2$ are known. For each of these reference points with the serial indices $I_3 \in \left[1, \frac{(N_3+1)(N_3+2)}{2}\right]$, one can calculate the three-digital indices $a_2\left(I_3\right)$, $b_2\left(I_3\right)$, and $c_2\left(I_3\right)$ relatively the small triangle $A_2B_2C_2$ with the serial index $J_2 \in \left[1, N_2^2\right]$ using Eq. 2.128 for N_3.

Thus, for each of these $\frac{(N_3+1)(N_3+2)}{2}$ reference points with the serial indices I_3 one can calculate new three-digital indices relatively the initial triangle ABC using Eqs. 2.146–2.148.

$$a\left(I_3, J_1, J_2\right) = a_2\left(I_3\right) a\left(a, J_1, J_2\right) + b_2\left(I_3\right) a\left(b, J_1, J_2\right) + c_2\left(I_3\right) a\left(c, J_1, J_2\right) \tag{2.146}$$

$$b\left(I_3, J_1, J_2\right) = a_2\left(I_3\right) b\left(a, J_1, J_2\right) + b_2\left(I_3\right) b\left(b, J_1, J_2\right) + c_2\left(I_3\right) b\left(c, J_1, J_2\right) \tag{2.147}$$

$$c\left(I_3, J_1, J_2\right) = a_2\left(I_3\right) c\left(a, J_1, J_2\right) + b_2\left(I_3\right) c\left(b, J_1, J_2\right) + c_2\left(I_3\right) c\left(c, J_1, J_2\right) \tag{2.148}$$

Step 3.2. Into each of the small triangles $A_2B_2C_2$ with the serial indices $J_2 \in \left[1, N_2^2\right]$, N_3^2 small triangles $A_3B_3C_3$ with the serial indices $J_3 \in \left[1, N_3^2\right]$ are known. The set of the three-digital indices:

$$a_2\left(a, J_3\right) \quad b_2\left(a, J_3\right) \quad c_2\left(a, J_3\right) \tag{2.149}$$

$$a_2\left(b, J_3\right) \quad b_2\left(b, J_3\right) \quad c_2\left(b, J_3\right) \tag{2.150}$$

$$a_2\left(c, J_3\right) \quad b_2\left(c, J_3\right) \quad c_2\left(c, J_3\right) \tag{2.151}$$

relatively the small triangle $A_2B_2C_2$ with the serial index $J_2 \in \left[1, N_2^2\right]$ for all of the three vertices A_3 (Eq. 2.149), B_3 (Eq. 2.150), and C_3 (Eq. 2.151) of each of the triangles $A_3B_3C_3$ with the serial indices $J_3 \in \left[1, N_3^2\right]$ independent on J_1 and J_2 are known and represented into the tables of the third type in Sect. 2.3.

Thus, one can find the set of the serial indices $I_3 \in \left[1, \frac{(N_3+1)(N_3+2)}{2}\right]$ for all of the three vertices A_3 (Eq. 2.152), B_3 (Eq. 2.153), and C_3 (Eq. 2.154) of each of the triangles $A_3B_3C_3$ with the serial indices $J_3 \in \left[1, N_3^2\right]$ using Eq. 2.129 for N_3.

$$I_3\left(a, J_3\right) = I_3\left(a_2\left(a, J_3\right), b_2\left(a, J_3\right), c_2\left(a, J_3\right)\right) \tag{2.152}$$

$$I_3\left(b, J_3\right) = I_3\left(a_2\left(b, J_3\right), b_2\left(b, J_3\right), c_2\left(b, J_3\right)\right) \tag{2.153}$$

$$I_3\left(c, J_3\right) = I_3\left(a_2\left(c, J_3\right), b_2\left(c, J_3\right), c_2\left(c, J_3\right)\right) \tag{2.154}$$

Therefore, using calculations carries out during Step 3.1 the set of the three-digital indices:

$$a\left(a, J_1, J_2, J_3\right) \quad b\left(a, J_1, J_2, J_3\right) \quad c\left(a, J_1, J_2, J_3\right) \tag{2.155}$$

$$a\left(b, J_1, J_2, J_3\right) \quad b\left(b, J_1, J_2, J_3\right) \quad c\left(b, J_1, J_2, J_3\right) \tag{2.156}$$

$$a\left(c, J_1, J_2, J_3\right) \quad b\left(c, J_1, J_2, J_3\right) \quad c\left(c, J_1, J_2, J_3\right) \tag{2.157}$$

relatively the initial triangle ABC for all of the three vertices A_3 (Eq. 2.155), B_3 (Eq. 2.156), and C_3 (Eq. 2.157) of each of the triangles $A_3B_3C_3$ with the serial indices $J_3 \in \left[1, N_3^2\right]$ are known. For example, using Eq. 2.146 one can find:

$$b\left(a, J_1, J_2, J_3\right) = b\left(I_3\left(a, J_3\right), J_1, J_2\right). \tag{2.158}$$

Step k.0. All steps starting with k are carried out for each of the small triangles $A_{k-1}B_{k-1}C_{k-1}$ with the serial indices $J_{k-1} \in \left[1, N_{k-1}^2\right]$.

Step k.1. For each of the small triangles $A_{k-1}B_{k-1}C_{k-1}$ with the serial indices $J_{k-1} \in \left[1, N_{k-1}^2\right]$, the local indices of $\frac{(N_k+1)(N_k+2)}{2}$ reference points into the small triangle $A_{k-1}B_{k-1}C_{k-1}$ are known. For each of these reference points with the serial indices $I_k \in \left[1, \frac{(N_k+1)(N_k+2)}{2}\right]$, one can calculate the three-digital indices $a_{k-1}\left(I_k\right)$, $b_{k-1}\left(I_k\right)$, and $c_{k-1}\left(I_k\right)$ relatively the small triangle $A_{k-1}B_{k-1}C_{k-1}$ with the serial index $J_{k-1} \in \left[1, N_{k-1}^2\right]$ using Eq. 2.128 for N_k.

Thus, for each of these $\frac{(N_k+1)(N_k+2)}{2}$ reference points with the serial indices I_k one can calculate new three-digital indices relatively the initial triangle ABC using Eqs. 2.159–2.161.

$$\begin{aligned} a\left(I_k, J_1, J_2, \ldots, J_{k-1}\right) = {} & a_{k-1}\left(I_k\right) a\left(a, J_1, J_2, \ldots, J_{k-1}\right) \\ & + b_{k-1}\left(I_k\right) a\left(b, J_1, J_2, \ldots, J_{k-1}\right) \\ & + c_{k-1}\left(I_k\right) a\left(c, J_1, J_2, \ldots, J_{k-1}\right) \end{aligned} \tag{2.159}$$

$$\begin{aligned} b\left(I_k, J_1, J_2, \ldots, J_{k-1}\right) = {} & a_{k-1}\left(I_k\right) b\left(a, J_1, J_2, \ldots, J_{k-1}\right) \\ & + b_{k-1}\left(I_k\right) b\left(b, J_1, J_2, \ldots, J_{k-1}\right) \\ & + c_{k-1}\left(I_k\right) b\left(c, J_1, J_2, \ldots, J_{k-1}\right) \end{aligned} \tag{2.160}$$

$$\begin{aligned} c\left(I_k, J_1, J_2, \ldots, J_{k-1}\right) = {} & a_{k-1}\left(I_k\right) c\left(a, J_1, J_2, \ldots, J_{k-1}\right) \\ & + b_{k-1}\left(I_k\right) c\left(b, J_1, J_2, \ldots, J_{k-1}\right) \\ & + c_{k-1}\left(I_k\right) c\left(c, J_1, J_2, \ldots, J_{k-1}\right) \end{aligned} \tag{2.161}$$

Step k.2. Into each of the small triangles $A_{k-1}B_{k-1}C_{k-1}$ with the serial indices $J_{k-1} \in \left[1, N_{k-1}^2\right]$, N_k^2 small triangles $A_kB_kC_k$ with the serial indices $J_k \in \left[1, N_k^2\right]$ are known. The set of the three-digital indices:

$$a_{k-1}\left(a, J_k\right) \quad b_{k-1}\left(a, J_k\right) \quad c_{k-1}\left(a, J_k\right) \tag{2.162}$$

$$a_{k-1}\left(b, J_k\right) \quad b_{k-1}\left(b, J_k\right) \quad c_{k-1}\left(b, J_k\right) \tag{2.163}$$

$$a_{k-1}(c, J_k) \quad b_{k-1}(c, J_k) \quad c_{k-1}(c, J_k) \tag{2.164}$$

relatively the small triangle $A_{k-1}B_{k-1}C_{k-1}$ with the serial index $J_{k-1} \in \left[1, N_{k-1}^2\right]$ for all of the three vertices A_k (Eq. 2.162), B_k (Eq. 2.163), and C_k (Eq. 2.164) of each of the triangles $A_k B_k C_k$ with the serial indices $J_k \in \left[1, N_k^2\right]$ independent on J_1, J_2,..., J_{k-1} are known and represented into the tables of the third type in Sect. 2.3.

Thus, one can find the set of the serial indices $I_k \in \left[1, \frac{(N_k+1)(N_k+2)}{2}\right]$ for all of the three vertices A_k (Eq. 2.165), B_k (Eq. 2.166), and C_k (Eq. 2.167) of each of the triangles $A_k B_k C_k$ with the serial indices $J_k \in \left[1, N_k^2\right]$ using Eq. 2.129 for N_k.

$$I_k(a, J_k) = I_k(a_{k-1}(a, J_k), b_{k-1}(a, J_k), c_{k-1}(a, J_k)) \tag{2.165}$$

$$I_k(b, J_k) = I_k(a_{k-1}(b, J_k), b_{k-1}(b, J_k), c_{k-1}(b, J_k)) \tag{2.166}$$

$$I_k(c, J_k) = I_k(a_{k-1}(c, J_k), b_{k-1}(c, J_k), c_{k-1}(c, J_k)) \tag{2.167}$$

Therefore, using calculations carries out during Step $k.1$ the set of the three-digital indices:

$$a(a, J_1, J_2, \ldots, J_k) \quad b(a, J_1, J_2, \ldots, J_k) \quad c(a, J_1, J_2, \ldots, J_k) \tag{2.168}$$

$$a(b, J_1, J_2, \ldots, J_k) \quad b(b, J_1, J_2, \ldots, J_k) \quad c(b, J_1, J_2, \ldots, J_k) \tag{2.169}$$

$$a(c, J_1, J_2, \ldots, J_k) \quad b(c, J_1, J_2, \ldots, J_k) \quad c(c, J_1, J_2, \ldots, J_k) \tag{2.170}$$

relatively the initial triangle ABC for all of the three vertices A_k (Eq. 2.168), B_k (Eq. 2.169), and C_k (Eq. 2.170) of each of the triangles $A_k B_k C_k$ with the serial indices $J_k \in \left[1, N_k^2\right]$ are known. For example, using Eq. 2.159 one can find:

$$b(a, J_1, J_2, \ldots, J_k) = b(I_k(a, J_k), J_1, J_2 \ldots, J_{k-1}). \tag{2.171}$$

Step $K-1.0$. All steps starting with $K-1$ are carried out for each of the small triangles $A_{K-2}B_{K-2}C_{K-2}$ with the serial indices $J_{K-2} \in \left[1, N_{K-2}^2\right]$.

Step $K-1.1$. For each of the small triangles $A_{K-2}B_{K-2}C_{K-2}$ with the serial indices $J_{K-2} \in \left[1, N_{K-2}^2\right]$, the local indices of $\frac{(N_{K-1}+1)(N_{K-1}+2)}{2}$ reference points into the small triangle $A_{K-2}B_{K-2}C_{K-2}$ are known. For each of these reference points with the serial indices $I_{K-1} \in \left[1, \frac{(N_{K-1}+1)(N_{K-1}+2)}{2}\right]$, one can calculate the three-digital indices $a_{K-2}(I_{K-1})$, $b_{K-2}(I_{K-1})$, and $c_{K-2}(I_{K-1})$ relatively the small triangle $A_{K-2}B_{K-2}C_{K-2}$ with the index $J_{K-2} \in \left[1, N_{K-2}^2\right]$ using Eq. 2.128 for N_{K-1}.

Thus, for each of these $\frac{(N_{K-1}+1)(N_{K-1}+2)}{2}$ reference points with the serial indices I_{K-1} one can calculate new three-digital indices relatively the initial triangle ABC using Eqs. 2.172–2.174.

$$\begin{aligned} a(I_K, J_1, J_2, .., J_{K-2}) = a_{K-2}(I_{K-1})\, a(a, J_1, J_2, \ldots, J_{K-2}) \\ + b_{K-2}(I_{K-1})\, a(b, J_1, J_2, \ldots, J_{K-2}) \\ + c_{K-2}(I_{K-1})\, a(c, J_1, J_2, \ldots, J_{K-2}) \end{aligned} \tag{2.172}$$

$$
\begin{aligned}
b\left(I_K, J_1, J_2, .., J_{K-2}\right) = {} & a_{K-2}\left(I_{K-1}\right) b\left(a, J_1, J_2, \ldots, J_{K-2}\right) \\
& + b_{K-2}\left(I_{K-1}\right) b\left(b, J_1, J_2, \ldots, J_{K-2}\right) \\
& + c_{K-2}\left(I_{K-1}\right) b\left(c, J_1, J_2, \ldots, J_{K-2}\right)
\end{aligned} \tag{2.173}
$$

$$
\begin{aligned}
c\left(I_K, J_1, J_2, .., J_{K-2}\right) = {} & a_{K-2}\left(I_{K-1}\right) c\left(a, J_1, J_2, \ldots, J_{K-2}\right) \\
& + b_{K-2}\left(I_{K-1}\right) c\left(b, J_1, J_2, \ldots., J_{K-2}\right) \\
& + c_{K-2}\left(I_{K-1}\right) c\left(c, J_1, J_2, \ldots, J_{K-2}\right)
\end{aligned} \tag{2.174}
$$

Step $K-1.2$. Into each of the small triangles $A_{K-2}B_{K-2}C_{K-2}$ with the serial indices $J_{K-2} \in \left[1, N_{K-2}^2\right]$, N_{K-1}^2 small triangles $A_{K-1}B_{K-1}C_{K-1}$ with the serial indices $J_{K-1} \in \left[1, N_{K-1}^2\right]$ are known. The set of the three-digital indices:

$$a_{K-2}\left(a, J_{K-1}\right) \quad b_{K-2}\left(a, J_{K-1}\right) \quad c_{K-2}\left(a, J_{K-1}\right) \tag{2.175}$$

$$a_{K-2}\left(b, J_{K-1}\right) \quad b_{K-2}\left(b, J_{K-1}\right) \quad c_{K-2}\left(b, J_{K-1}\right) \tag{2.176}$$

$$a_{K-2}\left(c, J_{K-1}\right) \quad b_{K-2}\left(c, J_{K-1}\right) \quad c_{K-2}\left(c, J_{K-1}\right) \tag{2.177}$$

relatively the small triangle $A_{K-2}B_{K-2}C_{K-2}$ with the serial index $J_{K-2} \in \left[1, N_{K-2}^2\right]$ for all of the three vertices A_{K-1} (Eq. 2.175), B_{K-1} (Eq. 2.176), and C_{K-1} (Eq. 2.177) of each of the triangles $A_{K-1}B_{K-1}C_{K-1}$ with the serial indices $J_{K-1} \in \left[1, N_{K-1}^2\right]$ independent on $J_1, J_2, \ldots, J_{K-2}$ are known and represented into the tables of the third type in Sect. 2.3.

Thus, one can find the set of the serial indices $I_{K-1} \in \left[1, \frac{(N_{K-1}+1)(N_{K-1}+2)}{2}\right]$ for all of the three vertices A_{K-1} (Eq. 2.178), B_{K-1} (Eq. 2.179), and C_{K-1} (Eq. 2.180) of each of the triangles $A_{K-1}B_{K-1}C_{K-1}$ with the serial indices $J_{K-1} \in \left[1, N_{K-1}^2\right]$ using Eq. 2.129 for N_{K-1}.

$$I_{K-1}\left(a, J_{K-1}\right) = I_{K-1}\left(a_{K-2}\left(a, J_{K-1}\right), b_{K-2}\left(a, J_{K-1}\right), c_{K-2}\left(a, J_{K-1}\right)\right) \tag{2.178}$$

$$I_{K-1}\left(b, J_{K-1}\right) = I_{K-1}\left(a_{K-2}\left(b, J_{K-1}\right), b_{K-2}\left(b, J_{K-1}\right), c_{K-2}\left(b, J_{K-1}\right)\right) \tag{2.179}$$

$$I_{K-1}\left(c, J_{K-1}\right) = I_{K-1}\left(a_{K-2}\left(c, J_{K-1}\right), b_{K-2}\left(c, J_{K-1}\right), c_{K-2}\left(c, J_{K-1}\right)\right) \tag{2.180}$$

Therefore, using calculations carries out during step $K-1.1$ the set of the three-digital indices:

$$a\left(a, J_1, J_2, \ldots, J_{K-1}\right) \quad b\left(a, J_1, J_2, \ldots, J_{K-1}\right) \quad c\left(a, J_1, J_2, \ldots, J_{K-1}\right) \tag{2.181}$$

$$a\left(b, J_1, J_2, \ldots, J_{K-1}\right) \quad b\left(b, J_1, J_2, \ldots, J_{K-1}\right) \quad c\left(b, J_1, J_2, \ldots, J_{K-1}\right) \tag{2.182}$$

$$a\,(c, J_1, J_2, \ldots, J_{K-1}) \quad b\,(c, J_1, J_2, \ldots, J_{K-1}) \quad c\,(c, J_1, J_2, \ldots, J_{K-1}) \tag{2.183}$$

relatively the initial triangle *ABC* for all of the three vertices A_{K-1} (Eq. 2.181), B_{K-1} (Eq. 2.182), and C_{K-1} (Eq. 2.183) of each of the triangles $A_{K-1}B_{K-1}C_{K-1}$ with the serial indices $J_{K-1} \in \left[1, N_{K-1}^2\right]$ are known. For example, using Eq. 2.172 one can find:

$$b\,(a, J_1, J_2, \ldots, J_{K-1}) = b\,(I_{K-1}\,(a, J_{K-1})\,, J_1, J_2 \ldots, J_{K-2})\,. \tag{2.184}$$

Step *K*.0. All steps *K* are carried out for each of the small triangles $A_{K-1}B_{K-1}C_{K-1}$ with the serial indices $J_{K-1} \in \left[1, N_{K-1}^2\right]$.

Step *K*.1. For each of the small triangles $A_{K-1}B_{K-1}C_{K-1}$ with the serial indices $J_{K-1} \in \left[1, N_{K-1}^2\right]$, the local indices of $\frac{(N_K+1)(N_K+2)}{2}$ reference points into the small triangle $A_{K-1}B_{K-1}C_{K-1}$ are known. For each of these $\frac{(N_K+1)(N_K+2)}{2}$ reference points with the three-digital indices $a_{K-1}b_{K-1}c_{K-1}$ relative the small triangle $A_{K-1}B_{K-1}C_{K-1}$ with the serial index J_{K-1}, one can calculate new three-digital indices relatively the initial triangle *ABC* using Eqs. 2.185–2.187.

$$\begin{aligned} a\,(a_{K-1}b_{K-1}c_{K-1}, J_1, J_2, \ldots, J_{K-1}) = {} & a_{K-1}a\,(a, J_1, J_2, \ldots, J_{K-1}) \\ & + b_{K-1}a\,(b, J_1, J_2, \ldots, J_{K-1}) \\ & + c_{K-1}a\,(c, J_1, J_2, \ldots, J_{K-1}) \end{aligned} \tag{2.185}$$

$$\begin{aligned} b\,(a_{K-1}b_{K-1}c_{K-1}, J_1, J_2, \ldots, J_{K-1}) = {} & a_{K-1}b\,(a, J_1, J_2, \ldots, J_{K-1}) \\ & + b_{K-1}b\,(b, J_1, J_2, \ldots, J_{K-1}) \\ & + c_{K-1}b\,(c, J_1, J_2, \ldots, J_{K-1}) \end{aligned} \tag{2.186}$$

$$\begin{aligned} c\,(a_{K-1}b_{K-1}c_{K-1},, J_1, J_2, \ldots, J_{K-1}) = {} & a_{K-1}c\,(a, J_1, J_2, \ldots, J_{K-1}) \\ & + b_{K-1}c\,(b, J_1, J_2, \ldots, J_{K-1}) \\ & + c_{K-1}c\,(c, J_1, J_2, \ldots, J_{K-1}) \end{aligned} \tag{2.187}$$

Note that the algorithms on Steps 1 and *K* differ from the algorithm on steps 2, 3, …, $K-1$.

Therefore, if the local three-digital index $a_{K-1}b_{K-1}c_{K-1}$ relative the triangle $A_{K-1}B_{K-1}C_{K-1}$ with the serial index J_{K-1} of the reference point under consideration, and the set of the indices of the small triangles $J_1, J_2, \ldots, J_{K-1}, J_k \in N_k^2$, in which this reference point under consideration lies are known, one can find the global three-digital index relative the initial triangle *ABC* using Eqs. 2.185–2.187.

Consider an example: $K = 2$, $N_1 = N_2 = 2$. The indices of the vertices of the triangle with the index $J_1 = 2$ are $110, 020, 011$ relative the initial triangle. Thus, the global index of the reference point with the local index 110 relative the small triangle with the index $J_1 = 2$ will be 130 and can be calculated using Eqs. 2.188–2.190.

$$a = 1 \cdot 1 + 1 \cdot 0 + 0 \cdot 0 = 1 \tag{2.188}$$

$$b = 1 \cdot 1 + 1 \cdot 2 + 0 \cdot 1 = 3 \tag{2.189}$$

$$c = 1 \cdot 0 + 1 \cdot 0 + 0 \cdot 1 = 0 \tag{2.190}$$

2.4.3 Hybrid Parabolic Linear Interpolation

In this Section, a hybrid parabolic linear interpolation method is proposed. This interpolation method combines the monotonicity of a piecewise-linear function in some small triangles with the second order of the quadratic function in other small triangles, where this does not lead to the formation of additional extremes that do not coincide with the reference points. The resulting function ought to remain the continuous, when passing from one triangle to another one. Such hybrid interpolation provides the numerical method with the property of monotonicity and avoids the non-physical oscillations of the solution.

The reference points 200, 020, 002, 110, 011, and 101 for the polynomial interpolation of the second order are used. The algorithm for restoring a monotone function in a grid triangle is as follows:

Step 1. Define the trial function using the normal polynomial interpolation of the second order.

Step 2. Determine the values in the centers of the edges as follows. If the trial function has not extremum on this edge, then take the value of the test function; otherwise use the linear interpolation on the given edge. Therefore, if there are values u_{A} and u_{B} into the vertices of the edge of the grid triangle and the value u_{AB} in the middle of this edge, then one can use the following inequalities. If $u_{\mathrm{AB}} \in [u_{\mathrm{A}}, u_{\mathrm{B}}]$ or $u_{\mathrm{AB}} \in [u_{\mathrm{B}}, u_{\mathrm{A}}]$, then $u_{\mathrm{AB}}^{\mathrm{NEW}} = u_{\mathrm{AB}}$. In another case, $u_{\mathrm{AB}}^{\mathrm{NEW}} = (u_{\mathrm{A}} + u_{\mathrm{B}})/2$.

This procedure should be performed for all the central-edge reference points 110, 011, and 101. If the central-edge reference point is 110, then the vertices of the edge are 200 and 020. If the central-edge reference point is 011, then the vertices of the edge are 020 and 002. If the central-edge reference point is 101, then the vertices of the edge are 200 and 002.

Step 3. Determine the final value of the interpolant using the normal polynomial interpolation of the second order with new values in the centers of the edges defined on Step 2.

2.4.4 Interpolation Using Min–Max Limiter

Algorithm for constructing an interpolant with a limiter on triangular grids based on interpolation by a polynomial of order N involves the following steps:

Step 1. The value of test function to the given point $\vec{r}$ using the polynomial interpolation of order N is determined. Let it be equal to $u_N\left(\vec{r}\right)$.

Step 2. The small triangle, in which the point $\vec{r}$ falls, should be determined.

Step 3. Then $u_N\left(\vec{r}\right)$ should be compared with the minimum m and maximum M values at the vertices of this small triangle, in which the point falls.

Step 3.1. If $m \le u_N\left(\vec{r}\right) \le M$, then the value of interpolant at the point $\vec{r}$ is taking equal to $u_N\left(\vec{r}\right)$.

Step 3.2. If $u_N\left(\vec{r}\right) < m$, then the value of interpolant at the point $\vec{r}$ is equal to m.

Step 3.3. If $u_N\left(\vec{r}\right) > M$, then the value of interpolant at the point $\vec{r}$ is equal to M.

The use of interpolation with a Min-Max limiter makes it possible to eliminate the non-physical oscillations of polynomials arising in the presence of discontinuities in the interpolated functions.

2.4.5 *Parabolic Interpolation on Reference Points for Interpolation of Fourth Order*

Consider a triangle ABC with the area S and 15 reference points 400, 040, 004, 310, 031, 103, 130, 013, 301, 220, 022, 202, 211, 121, and 112 into this triangle for $N = 4$. One can divide this triangle ABC into 4 small triangles with areas $S/4$ using a set of 6 reference points in the Table 2.17 for $N = 2$.

One can use the algorithm discussed in Sect. 2.4.1 in order to understand, in which of the 4 small triangles the point under consideration lies. Then one can apply the formulae for the parabolic interpolation discussed in Sect. 2.2.3 and use the Table 2.18 containing the congruence of three-digital indices for $N = 4$ and three-digital indices for $N = 2$ for each of the 4 small triangles.

Table 2.17 Congruence of three-digital indices of reference points for $N = 2$ in order to divide the initial triangle into 4 small triangles and initial three-digital indices for $N = 4$

Index to divide, $N = 2$	200	020	002	110	011	101
Initial index, $N = 4$	400	040	004	220	022	202

Table 2.18 Congruence of three-digital indices of reference points for $N = 2$ and initial three-digital indices for $N = 4$ for each of the 4 small triangles

Small triangle No	200	020	002	110	011	101
1	400	220	202	310	211	301
2	220	040	022	130	031	121
3	202	022	004	112	013	103
4	022	202	220	112	211	121

2.5 Conclusions

The analytical formulae for high-order interpolation techniques on the unstructured triangular grids, such as the polynomial interpolation, piecewise linear interpolation, hybrid parabolic linear interpolation, interpolation using Min-Max limiter, and parabolic interpolation on reference points for interpolation of fourth order are suggested in this chapter. The cases of order from 1 to 5 inclusively are considered. These interpolation techniques can be used during creation a new unstructured triangular or regular gird instead of previous one as an element of numerical method for finding 2D solutions on the unstructured triangular meshes, as well as during the visualization of some 2D field and images creation or transformation. Also, the hierarchical nested unstructured triangular grids and formulae for recalculation from local to global indices are discussed. This type of grids can be used also as an element of numerical method on the unstructured triangular grids and for the visualization, creation, and transformation of 2D images. These formulae for recalculation indices are used to decrease the amount of calculation and software operation time. In this chapter, one can find a vast amount of analytical expressions ready for use. These analytical expressions and tables help to achieve huge numerical modelling results in a case of the deficiency of hardware resources.

Acknowledgements This work has been performed at Non-state Educational Institution "Educational Scientific and Experimental Center of Moscow Institute of Physics and Technology" and supported by the Russian Science Foundation, grant no. 17-71-20088. This work has been carried out using computing resources of the federal collective usage center Complex for Simulation and Data Processing for Mega-science Facilities at NRC "Kurchatov Institute", http://ckp.nrcki.ru/.

References

1. Favorskaya AV, Muratov MV, Petrov IB, Sannikov IV (2014) Grid-characteristic method on unstructured tetrahedral meshes. Comput Math Math Phys 54(5):837–847
2. Kvasov IE, Leviant VB, Petrov IB (2016) Numerical study of wave propagation in porous media with the use of the grid-characteristic method. Comput Math MathPhys 56(9):1620–1630
3. Favorskaya AV, Petrov IB (2016) Wave responses from oil reservoirs in the Arctic shelf zone. Dokl Earth Sci 466(2):214–217
4. Khokhlov N, Yavich N, Malovichko M, Petrov I (2015) Solution of large-scale seismic modeling problems. Proc Comput Sci 66:191–199
5. Favorskaya AV, Petrov IB (2017) Numerical modeling of dynamic wave effects in rock masses. Dokl Math 95(3):287–290
6. Vassilevski YV, Beklemysheva KA, Grigoriev GK, Kazakov AO, Kulberg NS, Petrov IB, Salamatova VY, Vasyukov AV (2016) Transcranial ultrasound of cerebral vessels in silico: proof of concept. Russ J Numer Anal Math Model 31(5):317–328
7. Beklemysheva KA, Vasyukov AV, Petrov IB (2015) Numerical simulation of dynamic processes in biomechanics using the grid-characteristic method. Comput Math Math Phys 55(8):1346–1355
8. Beklemysheva KA, Danilov AA, Petrov IB, Salamatova VY, Vassilevski YV, Vasyukov AV (2015) Virtual blunt injury of human thorax: age-dependent response of vascular system. Russ J Numer Anal Math Model 30(5):259–268

9. Beklemysheva K, Vasyukov A, Ermakov A, Favorskaya A (2017) Numerical modeling of ultrasound beam forming in elastic medium. Proc Comput Sci 112:1488–1496
10. Petrov I, Vasyukov A, Beklemysheva K, Ermakov A, Favorskaya A (2016) Numerical modeling of non-destructive testing of composites. Proc Comput Sci 96:930–938
11. Ball JA, Gohberg I, Rodman L (1990) Interpolation of rational matrix functions. OT45, Birkhäuser Verlag, Basel, Boston, Berlin
12. Lama RK, Kwon GR (2015) New interpolation method based on combination of Discrete cosine transform and wavelet transform. In: IEEE International Conference Information Networking (ICOIN'2015), pp 363–366
13. Li J, Heap AD (2014) Spatial interpolation methods applied in the environmental sciences: a review. Environ Model Softw 53:173–189
14. Agarwal S, Khade S, Dandawate Y, Khandekar P (2015) Three dimensional image reconstruction using interpolation of distance and image registration. In: IEEE International Conference Computer, Communication and Control (IC4'2015), pp 1–5
15. Dong W, Zhang L, Lukac R, Shi G (2013) Sparse representation based image interpolation with nonlocal autoregressive modeling. IEEE Trans Image Process 22(4):1382–1394
16. Greco L, Cuomo M (2014) An implicit G1 multi patch B-spline interpolation for Kirchhoff-Love space rod. Comput Methods Appl Mech Eng 269:173–197
17. Hecht F, Kuate R (2014) An approximation of anisotropic metrics from higher order interpolation error for triangular mesh adaptation. J Comput Appl Math 258:99–115
18. Wen H, Zhang J, Meng Z, Guo S, Li F, Yang Y (2015) Harmonic estimation using symmetrical interpolation FFT based on triangular self-convolution window. IEEE Trans Ind Inf 11(1):16–26
19. Dell'Accio F, Di Tommaso F, Hormann K (2015) On the approximation order of triangular Shepard interpolation. IMA J Numer Anal 36(1):359–379
20. Wan H, Giorgetta MA, Zangl G, Restelli M, Majewski D, Bonaventura L, Frohlich K, Reinert D, Ripodas P, L. Kornblueh L, Forstner J (2013) The ICON-1.2 hydrostatic atmospheric dynamical core on triangular grids—part I: formulation and performance of the baseline version. Geosci Model Dev 6:735–763
21. Lee DT, Schachter BJ (1980) Two algorithms for constructing a Delaunay triangulation. Int J Comput Inform Sci 9(3):219–242

Chapter 3
Interpolation on Unstructured Tetrahedral Grids

Alena V. Favorskaya

Abstract Analytical expressions for polynomial interpolations of high-orders are discussed in the chapter. Also in this Chapter several approaches for hybrid interpolation are discussed. Using algorithms for the piecewise linear interpolation one can obtain the hierarchical nested unstructured tetrahedral grids. The recurrent formulae for recalculation of the local reference points' indices to the global reference points' indices are discussed in the chapter as well. An example of hybridization called hybrid parabolic—linear interpolation is considered. Another example of hybridization based on a limiter is considered as well. An example of approach based on both hierarchical nested unstructured grids and hybridization called parabolic interpolation on the reference points for interpolation of fourth order is offered in the chapter.

Keywords Interpolation · High-order interpolation · Unstructured grids
Tetrahedral grids · Polynomial interpolation · Piecewise interpolation
Hybrid interpolation · Nested meshes

A. V. Favorskaya (✉)
Non-state Educational Institution "Educational Scientific and Experimental Center of Moscow Institute of Physics and Technology", 9, Institutsky Pereulok St, Dolgoprudny, Moscow Region 141700, Russian Federation
e-mail: aleanera@yandex.ru

A. V. Favorskaya
Moscow Institute of Physics and Technology, Department of Computer Science and Numerical Mathematics, 9, Institutsky Pereulok St, Dolgoprudny, Moscow Region 141700, Russian Federation

A. V. Favorskaya
Scientific Research Institute for System Studies of the Russian Academy of Sciences, 36(1), Nahimovskij Ave, Dolgoprudny, Moscow 117218, Russian Federation

A. V. Favorskaya and I. B. Petrov (eds.), *Innovations in Wave Processes Modelling and Decision Making*, Smart Innovation, Systems and Technologies 90,
https://doi.org/10.1007/978-3-319-76201-2_3

3.1 Introduction

In this Chapter the interpolation on unstructured tetrahedral grids is discussed. The main applications of the interpolation are discussed in Chap. 2. The main aspects of using tetrahedral interpolation as an element of grid-characteristic method [1–10] are discussed in Sect. 2.1 as well.

Due to the question of interpolation is very difficult there are a lot of recent works in this area [11–16]. More detailed review of these works [11–16] is given in Chap. 2. A family of conforming mixed finite elements for linear elasticity on tetrahedral grids using Scott-Zhang interpolation operator is presented by Hu and Zhang [17]. A novel consecutive-interpolation 4-node tetrahedral element for heat transfer analysis was proposed by Nguyen et al. [18]. An interpolation operator on unstructured tetrahedral meshes that satisfies the properties of mass conservation was suggested by Alauzet [19]. A geometric representation of a tetrahedral mesh that is solely based on dihedral angles was presented by Paille et al. [20]. Nikitin et al. provide a description of numerical schemes with the use of adaptive moving grids for nonstationary problems of fluid and gas mechanics and mechanics of a deformable solid in the paper [21].

Tetrahedral grids are topologically qualitatively different from triangular ones. For example, the Delaunay triangulation algorithm [22] does not work for them. Therefore, the analytical expressions presented in this Chapter and Chap. 4 for interpolation on tetrahedral grids are so important.

Notice that in this Chapter only the case of scalar field $u\left(\vec{r}\right)$ is considered but one can use the same expressions for vector field $\vec{u}\left(\vec{r}\right)$ if apply these expressions to the components of this vector field $u_1\left(\vec{r}\right)$, $u_2\left(\vec{r}\right)$, $u_3\left(\vec{r}\right)$, $u_4\left(\vec{r}\right)$, ... one by one. Also notice that all formulae in this Chapter are true for tetrahedrons of arbitrary shape despite on in all figures in this Chapter regular tetrahedrons are drawn. The analogical formulae for triangles of arbitrary shapes are discussed in Chap. 2.

Chapter has the following structure. In Sect. 3.2, a polynomial interpolation on the unstructured tetrahedral grids is presented. In Sect. 3.3, several approaches for hybrid interpolation on the unstructured tetrahedral grids are discussed. Section 3.4 concludes the chapter.

3.2 Polynomial Interpolation

In this section, the analytical formulae for polynomial interpolation on the unstructured tetrahedral grids are considered. Using this type of interpolation, one can obtain the continuous piecewise polynomial field and continuous differentiable in each tetrahedron [1]. The method for obtaining these analytical formulae for any given polynomial degree and list the analytical formulae for degrees from 1 to 5 inclusive is considered.

In Sect. 3.2.1, one can find the description of a way to obtain the analytical expression for the polynomial interpolation. One can find these analytical expressions for the polynomial interpolation on the tetrahedral grids with degree from 1 to 5 inclusively in Sects. 3.2.2, 3.2.3, 3.2.4, 3.2.5, and 3.2.6, respectively.

3.2.1 Obtaining the Analytical Formulae

In order to determine a polynomial field with degree N, which depends on x, y, and z, the values at $\frac{(N+1)(N+2)(N+3)}{6}$ reference points should be known.

The following method of arranging reference points in the tetrahedron is suggested. The planes parallel to the faces of the tetrahedron $ABCD$, which divide each of its edges into N equal parts, are drawn within the tetrahedron. In Figs. 3.1, 3.2, 3.3, 3.4 and 3.5 one can see the faces of tetrahedron $ABCD$ and lines obtained, when the planes cross the corresponding face. On these faces, there are the numbered reference points. These planes divide the tetrahedron into $\frac{N(N^2+2)}{3}$ smaller tetrahedrons similar to it and $\frac{N(N-1)(N+1)}{6}$ octahedrons. The case of octahedron is discussed circumstantially in Sect. 3.3.1.

The interpolation polynomials for finding some scalar field $u\left(\vec{r}\right)$ in the tetrahedron can be written in the form of Eq. 3.1.

$$u\left(\vec{r}\right) = \sum_{i,j,k} u_{ijk} x^i y^j z^k \tag{3.1}$$

Note that values of the field in the reference points are known:

$$u_{abcd} = u\left(\vec{r}_{abcd}\right) = \sum_{i,j,k} u_{ijk} x^i_{abcd} y^j_{abcd} z^k_{abcd}. \tag{3.2}$$

One can write the solution of the system of linear Eq. 3.2 as Eq. 3.3.

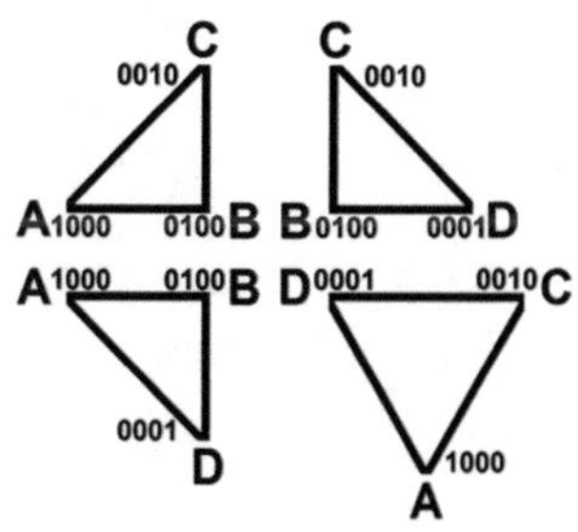

Fig. 3.1 Reference points in the tetrahedron for polynomial interpolation with order 1

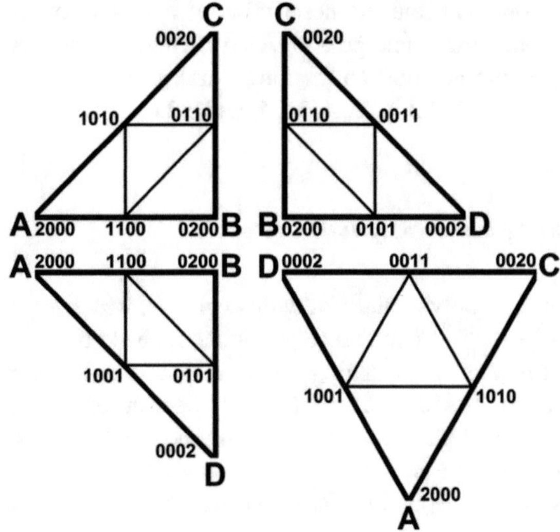

Fig. 3.2 Reference points in the tetrahedron for polynomial interpolation with order 2

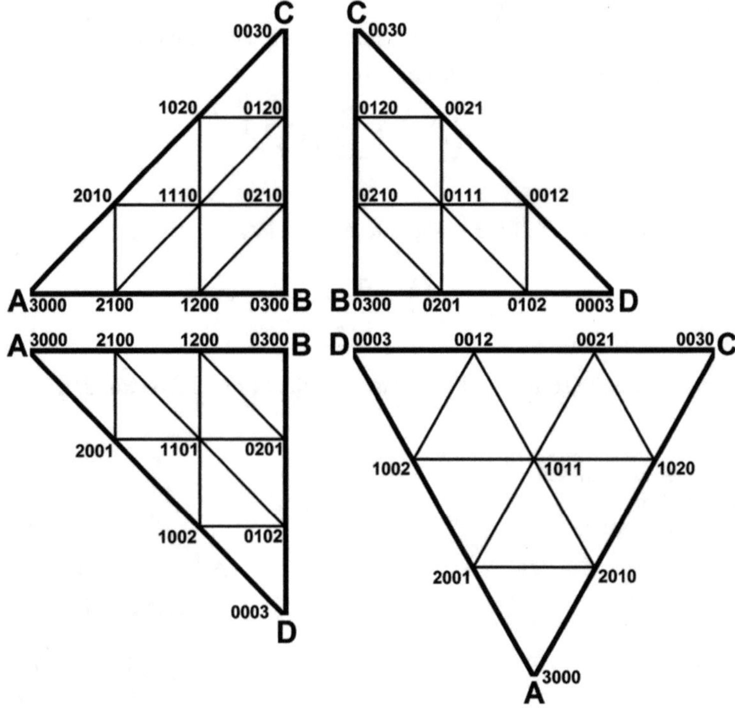

Fig. 3.3 Reference points in the tetrahedron for polynomial interpolation with order 3

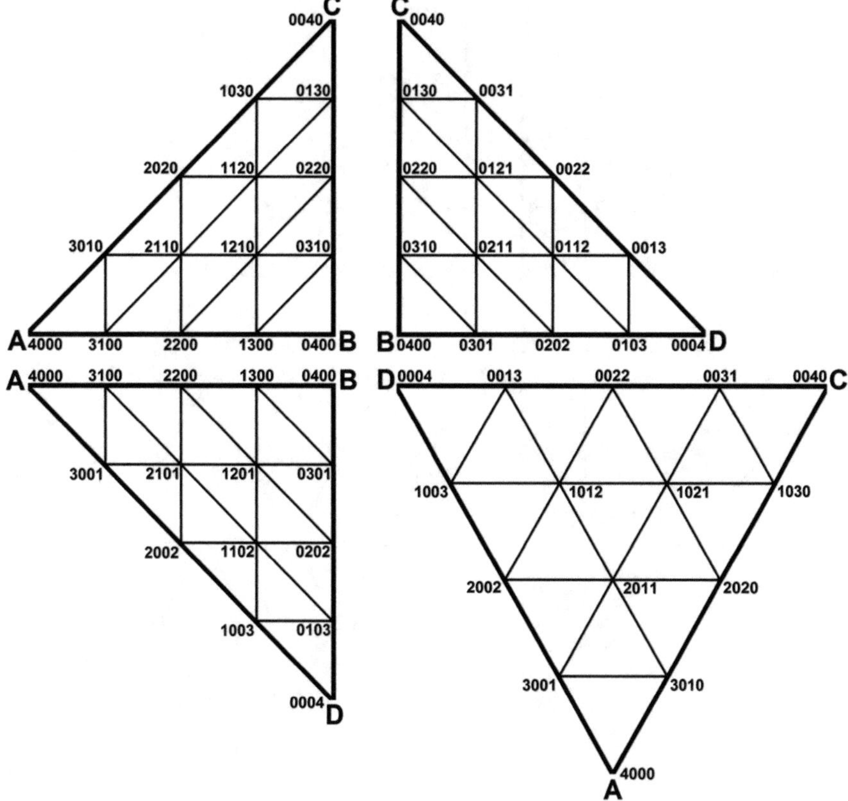

Fig. 3.4 Reference points in the tetrahedron for polynomial interpolation with order 4

$$u_{ijk} = \sum_{a,b,c,d} a_{abcd}^{ijk} u_{abcd} \tag{3.3}$$

Thus, one can determine the weights of the reference points $w_{abcd}\left(\vec{r}\right)$ as follows:

$$w_{abcd}\left(\vec{r}\right) = \sum_{i,j,k} a_{abcd}^{ijk} x^i y^j z^k \tag{3.4}$$

and write the Eq. 3.1 in more applicable form:

$$u\left(\vec{r}\right) = \sum_{a,b,c,d} w_{abcd}\left(\vec{r}\right) u_{abcd}. \tag{3.5}$$

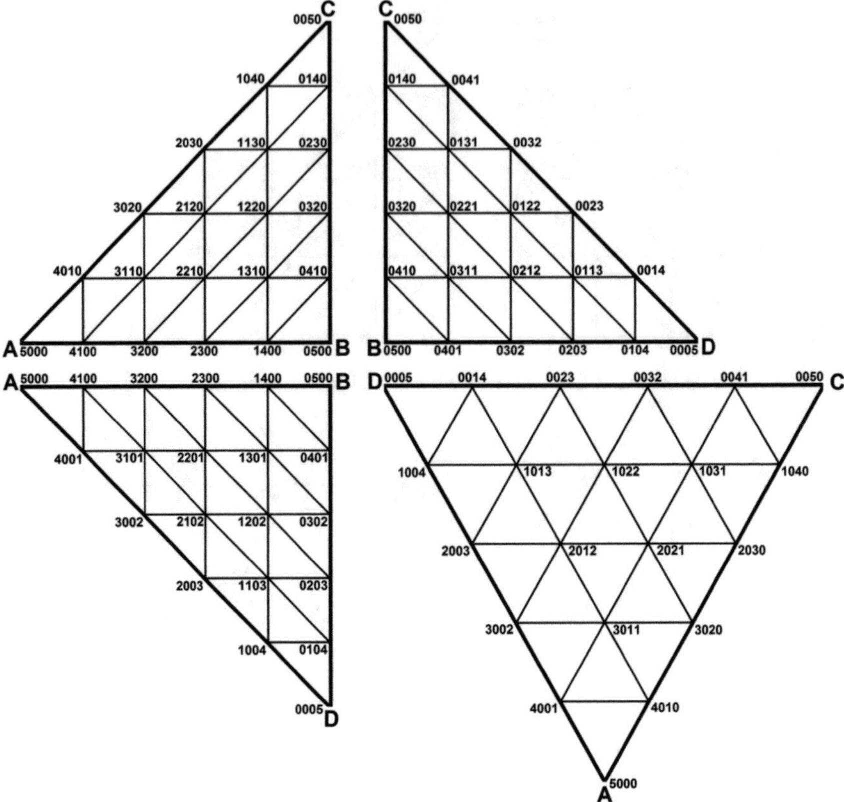

Fig. 3.5 Reference points in the tetrahedron for polynomial interpolation with order 5

The field in the point R with radius-vector $\vec{r}$ into the tetrahedron $ABCD$ should be found. Also, four tetrahedrons $BCDR$, $CDAR$, $DABR$, and $ABCR$ are considered. One can find volumes of these tetrahedrons using Eqs. 3.6–3.10.

$$V_A = \frac{1}{6}\left((\vec{r} - \vec{r}_C) \cdot \left[(\vec{r}_D - \vec{r}_C),\ (\vec{r}_B - \vec{r}_C)\right]\right) \tag{3.6}$$

$$V_B = \frac{1}{6}\left((\vec{r} - \vec{r}_D) \cdot \left[(\vec{r}_C - \vec{r}_D),\ (\vec{r}_A - \vec{r}_D)\right]\right) \tag{3.7}$$

$$V_C = \frac{1}{6}\left((\vec{r} - \vec{r}_B) \cdot \left[(\vec{r}_D - \vec{r}_B),\ (\vec{r}_A - \vec{r}_B)\right]\right) \tag{3.8}$$

$$V_D = \frac{1}{6}\left((\vec{r} - \vec{r}_C) \cdot \left[(\vec{r}_B - \vec{r}_C),\ (\vec{r}_A - \vec{r}_C)\right]\right) \tag{3.9}$$

$$V = \frac{1}{6}\left(\left(\vec{r}_C - \vec{r}_D\right) \cdot \left[\left(\vec{r}_A - \vec{r}_D\right),\ \left(\vec{r}_B - \vec{r}_D\right)\right]\right) \tag{3.10}$$

One can find the relative volumes using Eq. 3.11, where $T = A,\ B,\ C,\ D$.

$$v_T\left(\vec{r}\right) = \frac{V_T\left(\vec{r}\right)}{V} \tag{3.11}$$

The relative volumes have several properties. Their sum is equal to 1 (Eq. 3.12).

$$v_A\left(\vec{r}\right) + v_B\left(\vec{r}\right) + v_C\left(\vec{r}\right) + v_D\left(\vec{r}\right) = 1 \tag{3.12}$$

If the point R is one of the reference points with indices $abcd$, then for any b, any c, and any d the following expression is true:

$$v_A\left(\vec{r}_{abcd}\right) = \frac{a}{N}, \tag{3.13}$$

for any a, any c, and any d the following expression is true:

$$v_B\left(\vec{r}_{abcd}\right) = \frac{b}{N}, \tag{3.14}$$

for any a, any b, and any d the following expression is true:

$$v_C\left(\vec{r}_{abcd}\right) = \frac{c}{N}, \tag{3.15}$$

and for any a, any b, and any c the following expression is true:

$$v_D\left(\vec{r}_{abcd}\right) = \frac{d}{N}. \tag{3.16}$$

The weights $w_{abcd}\left(\vec{r}\right)$ of the reference points can be written by Eq. 3.17.

$$w_{abcd}\left(\vec{r}\right) = \frac{\prod_{i=1}^{N}\left(v_{T_i}\left(\vec{r}\right) - \frac{n_i}{N}\right)}{\prod_{i=1}^{N}\left(v_{T_i}\left(\vec{r}_{abcd}\right) - \frac{n_i}{N}\right)} \tag{3.17}$$

In Eq. 3.17, letter-indices $T_i = A,\ B,\ C,\ D$ and natural numbers n_i should be founded for the following expression to be true:

$$w_{abcd}\left(\vec{r}_{a'b'c'd'}\right) = \delta_{aa'}\delta_{bb'}\delta_{cc'}\delta_{dd'}. \tag{3.18}$$

3.2.2 Weights of Reference Points for N = 1

Consider some example of finding the letter-indices and natural numbers in Eq. 3.17 for the case of $N = 1$. In this case, 4 reference points: 1000, 0100, 0010, and 0001 are. These points are represented in Fig. 3.1.

Firstly the weight $w_{1000}\left(\vec{r}\right)$ using Eq. 3.19 should be found.

$$w_{1000}\left(\vec{r}\right) = \frac{v_{T_1}\left(\vec{r}\right) - n_1}{v_{T_1}\left(\vec{r}_{1000}\right) - n_1} \tag{3.19}$$

Let us consider $\vec{r} = \vec{r}_{0001}$:

$$0 = \frac{v_{T_1}\left(\vec{r}_{0001}\right) - n_1}{v_{T_1}\left(\vec{r}_{1000}\right) - n_1} = \frac{v_A\left(\vec{r}_{0001}\right) - n_1}{v_A\left(\vec{r}_{1000}\right) - n_1} = \frac{0 - n_1}{1 - n_1} = \frac{0 - 0}{1 - 0} = v_A\left(\vec{r}_{0001}\right). \tag{3.20}$$

Thus, the weight in a view of Eq. 3.21 should be found.

$$w_{1000}\left(\vec{r}\right) = v_A\left(\vec{r}\right) \tag{3.21}$$

Similarly, one can find the weights $w_{0100}\left(\vec{r}\right)$, $w_{0010}\left(\vec{r}\right)$, and $w_{0001}\left(\vec{r}\right)$ using Eqs. 3.22–3.24.

$$w_{0100}\left(\vec{r}\right) = v_B\left(\vec{r}\right) \tag{3.22}$$

$$w_{0010}\left(\vec{r}\right) = s_C\left(\vec{r}\right) \tag{3.23}$$

$$w_{0001}\left(\vec{r}\right) = s_D\left(\vec{r}\right) \tag{3.24}$$

Thus, the formula for linear interpolation in the tetrahedron can be written in the form of Eq. 3.25.

$$u\left(\vec{r}\right) = v_A\left(\vec{r}\right)u_{1000} + v_B\left(\vec{r}\right)u_{0100} + v_C\left(\vec{r}\right)u_{0010} + v_D\left(\vec{r}\right)u_{0001} \tag{3.25}$$

3.2.3 Weights of Reference Points for N = 2

In the case of parabolic interpolation, when $N = 2$, the weights of reference points represented in Fig. 3.2 one can find using Eqs. 3.26–3.35.

$$w_{2000}\left(\vec{r}\right) = v_A\left(\vec{r}\right)\left(2v_A\left(\vec{r}\right) - 1\right) \tag{3.26}$$

$$w_{0200}\left(\vec{r}\right) = v_B\left(\vec{r}\right)\left(2v_B\left(\vec{r}\right) - 1\right) \tag{3.27}$$

$$w_{0020}\left(\vec{r}\right) = v_C\left(\vec{r}\right)\left(2v_C\left(\vec{r}\right) - 1\right) \tag{3.28}$$

$$w_{0002}\left(\vec{r}\right) = \nu_D\left(\vec{r}\right)\left(2\nu_D\left(\vec{r}\right) - 1\right) \tag{3.29}$$

$$w_{1100}\left(\vec{r}\right) = 4\nu_A\left(\vec{r}\right)\nu_B\left(\vec{r}\right) \tag{3.30}$$

$$w_{0110}\left(\vec{r}\right) = 4\nu_B\left(\vec{r}\right)\nu_C\left(\vec{r}\right) \tag{3.31}$$

$$w_{0011}\left(\vec{r}\right) = 4\nu_C\left(\vec{r}\right)\nu_D\left(\vec{r}\right) \tag{3.32}$$

$$w_{1001}\left(\vec{r}\right) = 4\nu_D\left(\vec{r}\right)\nu_A\left(\vec{r}\right) \tag{3.33}$$

$$w_{1010}\left(\vec{r}\right) = 4\nu_A\left(\vec{r}\right)\nu_C\left(\vec{r}\right) \tag{3.34}$$

$$w_{0101}\left(\vec{r}\right) = 4\nu_B\left(\vec{r}\right)\nu_D\left(\vec{r}\right) \tag{3.35}$$

3.2.4 Weights of Reference Points for N = 3

In the case of $N = 3$, the weights of reference points represented in Fig. 3.3 can be defined using Eqs. 3.36–3.55.

$$w_{3000}\left(\vec{r}\right) = \frac{1}{2}\nu_A\left(\vec{r}\right)\left(3\nu_A\left(\vec{r}\right) - 1\right)\left(3\nu_A\left(\vec{r}\right) - 2\right) \tag{3.36}$$

$$w_{0300}\left(\vec{r}\right) = \frac{1}{2}\nu_B\left(\vec{r}\right)\left(3\nu_B\left(\vec{r}\right) - 1\right)\left(3\nu_B\left(\vec{r}\right) - 2\right) \tag{3.37}$$

$$w_{0030}\left(\vec{r}\right) = \frac{1}{2}\nu_C\left(\vec{r}\right)\left(3\nu_C\left(\vec{r}\right) - 1\right)\left(3\nu_C\left(\vec{r}\right) - 2\right) \tag{3.38}$$

$$w_{0003}\left(\vec{r}\right) = \frac{1}{2}\nu_D\left(\vec{r}\right)\left(3\nu_D\left(\vec{r}\right) - 1\right)\left(3\nu_D\left(\vec{r}\right) - 2\right) \tag{3.39}$$

$$w_{2100}\left(\vec{r}\right) = \frac{9}{2}\nu_A\left(\vec{r}\right)\nu_B\left(\vec{r}\right)\left(3\nu_A\left(\vec{r}\right) - 1\right) \tag{3.40}$$

$$w_{0210}\left(\vec{r}\right) = \frac{9}{2}\nu_B\left(\vec{r}\right)\nu_C\left(\vec{r}\right)\left(3\nu_B\left(\vec{r}\right) - 1\right) \tag{3.41}$$

$$w_{0021}\left(\vec{r}\right) = \frac{9}{2}\nu_C\left(\vec{r}\right)\nu_D\left(\vec{r}\right)\left(3\nu_C\left(\vec{r}\right) - 1\right) \tag{3.42}$$

$$w_{1002}\left(\vec{r}\right) = \frac{9}{2}\nu_D\left(\vec{r}\right)\nu_A\left(\vec{r}\right)\left(3\nu_D\left(\vec{r}\right) - 1\right) \tag{3.43}$$

$$w_{1200}\left(\vec{r}\right) = \frac{9}{2}\nu_B\left(\vec{r}\right)\nu_A\left(\vec{r}\right)\left(3\nu_B\left(\vec{r}\right) - 1\right) \tag{3.44}$$

$$w_{0120}\left(\vec{r}\right) = \frac{9}{2}\nu_C\left(\vec{r}\right)\nu_B\left(\vec{r}\right)\left(3\nu_C\left(\vec{r}\right) - 1\right) \tag{3.45}$$

$$w_{0012}\left(\vec{r}\right) = \frac{9}{2}\nu_D\left(\vec{r}\right)\nu_C\left(\vec{r}\right)\left(3\nu_D\left(\vec{r}\right) - 1\right) \tag{3.46}$$

$$w_{2001}\left(\vec{r}\right) = \frac{9}{2}\nu_A\left(\vec{r}\right)\nu_D\left(\vec{r}\right)\left(3\nu_A\left(\vec{r}\right) - 1\right) \tag{3.47}$$

$$w_{2010}\left(\vec{r}\right)=\frac{9}{2}\nu_A\left(\vec{r}\right)\nu_C\left(\vec{r}\right)\left(3\nu_A\left(\vec{r}\right)-1\right) \tag{3.48}$$

$$w_{0201}\left(\vec{r}\right)=\frac{9}{2}\nu_B\left(\vec{r}\right)\nu_D\left(\vec{r}\right)\left(3\nu_B\left(\vec{r}\right)-1\right) \tag{3.49}$$

$$w_{1020}\left(\vec{r}\right)=\frac{9}{2}\nu_C\left(\vec{r}\right)\nu_A\left(\vec{r}\right)\left(3\nu_C\left(\vec{r}\right)-1\right) \tag{3.50}$$

$$w_{0102}\left(\vec{r}\right)=\frac{9}{2}\nu_D\left(\vec{r}\right)\nu_B\left(\vec{r}\right)\left(3\nu_D\left(\vec{r}\right)-1\right) \tag{3.51}$$

$$w_{1110}\left(\vec{r}\right)=27\nu_A\left(\vec{r}\right)\nu_B\left(\vec{r}\right)\nu_C\left(\vec{r}\right) \tag{3.52}$$

$$w_{0111}\left(\vec{r}\right)=27\nu_B\left(\vec{r}\right)\nu_C\left(\vec{r}\right)\nu_D\left(\vec{r}\right) \tag{3.53}$$

$$w_{1011}\left(\vec{r}\right)=27\nu_C\left(\vec{r}\right)\nu_D\left(\vec{r}\right)\nu_A\left(\vec{r}\right) \tag{3.54}$$

$$w_{1101}\left(\vec{r}\right)=27\nu_D\left(\vec{r}\right)\nu_A\left(\vec{r}\right)\nu_B\left(\vec{r}\right) \tag{3.55}$$

3.2.5 Weights of Reference Points for N = 4

In the case of $N = 4$, the weights of reference points represented in Fig. 3.4 are determined using Eqs. 3.56–3.90.

$$w_{4000}\left(\vec{r}\right)=\frac{1}{3}\nu_A\left(\vec{r}\right)\left(4\nu_A\left(\vec{r}\right)-1\right)\left(2\nu_A\left(\vec{r}\right)-1\right)\left(4\nu_A\left(\vec{r}\right)-3\right) \tag{3.56}$$

$$w_{0400}\left(\vec{r}\right)=\frac{1}{3}\nu_B\left(\vec{r}\right)\left(4\nu_B\left(\vec{r}\right)-1\right)\left(2\nu_B\left(\vec{r}\right)-1\right)\left(4\nu_B\left(\vec{r}\right)-3\right) \tag{3.57}$$

$$w_{0040}\left(\vec{r}\right)=\frac{1}{3}\nu_C\left(\vec{r}\right)\left(4\nu_C\left(\vec{r}\right)-1\right)\left(2\nu_C\left(\vec{r}\right)-1\right)\left(4\nu_C\left(\vec{r}\right)-3\right) \tag{3.58}$$

$$w_{0004}\left(\vec{r}\right)=\frac{1}{3}\nu_D\left(\vec{r}\right)\left(4\nu_D\left(\vec{r}\right)-1\right)\left(2\nu_D\left(\vec{r}\right)-1\right)\left(4\nu_D\left(\vec{r}\right)-3\right) \tag{3.59}$$

$$w_{3100}\left(\vec{r}\right)=\frac{16}{3}\nu_A\left(\vec{r}\right)\nu_B\left(\vec{r}\right)\left(4\nu_A\left(\vec{r}\right)-1\right)\left(2\nu_A\left(\vec{r}\right)-1\right) \tag{3.60}$$

$$w_{0310}\left(\vec{r}\right)=\frac{16}{3}\nu_B\left(\vec{r}\right)\nu_C\left(\vec{r}\right)\left(4\nu_B\left(\vec{r}\right)-1\right)\left(2\nu_B\left(\vec{r}\right)-1\right) \tag{3.61}$$

$$w_{0031}\left(\vec{r}\right)=\frac{16}{3}\nu_C\left(\vec{r}\right)\nu_D\left(\vec{r}\right)\left(4\nu_C\left(\vec{r}\right)-1\right)\left(2\nu_C\left(\vec{r}\right)-1\right) \tag{3.62}$$

$$w_{1003}\left(\vec{r}\right)=\frac{16}{3}\nu_D\left(\vec{r}\right)\nu_A\left(\vec{r}\right)\left(4\nu_D\left(\vec{r}\right)-1\right)\left(2\nu_D\left(\vec{r}\right)-1\right) \tag{3.63}$$

$$w_{1300}\left(\vec{r}\right)=\frac{16}{3}\nu_B\left(\vec{r}\right)\nu_A\left(\vec{r}\right)\left(4\nu_B\left(\vec{r}\right)-1\right)\left(2\nu_B\left(\vec{r}\right)-1\right) \tag{3.64}$$

$$w_{0130}\left(\vec{r}\right)=\frac{16}{3}\nu_C\left(\vec{r}\right)\nu_B\left(\vec{r}\right)\left(4\nu_C\left(\vec{r}\right)-1\right)\left(2\nu_C\left(\vec{r}\right)-1\right) \tag{3.65}$$

$$w_{0013}\left(\vec{r}\right) = \frac{16}{3}\nu_D\left(\vec{r}\right)\nu_C\left(\vec{r}\right)\left(4\nu_D\left(\vec{r}\right)-1\right)\left(2\nu_D\left(\vec{r}\right)-1\right) \tag{3.66}$$

$$w_{3001}\left(\vec{r}\right) = \frac{16}{3}\nu_A\left(\vec{r}\right)\nu_D\left(\vec{r}\right)\left(4\nu_A\left(\vec{r}\right)-1\right)\left(2\nu_A\left(\vec{r}\right)-1\right) \tag{3.67}$$

$$w_{3010}\left(\vec{r}\right) = \frac{16}{3}\nu_A\left(\vec{r}\right)\nu_C\left(\vec{r}\right)\left(4\nu_A\left(\vec{r}\right)-1\right)\left(2\nu_A\left(\vec{r}\right)-1\right) \tag{3.68}$$

$$w_{0301}\left(\vec{r}\right) = \frac{16}{3}\nu_B\left(\vec{r}\right)\nu_D\left(\vec{r}\right)\left(4\nu_B\left(\vec{r}\right)-1\right)\left(2\nu_D\left(\vec{r}\right)-1\right) \tag{3.69}$$

$$w_{1030}\left(\vec{r}\right) = \frac{16}{3}\nu_C\left(\vec{r}\right)\nu_A\left(\vec{r}\right)\left(4\nu_C\left(\vec{r}\right)-1\right)\left(2\nu_A\left(\vec{r}\right)-1\right) \tag{3.70}$$

$$w_{0103}\left(\vec{r}\right) = \frac{16}{3}\nu_D\left(\vec{r}\right)\nu_B\left(\vec{r}\right)\left(4\nu_D\left(\vec{r}\right)-1\right)\left(2\nu_B\left(\vec{r}\right)-1\right) \tag{3.71}$$

$$w_{2200}\left(\vec{r}\right) = 4\nu_A\left(\vec{r}\right)\nu_B\left(\vec{r}\right)\left(4\nu_A\left(\vec{r}\right)-1\right)\left(4\nu_B\left(\vec{r}\right)-1\right) \tag{3.72}$$

$$w_{0220}\left(\vec{r}\right) = 4\nu_B\left(\vec{r}\right)\nu_C\left(\vec{r}\right)\left(4\nu_B\left(\vec{r}\right)-1\right)\left(4\nu_C\left(\vec{r}\right)-1\right) \tag{3.73}$$

$$w_{0022}\left(\vec{r}\right) = 4\nu_C\left(\vec{r}\right)\nu_D\left(\vec{r}\right)\left(4\nu_C\left(\vec{r}\right)-1\right)\left(4\nu_D\left(\vec{r}\right)-1\right) \tag{3.74}$$

$$w_{2002}\left(\vec{r}\right) = 4\nu_D\left(\vec{r}\right)\nu_A\left(\vec{r}\right)\left(4\nu_D\left(\vec{r}\right)-1\right)\left(4\nu_A\left(\vec{r}\right)-1\right) \tag{3.75}$$

$$w_{2020}\left(\vec{r}\right) = 4\nu_A\left(\vec{r}\right)\nu_C\left(\vec{r}\right)\left(4\nu_A\left(\vec{r}\right)-1\right)\left(4\nu_C\left(\vec{r}\right)-1\right) \tag{3.76}$$

$$w_{0202}\left(\vec{r}\right) = 4\nu_B\left(\vec{r}\right)\nu_D\left(\vec{r}\right)\left(4\nu_B\left(\vec{r}\right)-1\right)\left(4\nu_D\left(\vec{r}\right)-1\right) \tag{3.77}$$

$$w_{2110}\left(\vec{r}\right) = 32\nu_A\left(\vec{r}\right)\nu_B\left(\vec{r}\right)\nu_C\left(\vec{r}\right)\left(4\nu_A\left(\vec{r}\right)-1\right) \tag{3.78}$$

$$w_{0211}\left(\vec{r}\right) = 32\nu_B\left(\vec{r}\right)\nu_C\left(\vec{r}\right)\nu_D\left(\vec{r}\right)\left(4\nu_B\left(\vec{r}\right)-1\right) \tag{3.79}$$

$$w_{1021}\left(\vec{r}\right) = 32\nu_C\left(\vec{r}\right)\nu_D\left(\vec{r}\right)\nu_A\left(\vec{r}\right)\left(4\nu_C\left(\vec{r}\right)-1\right) \tag{3.80}$$

$$w_{1102}\left(\vec{r}\right) = 32\nu_D\left(\vec{r}\right)\nu_A\left(\vec{r}\right)\nu_B\left(\vec{r}\right)\left(4\nu_D\left(\vec{r}\right)-1\right) \tag{3.81}$$

$$w_{1210}\left(\vec{r}\right) = 32\nu_B\left(\vec{r}\right)\nu_A\left(\vec{r}\right)\nu_C\left(\vec{r}\right)\left(4\nu_B\left(\vec{r}\right)-1\right) \tag{3.82}$$

$$w_{0121}\left(\vec{r}\right) = 32\nu_C\left(\vec{r}\right)\nu_B\left(\vec{r}\right)\nu_D\left(\vec{r}\right)\left(4\nu_C\left(\vec{r}\right)-1\right) \tag{3.83}$$

$$w_{1012}\left(\vec{r}\right) = 32\nu_D\left(\vec{r}\right)\nu_C\left(\vec{r}\right)\nu_A\left(\vec{r}\right)\left(4\nu_D\left(\vec{r}\right)-1\right) \tag{3.84}$$

$$w_{2101}\left(\vec{r}\right) = 32\nu_A\left(\vec{r}\right)\nu_D\left(\vec{r}\right)\nu_B\left(\vec{r}\right)\left(4\nu_A\left(\vec{r}\right)-1\right) \tag{3.85}$$

$$w_{1120}\left(\vec{r}\right) = 32\nu_C\left(\vec{r}\right)\nu_A\left(\vec{r}\right)\nu_B\left(\vec{r}\right)\left(4\nu_C\left(\vec{r}\right)-1\right) \tag{3.86}$$

$$w_{0112}\left(\vec{r}\right) = 32\nu_D\left(\vec{r}\right)\nu_B\left(\vec{r}\right)\nu_C\left(\vec{r}\right)\left(4\nu_D\left(\vec{r}\right)-1\right) \tag{3.87}$$

$$w_{2011}\left(\vec{r}\right) = 32\nu_A\left(\vec{r}\right)\nu_C\left(\vec{r}\right)\nu_D\left(\vec{r}\right)\left(4\nu_A\left(\vec{r}\right)-1\right) \tag{3.88}$$

$$w_{1201}\left(\vec{r}\right) = 32\nu_B\left(\vec{r}\right)\nu_D\left(\vec{r}\right)\nu_A\left(\vec{r}\right)\left(4\nu_B\left(\vec{r}\right)-1\right), \tag{3.89}$$

$$w_{1111}\left(\vec{r}\right) = 256\nu_A\left(\vec{r}\right)\nu_B\left(\vec{r}\right)\nu_C\left(\vec{r}\right)\nu_D\left(\vec{r}\right) \tag{3.90}$$

Note that the reference point 1111 is not drawn in Fig. 3.4 because this reference point does not lie on any of the faces of the tetrahedron *ABCD*.

3.2.6 Weights of Reference Points for N = 5

In the case of $N = 5$, the weights of reference points represented in Fig. 3.5 are computed using Eqs. 3.91–3.146.

$$w_{5000}\left(\vec{r}\right) = \frac{1}{24}\nu_A\left(\vec{r}\right)\left(5\nu_A\left(\vec{r}\right)-1\right)\left(5\nu_A\left(\vec{r}\right)-2\right)\left(5\nu_A\left(\vec{r}\right)-3\right)\left(5\nu_A\left(\vec{r}\right)-4\right) \tag{3.91}$$

$$w_{0500}\left(\vec{r}\right) = \frac{1}{24}\nu_B\left(\vec{r}\right)\left(5\nu_B\left(\vec{r}\right)-1\right)\left(5\nu_B\left(\vec{r}\right)-2\right)\left(5\nu_B\left(\vec{r}\right)-3\right)\left(5\nu_B\left(\vec{r}\right)-4\right) \tag{3.92}$$

$$w_{0050}\left(\vec{r}\right) = \frac{1}{24}\nu_C\left(\vec{r}\right)\left(5\nu_C\left(\vec{r}\right)-1\right)\left(5\nu_C\left(\vec{r}\right)-2\right)\left(5\nu_C\left(\vec{r}\right)-3\right)\left(5\nu_C\left(\vec{r}\right)-4\right) \tag{3.93}$$

$$w_{0005}\left(\vec{r}\right) = \frac{1}{24}\nu_D\left(\vec{r}\right)\left(5\nu_D\left(\vec{r}\right)-1\right)\left(5\nu_D\left(\vec{r}\right)-2\right)\left(5\nu_D\left(\vec{r}\right)-3\right)\left(5\nu_D\left(\vec{r}\right)-4\right) \tag{3.94}$$

$$w_{4100}\left(\vec{r}\right) = \frac{25}{24}\nu_A\left(\vec{r}\right)\nu_B\left(\vec{r}\right)\left(5\nu_A\left(\vec{r}\right)-1\right)\left(5\nu_A\left(\vec{r}\right)-2\right)\left(5\nu_A\left(\vec{r}\right)-3\right) \tag{3.95}$$

$$w_{0410}\left(\vec{r}\right) = \frac{25}{24}\nu_B\left(\vec{r}\right)\nu_C\left(\vec{r}\right)\left(5\nu_B\left(\vec{r}\right)-1\right)\left(5\nu_B\left(\vec{r}\right)-2\right)\left(5\nu_B\left(\vec{r}\right)-3\right) \tag{3.96}$$

$$w_{0041}\left(\vec{r}\right) = \frac{25}{24}\nu_C\left(\vec{r}\right)\nu_D\left(\vec{r}\right)\left(5\nu_C\left(\vec{r}\right)-1\right)\left(5\nu_C\left(\vec{r}\right)-2\right)\left(5\nu_C\left(\vec{r}\right)-3\right) \tag{3.97}$$

$$w_{1004}\left(\vec{r}\right) = \frac{25}{24}\nu_D\left(\vec{r}\right)\nu_A\left(\vec{r}\right)\left(5\nu_D\left(\vec{r}\right)-1\right)\left(5\nu_D\left(\vec{r}\right)-2\right)\left(5\nu_D\left(\vec{r}\right)-3\right) \tag{3.98}$$

$$w_{1400}\left(\vec{r}\right) = \frac{25}{24}\nu_B\left(\vec{r}\right)\nu_A\left(\vec{r}\right)\left(5\nu_B\left(\vec{r}\right)-1\right)\left(5\nu_B\left(\vec{r}\right)-2\right)\left(5\nu_B\left(\vec{r}\right)-3\right) \tag{3.99}$$

$$w_{0140}\left(\vec{r}\right) = \frac{25}{24}\nu_C\left(\vec{r}\right)\nu_B\left(\vec{r}\right)\left(5\nu_C\left(\vec{r}\right)-1\right)\left(5\nu_C\left(\vec{r}\right)-2\right)\left(5\nu_C\left(\vec{r}\right)-3\right) \tag{3.100}$$

$$w_{0014}\left(\vec{r}\right) = \frac{25}{24}\nu_D\left(\vec{r}\right)\nu_C\left(\vec{r}\right)\left(5\nu_D\left(\vec{r}\right)-1\right)\left(5\nu_D\left(\vec{r}\right)-2\right)\left(5\nu_D\left(\vec{r}\right)-3\right) \tag{3.101}$$

$$w_{4001}\left(\vec{r}\right) = \frac{25}{24}\nu_A\left(\vec{r}\right)\nu_D\left(\vec{r}\right)\left(5\nu_A\left(\vec{r}\right)-1\right)\left(5\nu_A\left(\vec{r}\right)-2\right)\left(5\nu_A\left(\vec{r}\right)-3\right) \tag{3.102}$$

$$w_{4010}\left(\vec{r}\right) = \frac{25}{24}\nu_A\left(\vec{r}\right)\nu_C\left(\vec{r}\right)\left(5\nu_A\left(\vec{r}\right)-1\right)\left(5\nu_A\left(\vec{r}\right)-2\right)\left(5\nu_A\left(\vec{r}\right)-3\right) \tag{3.103}$$

$$w_{0401}\left(\vec{r}\right) = \frac{25}{24}\nu_B\left(\vec{r}\right)\nu_D\left(\vec{r}\right)\left(5\nu_B\left(\vec{r}\right)-1\right)\left(5\nu_B\left(\vec{r}\right)-2\right)\left(5\nu_B\left(\vec{r}\right)-3\right) \tag{3.104}$$

$$w_{1040}\left(\vec{r}\right) = \frac{25}{24}\nu_C\left(\vec{r}\right)\nu_A\left(\vec{r}\right)\left(5\nu_C\left(\vec{r}\right)-1\right)\left(5\nu_C\left(\vec{r}\right)-2\right)\left(5\nu_C\left(\vec{r}\right)-3\right) \tag{3.105}$$

$$w_{0104}\left(\vec{r}\right) = \frac{25}{24}\nu_D\left(\vec{r}\right)\nu_B\left(\vec{r}\right)\left(5\nu_D\left(\vec{r}\right)-1\right)\left(5\nu_D\left(\vec{r}\right)-2\right)\left(5\nu_D\left(\vec{r}\right)-3\right) \tag{3.106}$$

$$w_{3200}\left(\vec{r}\right) = \frac{25}{12}\nu_A\left(\vec{r}\right)\nu_B\left(\vec{r}\right)\left(5\nu_A\left(\vec{r}\right)-1\right)\left(5\nu_B\left(\vec{r}\right)-1\right)\left(5\nu_A\left(\vec{r}\right)-2\right) \tag{3.107}$$

$$w_{0320}\left(\vec{r}\right) = \frac{25}{12}\nu_B\left(\vec{r}\right)\nu_C\left(\vec{r}\right)\left(5\nu_B\left(\vec{r}\right)-1\right)\left(5\nu_C\left(\vec{r}\right)-1\right)\left(5\nu_B\left(\vec{r}\right)-2\right) \tag{3.108}$$

$$w_{0032}\left(\vec{r}\right) = \frac{25}{12}\nu_C\left(\vec{r}\right)\nu_D\left(\vec{r}\right)\left(5\nu_C\left(\vec{r}\right)-1\right)\left(5\nu_D\left(\vec{r}\right)-1\right)\left(5\nu_C\left(\vec{r}\right)-2\right) \tag{3.109}$$

$$w_{2003}\left(\vec{r}\right) = \frac{25}{12}\nu_D\left(\vec{r}\right)\nu_A\left(\vec{r}\right)\left(5\nu_D\left(\vec{r}\right)-1\right)\left(5\nu_A\left(\vec{r}\right)-1\right)\left(5\nu_D\left(\vec{r}\right)-2\right) \tag{3.110}$$

$$w_{2300}\left(\vec{r}\right) = \frac{25}{12}\nu_B\left(\vec{r}\right)\nu_A\left(\vec{r}\right)\left(5\nu_B\left(\vec{r}\right)-1\right)\left(5\nu_A\left(\vec{r}\right)-1\right)\left(5\nu_B\left(\vec{r}\right)-2\right) \tag{3.111}$$

$$w_{0230}\left(\vec{r}\right) = \frac{25}{12}\nu_C\left(\vec{r}\right)\nu_B\left(\vec{r}\right)\left(5\nu_C\left(\vec{r}\right)-1\right)\left(5\nu_B\left(\vec{r}\right)-1\right)\left(5\nu_C\left(\vec{r}\right)-2\right) \tag{3.112}$$

$$w_{0023}\left(\vec{r}\right) = \frac{25}{12}\nu_D\left(\vec{r}\right)\nu_C\left(\vec{r}\right)\left(5\nu_D\left(\vec{r}\right)-1\right)\left(5\nu_C\left(\vec{r}\right)-1\right)\left(5\nu_D\left(\vec{r}\right)-2\right) \tag{3.113}$$

$$w_{3002}\left(\vec{r}\right) = \frac{25}{12}\nu_A\left(\vec{r}\right)\nu_D\left(\vec{r}\right)\left(5\nu_A\left(\vec{r}\right)-1\right)\left(5\nu_D\left(\vec{r}\right)-1\right)\left(5\nu_A\left(\vec{r}\right)-2\right) \tag{3.114}$$

$$w_{3020}\left(\vec{r}\right) = \frac{25}{12}\nu_A\left(\vec{r}\right)\nu_C\left(\vec{r}\right)\left(5\nu_A\left(\vec{r}\right)-1\right)\left(5\nu_C\left(\vec{r}\right)-1\right)\left(5\nu_A\left(\vec{r}\right)-2\right) \tag{3.115}$$

$$w_{0302}\left(\vec{r}\right) = \frac{25}{12}\nu_B\left(\vec{r}\right)\nu_D\left(\vec{r}\right)\left(5\nu_B\left(\vec{r}\right)-1\right)\left(5\nu_D\left(\vec{r}\right)-1\right)\left(5\nu_B\left(\vec{r}\right)-2\right) \tag{3.116}$$

$$w_{2030}\left(\vec{r}\right) = \frac{25}{12}\nu_C\left(\vec{r}\right)\nu_A\left(\vec{r}\right)\left(5\nu_C\left(\vec{r}\right)-1\right)\left(5\nu_A\left(\vec{r}\right)-1\right)\left(5\nu_C\left(\vec{r}\right)-2\right) \tag{3.117}$$

$$w_{0203}\left(\vec{r}\right) = \frac{25}{12}\nu_D\left(\vec{r}\right)\nu_B\left(\vec{r}\right)\left(5\nu_D\left(\vec{r}\right)-1\right)\left(5\nu_B\left(\vec{r}\right)-1\right)\left(5\nu_D\left(\vec{r}\right)-2\right) \tag{3.118}$$

$$w_{3110}\left(\vec{r}\right) = \frac{125}{6}\nu_A\left(\vec{r}\right)\nu_B\left(\vec{r}\right)\nu_C\left(\vec{r}\right)\left(5\nu_A\left(\vec{r}\right)-1\right)\left(5\nu_A\left(\vec{r}\right)-2\right) \tag{3.119}$$

$$w_{0311}\left(\vec{r}\right) = \frac{125}{6}\nu_B\left(\vec{r}\right)\nu_C\left(\vec{r}\right)\nu_D\left(\vec{r}\right)\left(5\nu_B\left(\vec{r}\right)-1\right)\left(5\nu_B\left(\vec{r}\right)-2\right) \tag{3.120}$$

$$w_{1031}\left(\vec{r}\right) = \frac{125}{6}\nu_C\left(\vec{r}\right)\nu_D\left(\vec{r}\right)\nu_A\left(\vec{r}\right)\left(5\nu_C\left(\vec{r}\right)-1\right)\left(5\nu_C\left(\vec{r}\right)-2\right) \tag{3.121}$$

$$w_{1103}\left(\vec{r}\right) = \frac{125}{6}\nu_D\left(\vec{r}\right)\nu_A\left(\vec{r}\right)\nu_B\left(\vec{r}\right)\left(5\nu_D\left(\vec{r}\right)-1\right)\left(5\nu_D\left(\vec{r}\right)-2\right) \tag{3.122}$$

$$w_{1310}\left(\vec{r}\right) = \frac{125}{6}\nu_B\left(\vec{r}\right)\nu_A\left(\vec{r}\right)\nu_C\left(\vec{r}\right)\left(5\nu_B\left(\vec{r}\right)-1\right)\left(5\nu_B\left(\vec{r}\right)-2\right) \tag{3.123}$$

$$w_{0131}\left(\vec{r}\right) = \frac{125}{6}\nu_C\left(\vec{r}\right)\nu_B\left(\vec{r}\right)\nu_D\left(\vec{r}\right)\left(5\nu_C\left(\vec{r}\right)-1\right)\left(5\nu_C\left(\vec{r}\right)-2\right) \tag{3.124}$$

$$w_{1013}\left(\vec{r}\right) = \frac{125}{6}\nu_D\left(\vec{r}\right)\nu_C\left(\vec{r}\right)\nu_A\left(\vec{r}\right)\left(5\nu_D\left(\vec{r}\right)-1\right)\left(5\nu_D\left(\vec{r}\right)-2\right) \tag{3.125}$$

$$w_{3101}\left(\vec{r}\right) = \frac{125}{6}\nu_A\left(\vec{r}\right)\nu_D\left(\vec{r}\right)\nu_B\left(\vec{r}\right)\left(5\nu_A\left(\vec{r}\right)-1\right)\left(5\nu_A\left(\vec{r}\right)-2\right) \tag{3.126}$$

$$w_{1130}\left(\vec{r}\right) = \frac{125}{6}\nu_C\left(\vec{r}\right)\nu_A\left(\vec{r}\right)\nu_B\left(\vec{r}\right)\left(5\nu_C\left(\vec{r}\right)-1\right)\left(5\nu_C\left(\vec{r}\right)-2\right) \tag{3.127}$$

$$w_{0113}\left(\vec{r}\right) = \frac{125}{6}\nu_D\left(\vec{r}\right)\nu_B\left(\vec{r}\right)\nu_C\left(\vec{r}\right)\left(5\nu_D\left(\vec{r}\right)-1\right)\left(5\nu_D\left(\vec{r}\right)-2\right) \tag{3.128}$$

$$w_{3011}\left(\vec{r}\right) = \frac{125}{6}\nu_A\left(\vec{r}\right)\nu_C\left(\vec{r}\right)\nu_D\left(\vec{r}\right)\left(5\nu_A\left(\vec{r}\right)-1\right)\left(5\nu_A\left(\vec{r}\right)-2\right) \tag{3.129}$$

$$w_{1301}\left(\vec{r}\right) = \frac{125}{6}\nu_B\left(\vec{r}\right)\nu_D\left(\vec{r}\right)\nu_A\left(\vec{r}\right)\left(5\nu_B\left(\vec{r}\right)-1\right)\left(5\nu_B\left(\vec{r}\right)-2\right) \tag{3.130}$$

$$w_{2210}\left(\vec{r}\right) = \frac{125}{4}\nu_A\left(\vec{r}\right)\nu_B\left(\vec{r}\right)\nu_C\left(\vec{r}\right)\left(5\nu_A\left(\vec{r}\right)-1\right)\left(5\nu_B\left(\vec{r}\right)-1\right) \quad (3.131)$$

$$w_{0221}\left(\vec{r}\right) = \frac{125}{4}\nu_B\left(\vec{r}\right)\nu_C\left(\vec{r}\right)\nu_D\left(\vec{r}\right)\left(5\nu_B\left(\vec{r}\right)-1\right)\left(5\nu_C\left(\vec{r}\right)-1\right) \quad (3.132)$$

$$w_{1022}\left(\vec{r}\right) = \frac{125}{4}\nu_C\left(\vec{r}\right)\nu_D\left(\vec{r}\right)\nu_A\left(\vec{r}\right)\left(5\nu_C\left(\vec{r}\right)-1\right)\left(5\nu_D\left(\vec{r}\right)-1\right) \quad (3.133)$$

$$w_{2102}\left(\vec{r}\right) = \frac{125}{4}\nu_D\left(\vec{r}\right)\nu_A\left(\vec{r}\right)\nu_B\left(\vec{r}\right)\left(5\nu_D\left(\vec{r}\right)-1\right)\left(5\nu_A\left(\vec{r}\right)-1\right) \quad (3.134)$$

$$w_{1220}\left(\vec{r}\right) = \frac{125}{4}\nu_B\left(\vec{r}\right)\nu_C\left(\vec{r}\right)\nu_A\left(\vec{r}\right)\left(5\nu_B\left(\vec{r}\right)-1\right)\left(5\nu_C\left(\vec{r}\right)-1\right) \quad (3.135)$$

$$w_{0122}\left(\vec{r}\right) = \frac{125}{4}\nu_C\left(\vec{r}\right)\nu_D\left(\vec{r}\right)\nu_B\left(\vec{r}\right)\left(5\nu_C\left(\vec{r}\right)-1\right)\left(5\nu_D\left(\vec{r}\right)-1\right) \quad (3.136)$$

$$w_{2012}\left(\vec{r}\right) = \frac{125}{4}\nu_D\left(\vec{r}\right)\nu_A\left(\vec{r}\right)\nu_C\left(\vec{r}\right)\left(5\nu_D\left(\vec{r}\right)-1\right)\left(5\nu_A\left(\vec{r}\right)-1\right) \quad (3.137)$$

$$w_{2201}\left(\vec{r}\right) = \frac{125}{4}\nu_A\left(\vec{r}\right)\nu_B\left(\vec{r}\right)\nu_D\left(\vec{r}\right)\left(5\nu_A\left(\vec{r}\right)-1\right)\left(5\nu_B\left(\vec{r}\right)-1\right) \quad (3.138)$$

$$w_{2120}\left(\vec{r}\right) = \frac{125}{4}\nu_A\left(\vec{r}\right)\nu_C\left(\vec{r}\right)\nu_B\left(\vec{r}\right)\left(5\nu_A\left(\vec{r}\right)-1\right)\left(5\nu_C\left(\vec{r}\right)-1\right) \quad (3.139)$$

$$w_{0212}\left(\vec{r}\right) = \frac{125}{4}\nu_B\left(\vec{r}\right)\nu_D\left(\vec{r}\right)\nu_C\left(\vec{r}\right)\left(5\nu_B\left(\vec{r}\right)-1\right)\left(5\nu_D\left(\vec{r}\right)-1\right) \quad (3.140)$$

$$w_{2021}\left(\vec{r}\right) = \frac{125}{4}\nu_C\left(\vec{r}\right)\nu_A\left(\vec{r}\right)\nu_D\left(\vec{r}\right)\left(5\nu_C\left(\vec{r}\right)-1\right)\left(5\nu_A\left(\vec{r}\right)-1\right) \quad (3.141)$$

$$w_{1202}\left(\vec{r}\right) = \frac{125}{4}\nu_D\left(\vec{r}\right)\nu_B\left(\vec{r}\right)\nu_A\left(\vec{r}\right)\left(5\nu_D\left(\vec{r}\right)-1\right)\left(5\nu_B\left(\vec{r}\right)-1\right) \quad (3.142)$$

$$w_{2111}\left(\vec{r}\right) = \frac{625}{2}\nu_A\left(\vec{r}\right)\nu_B\left(\vec{r}\right)\nu_C\left(\vec{r}\right)\nu_D\left(\vec{r}\right)\left(5\nu_A\left(\vec{r}\right)-1\right) \quad (3.143)$$

$$w_{1211}\left(\vec{r}\right) = \frac{625}{2}\nu_B\left(\vec{r}\right)\nu_C\left(\vec{r}\right)\nu_D\left(\vec{r}\right)\nu_A\left(\vec{r}\right)\left(5\nu_B\left(\vec{r}\right)-1\right) \quad (3.144)$$

$$w_{1121}\left(\vec{r}\right) = \frac{625}{2}\nu_C\left(\vec{r}\right)\nu_D\left(\vec{r}\right)\nu_A\left(\vec{r}\right)\nu_B\left(\vec{r}\right)\left(5\nu_C\left(\vec{r}\right)-1\right) \quad (3.145)$$

$$w_{1112}\left(\vec{r}\right) = \frac{625}{2}\nu_D\left(\vec{r}\right)\nu_A\left(\vec{r}\right)\nu_B\left(\vec{r}\right)\nu_C\left(\vec{r}\right)\left(5\nu_D\left(\vec{r}\right)-1\right) \quad (3.146)$$

Note that the reference points 2111, 1211, 1121, and 1112 are not drawn in Fig. 3.5 because these reference points do not lie on any of the faces of the tetrahedron *ABCD*.

3.3 Several Approaches for Hybrid Interpolation

Consider several approaches for hybrid interpolation on the unstructured tetrahedral grids [1]. Using algorithms for the piecewise linear interpolation discussed in Chap. 4, one can obtain the hierarchical nested unstructured tetrahedral grids described in Sect. 3.3.1. In Sect. 3.3.2, the recurrent formulae for recalculation of the local reference points' indices to the global reference points' indices are discussed. In Sect. 3.3.3, an example of hybridization called hybrid parabolic—linear interpolation is considered. In Sect. 3.3.4, another example of hybridization based a limiter is discussed. In Sect. 3.3.5, an example of approach based on both hierarchical nested unstructured grids and hybridization called parabolic interpolation on the reference points for interpolation of fourth order is offered.

3.3.1 Hierarchical Nested Unstructured Tetrahedral Grids

Using algorithms for the piecewise linear interpolation discussed in Sect. 3.3, one can obtain hybrid hierarchical unstructured grids. The use of these types of grids allows to avoid spending resources like execution time of software during building the detailed unstructured tetrahedral grid. This result is achieved due to the analytical expressions given in Chap. 4 and allows to understand, in which tetrahedron the point under consideration lies without using the search algorithms.

One can use the hierarchical nested unstructured tetrahedral grids applying the following algorithm for the point under consideration $\vec{r}$:

Step 0 Firstly, the relative volumes $v_A\left(\vec{r}\right)$, $v_B\left(\vec{r}\right)$, $v_C\left(\vec{r}\right)$, and $v_D\left(\vec{r}\right)$ corresponding the big tetrahedron *ABCD* should be calculated using Eqs. 3.6–3.11 given in Sect. 3.2.1.

Step 1 For the first level of hybridization with N_1 using algorithms discussed in Chap. 4, one can find the reference points associated with the big tetrahedron *ABCD* forming the small tetrahedron $A_1B_1C_1D_1$ and calculate corresponding relative volumes $v_{A,1}\left(v_A\right)$, $v_{B,1}\left(v_B\right)$, $v_{C,1}\left(v_C\right)$, and $v_{D,1}\left(v_D\right)$.

Step 2 For the second level of hybridization with N_2 using algorithms discussed in Chap. 4, one can find the reference points associated with the tetrahedron $A_1B_1C_1D_1$ forming the small tetrahedron $A_2B_2C_2D_2$ and calculate corresponding relative volumes $v_{A,2}\left(v_{A,1}\right)$, $v_{B,2}\left(v_{B,1}\right)$, $v_{C,2}\left(v_{C,1}\right)$, and $v_{D,2}\left(v_{D,1}\right)$.

Step 3 For the third level of hybridization with N_3 using algorithms discussed in Chap. 4, one can find the reference points associated with the tetrahedron $A_2B_2C_2D_2$ forming the small tetrahedron $A_3B_3C_3D_3$ and calculate corresponding relative volumes $v_{A,3}\left(v_{A,2}\right)$, $v_{B,3}\left(v_{B,2}\right)$, $v_{C,3}\left(v_{C,2}\right)$, and $v_{D,3}\left(v_{D,2}\right)$.

Step k For the k level of hybridization with N_k using algorithms discussed in Chap. 4, one can find the reference points associated with the tetrahedron $A_{k-1}B_{k-1}C_{k-1}D_{k-1}$ forming the small tetrahedron $A_kB_kC_kD_k$ and calculate corresponding relative volumes $v_{A,k}\left(v_{A,k-1}\right)$, $v_{B,k}\left(v_{B,k-1}\right)$, $v_{C,k}\left(v_{C,k-1}\right)$, and $v_{D,k}\left(v_{D,k-1}\right)$.

Note that one can find coordinates of the reference point *abcd* associated with the tetrahedron *ABCD* using Eq. 3.147.

$$\vec{r}_{abcd} = \frac{a}{N}\left(\vec{r}_A - \vec{r}_D\right) + \frac{b}{N}\left(\vec{r}_B - \vec{r}_D\right) + \frac{c}{N}\left(\vec{r}_C - \vec{r}_D\right) + \vec{r}_D \qquad (3.147)$$

3.3.2 Formulae for Recalculation from Local to Global Indices

In order to diminish the amount of calculations, the values of radius-vectors corresponding to the reference point into the big tetrahedron $ABCD$ in the hierarchical nested unstructured tetrahedral grid are defined using the recurrent formulae for recalculation from local reference points' indices to global reference points' indices. One can use these recurrent formulae before calculations and save the non-recurrent formulae for recalculation from local to global reference points' indices as a table and then use Eq. 3.149 with global four-digital index $abcd$ and $N = N_1 N_2 \dots N_K$.

There are $\frac{(N+1)(N+2)(N+3)}{6}$ reference points into the tetrahedron if the appropriate order is equal to N. One can determine the rule for recalculation from serial reference points' index $I \in \left[1, \frac{(N+1)(N+2)(N+3)}{6}\right]$ into four-digital index $abcd$ introduced in Sect. 3.2 (Eq. 3.148) and vice versa (Eq. 3.149).

$$I \in \left[1, \frac{(N+1)(N+2)(N+3)}{6}\right] \mapsto \{a(I), b(I), c(I), d(I)\}$$

$$\{a \in [0, N], b \in [0, N], c \in [0, N], d \in [0, N], \tag{3.148}$$

$$a + b + c + d = N\} \mapsto I(a, b, c, d) \tag{3.149}$$

The steps of this algorithm are mentioned below.

Step 1 N_1^3 small tetrahedrons $A_1 B_1 C_1 D_1$ with the serial indices $J_1 \in \left[1, N_1^3\right]$ are known into the initial tetrahedron $ABCD$. The set of the four-digital indices:

$$a(a, J_1)\ b(a, J_1)\ c(a, J_1)\ d(a, J_1) \tag{3.150}$$

$$a(b, J_1)\ b(b, J_1)\ c(b, J_1)\ c(b, J_1) \tag{3.151}$$

$$a(c, J_1)\ b(c, J_1)\ c(c, J_1)\ d(c, J_1) \tag{3.152}$$

$$a(d, J_1)\ b(d, J_1)\ c(d, J_1)\ d(d, J_1) \tag{3.153}$$

relatively the initial tetrahedron $ABCD$ for all of the four vertices A_1 (Eq. 3.150), B_1 (Eq. (3.151), C_1 (Eq. 3.152), and D_1 (Eq. 3.153) of each of the tetrahedrons $A_1 B_1 C_1 D_1$ with the serial indices $J_1 \in \left[1, N_1^3\right]$ are known and represented into the tables of the third, fifth, and sixth types in Chap. 4.

Step 2.0 All steps starting with 2 are carried out for each of the small tetrahedrons $A_1 B_1 C_1 D_1$ with the serial indices $J_1 \in \left[1, N_1^3\right]$.

Step 2.1 For each of the small tetrahedrons $A_1 B_1 C_1 D_1$ with the serial indices $J_1 \in \left[1, N_1^3\right]$, the local indices of $\frac{(N_2+1)(N_2+2)(N_2+3)}{6}$ reference points into the small tetrahedron $A_1 B_1 C_1 D_1$ are known. For each of these reference points with the serial indices $I_2 \in \left[1, \frac{(N_2+1)(N_2+2)(N_2+3)}{6}\right]$, one can calculate the four-digital indices $a_1(I_2)$, $b_1(I_2)$, $c_1(I_2)$, and $d_1(I_2)$ relatively the small

tetrahedron $A_1B_1C_1D_1$ with the serial index $J_1 \in \left[1, N_1^3\right]$ using Eq. 3.148 for N_2.
Thus, for each of these $\frac{(N_2+1)(N_2+2)(N_2+3)}{6}$ reference points with the serial indices I_2, one can calculate new four-digital indices relatively the initial tetrahedron $ABCD$ using Eqs. 1.154–1.157.

$$a(I_2, J_1) = a_1(I_2)\, a(a, J_1) + b_1(I_2)\, a(b, J_1) + \\ + c_1(I_2)\, a(c, J_1) + d_1(I_2)\, a(d, J_1) \tag{3.154}$$

$$b(I_2, J_1) = a_1(I_2)\, b(a, J_1) + b_1(I_2)\, b(b, J_1) + \\ + c_1(I_2)\, b(c, J_1) + d_1(I_2)\, b(d, J_1) \tag{3.155}$$

$$c(I_2, J_1) = a_1(I_2)\, c(a, J_1) + b_1(I_2)\, c(b, J_1) + \\ + c_1(I_2)\, c(c, J_1) + d_1(I_2)\, c(d, J_1) \tag{3.156}$$

$$d(I_2, J_1) = a_1(I_2)\, d(a, J_1) + b_1(I_2)\, d(b, J_1) + \\ + c_1(I_2)\, d(c, J_1) + d_1(I_2)\, d(d, J_1) \tag{3.157}$$

Step 2.2 Into each of the small tetrahedrons $A_1B_1C_1D_1$ with the serial indices $J_1 \in \left[1, N_1^3\right]$, N_2^3 small tetrahedrons $A_2B_2C_2D_2$ with the serial indices $J_2 \in \left[1, N_2^3\right]$ are known.

The set of the four digital indices:

$$a_1(a, J_2)\ b_1(a, J_2)\ b_1(a, J_2)\ d_1(a, J_2) \tag{3.158}$$

$$a_1(b, J_2)\ b_1(b, J_2)\ b_1(b, J_2)\ d_1(b, J_2) \tag{3.159}$$

$$a_1(c, J_2)\ b_1(c, J_2)\ c_1(c, J_2)\ d_1(c, J_2) \tag{3.160}$$

$$a_1(d, J_2)\ b_1(d, J_2)\ c_1(d, J_2)\ d_1(d, J_2) \tag{3.161}$$

relatively the small tetrahedron $A_1B_1C_1D_1$ with the serial index $J_1 \in \left[1, N_1^3\right]$ for all of the four vertices A_2 (Eq. 3.158), B_2 (Eq. 3.159), C_2 (Eq. 3.160), and D_2 (Eq. 3.161) of each of the tetrahedrons $A_2B_2C_2D_2$ with the serial indices $J_2 \in \left[1, N_2^3\right]$ independent on J_1 are known and represented into the tables of the third, fifth, and sixth types in Chap. 4.

Thus, one can find the set of the serial indices $I_2 \in \left[1, \frac{(N_2+1)(N_2+2)(N_2+3)}{6}\right]$ for all of the four vertices A_2 (Eq. 3.162), B_2 (Eq. 3.163), C_2 (Eq. 3.164), and D_2 (Eq. 3.165) of each of the tetrahedrons $A_2B_2C_2D_2$ with the serial indices $J_2 \in \left[1, N_2^3\right]$ using equation (Eq. 3.149) for N_2.

$$I_2(a, J_2) = I_2(a_1(a, J_2), b_1(a, J_2), c_1(a, J_2), d_1(a, J_2)) \tag{3.162}$$

$$I_2(b, J_2) = I_2(a_1(b, J_2), b_1(b, J_2), c_1(b, J_2), d_1(b, J_2)) \tag{3.163}$$

$$I_2(c, J_2) = I_2(a_1(c, J_2), b_1(c, J_2), c_1(c, J_2), d_1(c, J_2)) \tag{3.164}$$

$$I_2(d, J_2) = I_2(a_1(d, J_2), b_1(d, J_2), c_1(d, J_2), d_1(d, J_2)) \tag{3.165}$$

Therefore, using calculations carries out during Step 2.1 the set of the four-digital indices:

$$a\left(a, J_1, J_2\right) b\left(a, J_1, J_2\right) c\left(a, J_1, J_2\right) d\left(a, J_1, J_2\right) \tag{3.166}$$

$$a\left(b, J_1, J_2\right) b\left(b, J_1, J_2\right) c\left(b, J_1, J_2\right) d\left(b, J_1, J_2\right) \tag{3.167}$$

$$a\left(c, J_1, J_2\right) b\left(c, J_1, J_2\right) c\left(c, J_1, J_2\right) d\left(c, J_1, J_2\right) \tag{3.168}$$

$$a\left(d, J_1, J_2\right) b\left(d, J_1, J_2\right) c\left(d, J_1, J_2\right) d\left(d, J_1, J_2\right) \tag{3.169}$$

relatively the initial tetrahedron $ABCD$ for all of the four vertices A_2 (Eq. 3.166), B_2 (Eq. 3.167), C_2 (Eq. 3.168), and (Eq. 3.169) of each of the tetrahedrons $A_2B_2C_2D_2$ with the serial indices $J_2 \in \left[1, N_2^3\right]$ are known. For example, using Eq. 3.162 one can find:

$$b\left(a, J_1, J_2\right) = b\left(I_2\left(a, J_2\right), J_1\right). \tag{3.170}$$

Step 3.0 All steps starting with Step 3 are carried out for each of the small tetrahedrons $A_2B_2C_2D_2$ with the serial indices $J_2 \in \left[1, N_2^3\right]$.

Step 3.1 For each of the small tetrahedrons $A_2B_2C_2D_2$ with the serial indices $J_2 \in \left[1, N_2^3\right]$, the local indices of $\frac{(N_3+1)(N_3+2)(N_3+3)}{6}$ reference points into the small tetrahedron $A_2B_2C_2D_2$ are known. For each of these reference points with the serial indices $I_3 \in \left[1, \frac{(N_3+1)(N_3+2)(N_3+3)}{6}\right]$, one can calculate the four-digital indices $a_2\left(I_3\right)$, $b_2\left(I_3\right)$, $c_2\left(I_3\right)$, and $d_2\left(I_3\right)$ relatively the small tetrahedron $A_2B_2C_2D_2$ with the serial index $J_2 \in \left[1, N_2^3\right]$ using Eq. 3.148 for N_3.

Thus, for each of these $\frac{(N_3+1)(N_3+2)(N_3+3)}{6}$ reference points with the serial indices I_3, one can calculate new four-digital indices relatively the initial tetrahedron $ABCD$ using Eqs. 3.171–3.174.

$$\begin{aligned} a\left(I_3, J_1, J_2\right) = {} & a_2\left(I_3\right) a\left(a, J_1, J_2\right) + b_2\left(I_3\right) a\left(b, J_1, J_2\right) + \\ & + c_2\left(I_3\right) a\left(c, J_1, J_2\right) + d_2\left(I_3\right) a\left(d, J_1, J_2\right) \end{aligned} \tag{3.171}$$

$$\begin{aligned} b\left(I_3, J_1, J_2\right) = {} & a_2\left(I_3\right) b\left(a, J_1, J_2\right) + b_2\left(I_3\right) b\left(b, J_1, J_2\right) + \\ & + c_2\left(I_3\right) b\left(c, J_1, J_2\right) + d_2\left(I_3\right) b\left(d, J_1, J_2\right) \end{aligned} \tag{3.172}$$

$$\begin{aligned} c\left(I_3, J_1, J_2\right) = {} & a_2\left(I_3\right) c\left(a, J_1, J_2\right) + b_2\left(I_3\right) c\left(b, J_1, J_2\right) + \\ & + c_2\left(I_3\right) c\left(c, J_1, J_2\right) + d_2\left(I_3\right) c\left(d, J_1, J_2\right) \end{aligned} \tag{3.173}$$

$$\begin{aligned} d\left(I_3, J_1, J_2\right) = {} & a_2\left(I_3\right) d\left(a, J_1, J_2\right) + b_2\left(I_3\right) d\left(b, J_1, J_2\right) + \\ & + c_2\left(I_3\right) d\left(c, J_1, J_2\right) + d_2\left(I_3\right) d\left(d, J_1, J_2\right) \end{aligned} \tag{3.174}$$

Step 3.2 Into each of the small tetrahedrons $A_2B_2C_2D_2$ with the serial indices $J_2 \in \left[1, N_2^3\right]$, N_3^3 small tetrahedrons $A_3B_3C_3D_3$ with the serial indices $J_3 \in \left[1, N_3^3\right]$ are known.

The set of the four-digital indices:

$$a_2(a, J_3)\ b_2(a, J_3)\ c_2(a, J_3)\ d_2(a, J_3) \tag{3.175}$$

$$a_2(b, J_3)\ b_2(b, J_3)\ c_2(b, J_3)\ d_2(b, J_3) \tag{3.176}$$

$$a_2(c, J_3)\ b_2(c, J_3)\ c_2(c, J_3)\ d_2(c, J_3) \tag{3.177}$$

$$a_2(d, J_3)\ b_2(d, J_3)\ c_2(d, J_3)\ d_2(d, J_3) \tag{3.178}$$

relatively the small tetrahedron $A_2B_2C_2D_2$ with the serial index $J_2 \in \left[1, N_2^3\right]$ for all of the four vertices A_3 (Eq. 3.175), B_3 (Eq. 3.176), C_3 (Eq. 3.177), and D_3 (Eq. 3.178) of each of the tetrahedrons $A_3B_3C_3D_3$ with the serial indices $J_3 \in \left[1, N_3^3\right]$ independent on J_1, J_2 are known and represented into the tables of the third, fifth, and sixth types in Chap. 4.

Thus, one can find the set of the serial indices $I_3 \in \left[1, \frac{(N_3+1)(N_3+2)(N_3+3)}{6}\right]$ for all of the four vertices A_3 (Eq. 3.179), B_3 (Eq. 3.180), C_3 (Eq. 3.181), and D_3 (Eq. 3.182) of each of the tetrahedrons $A_3B_3C_3D_3$ with the serial indices $J_3 \in \left[1, N_3^3\right]$ using Eq. 3.149 for N_3.

$$I_3(a, J_3) = I_3(a_2(a, J_3), b_2(a, J_3), c_2(a, J_3), d_2(a, J_3)) \tag{3.179}$$

$$I_3(b, J_3) = I_3(a_2(b, J_3), b_2(b, J_3), c_2(b, J_3), d_2(b, J_3)) \tag{3.180}$$

$$I_3(c, J_3) = I_3(a_2(c, J_3), b_2(c, J_3), c_2(c, J_3), d_2(c, J_3)) \tag{3.181}$$

$$I_3(d, J_3) = I_3(a_2(d, J_3), b_2(d, J_3), c_2(d, J_3), d_2(d, J_3)) \tag{3.182}$$

Therefore, using calculations carries out during Step 3.1 the set of the four-digital indices:

$$a(a, J_1, J_2, J_3)\ b(a, J_1, J_2, J_3)\ c(a, J_1, J_2, J_3)\ d(a, J_1, J_2, J_3) \tag{3.183}$$

$$a(b, J_1, J_2, J_3)\ b(b, J_1, J_2, J_3)\ c(b, J_1, J_2, J_3)\ d(b, J_1, J_2, J_3) \tag{3.184}$$

$$a(c, J_1, J_2, J_3)\ b(c, J_1, J_2, J_3)\ c(c, J_1, J_2, J_3)\ d(c, J_1, J_2, J_3) \tag{3.185}$$

$$a(d, J_1, J_2, J_3)\ b(d, J_1, J_2, J_3)\ c(d, J_1, J_2, J_3)\ d(d, J_1, J_2, J_3) \tag{3.186}$$

relatively the initial tetrahedron $ABCD$ for all of the four vertices A_3 (Eq. 3.183), B_3 (Eq. 3.184), C_3 (Eq. 3.185), and D_3 (Eq. 3.186) of each of the tetrahedrons $A_3B_3C_3D_3$ with the serial indices $J_3 \in \left[1, N_3^3\right]$ are known. For example, using Eq. 3.171 one can find:

$$b(a, J_1, J_2, J_3) = b(I_3(a, J_3), J_1, J_2). \tag{3.187}$$

Step k.0 All steps starting with Step k are carried out for each of the small tetrahedrons $A_{k-1}B_{k-1}C_{k-1}D_{k-1}$ with the serial indices $J_{k-1} \in \left[1, N_{k-1}^3\right]$.

Step k.1 For each of the small tetrahedrons $A_{k-1}B_{k-1}C_{k-1}D_{k-1}$ with the serial indices $J_{k-1} \in \left[1, N_{k-1}^3\right]$, the local indices of $\frac{(N_k+1)(N_k+2)(N_k+3)}{6}$ reference points into the small tetrahedron $A_{k-1}B_{k-1}C_{k-1}D_{k-1}$ are known.

For each of these reference points with the serial indices $I_k \in \left[1, \frac{(N_k+1)(N_k+2)(N_k+3)}{6}\right]$, one can calculate the four-digital indices $a_{k-1}(I_k)$, $b_{k-1}(I_k)$, $c_{k-1}(I_k)$, and $d_{k-1}(I_k)$ relatively the small tetrahedron $A_{k-1}B_{k-1}C_{k-1}D_{k-1}$ with the serial index $J_{k-1} \in \left[1, N_{k-1}^3\right]$ using Eq. 3.148 for N_k.

Thus, for each of these $\frac{(N_k+1)(N_k+2)(N_k+3)}{6}$ reference points with the serial indices I_k one can calculate new four-digital indices relatively the initial tetrahedron $ABCD$ using Eqs. 3.188–3.191.

$$a(I_k, J_1, \ldots, J_{k-1}) = a_{k-1}(I_k)\, a(a, J_1, \ldots, J_{k-1}) + b_{k-1}(I_k)\, a(b, J_1, \ldots, J_{k-1}) + \\ + c_{k-1}(I_k)\, a(c, J_1, \ldots, J_{k-1}) + d_{k-1}(I_k)\, a(d, J_1, \ldots, J_{k-1}) \quad (3.188)$$

$$b(I_k, J_1, \ldots, J_{k-1}) = a_{k-1}(I_k)\, b(a, J_1, \ldots, J_{k-1}) + b_{k-1}(I_k)\, b(b, J_1, \ldots, J_{k-1}) + \\ + c_{k-1}(I_k)\, b(c, J_1, \ldots, J_{k-1}) + d_{k-1}(I_k)\, b(d, J_1, \ldots, J_{k-1}) \quad (3.189)$$

$$c(I_k, J_1, \ldots, J_{k-1}) = a_{k-1}(I_k)\, c(a, J_1, \ldots, J_{k-1}) + b_{k-1}(I_k)\, c(b, J_1, \ldots, J_{k-1}) + \\ + c_{k-1}(I_k)\, c(c, J_1, \ldots, J_{k-1}) + d_{k-1}(I_k)\, c(d, J_1, \ldots, J_{k-1}) \quad (3.190)$$

$$d(I_k, J_1, \ldots, J_{k-1}) = a_{k-1}(I_k)\, d(a, J_1, \ldots, J_{k-1}) + b_{k-1}(I_k)\, d(b, J_1, \ldots, J_{k-1}) + \\ + c_{k-1}(I_k)\, d(c, J_1, \ldots, J_{k-1}) + d_{k-1}(I_k)\, d(d, J_1, \ldots, J_{k-1}) \quad (3.191)$$

Step k.2 Into each of the small tetrahedrons $A_{k-1}B_{k-1}C_{k-1}D_{k-1}$ with the serial indices $J_{k-1} \in \left[1, N_{k-1}^3\right]$, N_k^3 small tetrahedrons $A_kB_kC_kD_k$ with the serial indices $J_k \in \left[1, N_k^3\right]$ are known. The set of the four-digital indices:

$$a_{k-1}(a, J_k)\; b_{k-1}(a, J_k)\; c_{k-1}(a, J_k)\; d_{k-1}(a, J_k) \quad (3.192)$$

$$a_{k-1}(b, J_k)\; b_{k-1}(b, J_k)\; c_{k-1}(b, J_k)\; d_{k-1}(b, J_k) \quad (3.193)$$

$$a_{k-1}(c, J_k)\; b_{k-1}(c, J_k)\; c_{k-1}(c, J_k)\; d_{k-1}(c, J_k) \quad (3.194)$$

$$a_{k-1}(d, J_k)\; b_{k-1}(d, J_k)\; c_{k-1}(d, J_k)\; d_{k-1}(d, J_k) \quad (3.195)$$

relatively the small tetrahedron $A_{k-1}B_{k-1}C_{k-1}D_{k-1}$ with the serial index $J_{k-1} \in \left[1, N_{k-1}^3\right]$ for all of the four vertices A_k (Eq. 3.192), B_k (Eq. 3.193), C_k (Eq. 3.194), and D_k (Eq. 3.195) of each of the tetrahedrons $A_kB_kC_kD_k$ with the serial indices $J_k \in \left[1, N_k^3\right]$ independent on $J_1, J_2, \ldots, J_{k-1}$ are known and represented into the tables of the third, fifth, and sixth types in Chap. 4.

Thus, one can find the set of the serial indices $I_k \in \left[1, \frac{(N_k+1)(N_k+2)(N_k+3)}{6}\right]$ for all of the four vertices A_k (Eq. 3.196), B_k (Eq. 3.197), C_k (Eq. 3.198), and D_k (Eq. 3.199) of each of the tetrahedrons $A_kB_kC_kD_k$ with the serial indices $J_k \in \left[1, N_k^3\right]$ using Eq. 3.148 for N_k.

$$I_k(a, J_k) = I_k\left(a_{k-1}(a, J_k), b_{k-1}(a, J_k), c_{k-1}(a, J_k), d_{k-1}(a, J_k)\right) \quad (3.196)$$

$$I_k(b, J_k) = I_k\left(a_{k-1}(b, J_k), b_{k-1}(b, J_k), c_{k-1}(b, J_k), d_{k-1}(b, J_k)\right) \quad (3.197)$$

$$I_k(c, J_k) = I_k(a_{k-1}(c, J_k), b_{k-1}(c, J_k), c_{k-1}(c, J_k), d_{k-1}(c, J_k)) \quad (3.198)$$

$$I_k(d, J_k) = I_k(a_{k-1}(d, J_k), b_{k-1}(d, J_k), c_{k-1}(d, J_k), d_{k-1}(d, J_k)) \quad (3.199)$$

Therefore, using calculations carries out during Step $k.1$ the set of the four-digital indices:

$$a(a, J_1, \ldots, J_k)\ b(a, J_1, \ldots, J_k)\ c(a, J_1, \ldots, J_k)\ d(a, J_1, \ldots, J_k) \quad (3.200)$$

$$a(b, J_1, \ldots, J_k)\ b(b, J_1, \ldots, J_k)\ c(b, J_1, \ldots, J_k)\ d(b, J_1, \ldots, J_k) \quad (3.201)$$

$$a(c, J_1, \ldots, J_k)\ b(c, J_1, \ldots, J_k)\ c(c, J_1, \ldots, J_k)\ d(c, J_1, \ldots, J_k) \quad (3.202)$$

$$a(d, J_1, \ldots, J_k)\ b(d, J_1, \ldots, J_k)\ c(d, J_1, \ldots, J_k)\ d(d, J_1, \ldots, J_k) \quad (3.203)$$

relatively the initial tetrahedron *ABCD* for all of the four vertices A_k (Eq. 3.200), B_k (Eq. 3.201), C_k (Eq. 3.202), and D_k (Eq. 3.203) of each of the tetrahedrons $A_k B_k C_k D_k$ with the serial indices $J_k \in \left[1, N_k^3\right]$ are known. For example, using Eq. 3.188 one can find:

$$b(a, J_1, J_2, \ldots, J_k) = b(I_k(a, J_k), J_1, J_2 \ldots, J_{k-1}). \quad (3.204)$$

Step K−1.0 All steps starting with Step $K-1$ are carried out for each of the small tetrahedrons $A_{K-2}B_{K-2}C_{K-2}D_{K-2}$ with the serial indices $J_{K-2} \in \left[1, N_{K-2}^3\right]$.

Step K−1.1 For each of the small tetrahedrons $A_{K-2}B_{K-2}C_{K-2}D_{K-2}$ with the serial indices $J_{K-2} \in \left[1, N_{K-2}^3\right]$, the local indices of $\frac{(N_{K-1}+1)(N_{K-1}+2)(N_{K-1}+3)}{6}$ reference points into the small tetrahedron $A_{K-2}B_{K-2}C_{K-2}D_{K-2}$ are known. For each of these reference points with the serial indices $I_{K-1} \in \left[1, \frac{(N_{K-1}+1)(N_{K-1}+2)(N_{K-1}+3)}{6}\right]$, one can calculate the four-digital indices $a_{K-2}(I_{K-1})$, $b_{K-2}(I_{K-1})$, $c_{K-2}(I_{K-1})$, and $d_{K-2}(I_{K-1})$ relatively the small tetrahedron $A_{K-2}B_{K-2}C_{K-2}D_{K-2}$ with the index $J_{K-2} \in \left[1, N_{K-2}^3\right]$ using Eq. 3.148 for N_{K-1}.

Thus, for each of these $\frac{(N_{K-1}+1)(N_{K-1}+2)(N_{K-1}+3)}{6}$ reference points with the serial indices I_{K-1} one can calculate new four-digital indices relatively the initial tetrahedron *ABCD* using Eqs. 3.205–3.208.

$$\begin{aligned} a(I_{K-1}, J_1, \ldots, J_{K-2}) = {} & a_{K-2}(I_{K-1})\, a(a, J_1, \ldots, J_{K-2}) \\ & + b_{K-2}(I_{K-1})\, a(b, J_1, \ldots, J_{K-2}) \\ & + c_{K-2}(I_{K-1})\, a(c, J_1, \ldots, J_{K-2}) \\ & + d_{K-2}(I_{K-1})\, a(d, J_1, \ldots, J_{K-2}) \end{aligned} \quad (3.205)$$

$$
\begin{aligned}
b\left(I_{K-1}, J_1, \ldots, J_{K-2}\right) = {} & a_{K-2}\left(I_{K-1}\right) b\left(a, J_1, \ldots, J_{K-2}\right) \\
& + b_{K-2}\left(I_{K-1}\right) b\left(b, J_1, \ldots, J_{K-2}\right) \\
& + c_{K-2}\left(I_{K-1}\right) b\left(c, J_1, \ldots, J_{K-2}\right) \\
& + d_{K-2}\left(I_{K-1}\right) b\left(d, J_1, \ldots, J_{K-2}\right)
\end{aligned} \tag{3.206}
$$

$$
\begin{aligned}
c\left(I_{K-1}, J_1, \ldots, J_{K-2}\right) = {} & a_{K-2}\left(I_{K-1}\right) c\left(a, J_1, \ldots, J_{K-2}\right) \\
& + b_{K-2}\left(I_{K-1}\right) c\left(b, J_1, \ldots, J_{K-2}\right) \\
& + c_{K-2}\left(I_{K-1}\right) c\left(c, J_1, \ldots, J_{K-2}\right) \\
& + d_{K-2}\left(I_{K-1}\right) c\left(d, J_1, \ldots, J_{K-2}\right)
\end{aligned} \tag{3.207}
$$

$$
\begin{aligned}
d\left(I_{K-1}, J_1, \ldots, J_{K-2}\right) = {} & a_{K-2}\left(I_{K-1}\right) d\left(a, J_1, \ldots, J_{K-2}\right) \\
& + b_{K-2}\left(I_{K-1}\right) d\left(b, J_1, \ldots, J_{K-2}\right) \\
& + c_{K-2}\left(I_{K-1}\right) d\left(c, J_1, \ldots, J_{K-2}\right) \\
& + d_{K-2}\left(I_{K-1}\right) d\left(d, J_1, \ldots, J_{K-2}\right)
\end{aligned} \tag{3.208}
$$

Step K−1.2 Into each of the small tetrahedrons $A_{K-2}B_{K-2}C_{K-2}D_{K-2}$ with the serial indices $J_{K-2} \in \left[1, N_{K-2}^3\right]$, N_{K-1}^3 small tetrahedrons $A_{K-1}B_{K-1}C_{K-1}D_{K-1}$ with the serial indices $J_{K-1} \in \left[1, N_{K-1}^3\right]$ are known. The set of the four-digital indices:

$$a_{K-2}\left(a, J_{K-1}\right) b_{K-2}\left(a, J_{K-1}\right) c_{K-2}\left(a, J_{K-1}\right) d_{K-2}\left(a, J_{K-1}\right) \tag{3.209}$$

$$a_{K-2}\left(b, J_{K-1}\right) b_{K-2}\left(b, J_{K-1}\right) c_{K-2}\left(b, J_{K-1}\right) d_{K-2}\left(b, J_{K-1}\right) \tag{3.210}$$

$$a_{K-2}\left(c, J_{K-1}\right) b_{K-2}\left(c, J_{K-1}\right) c_{K-2}\left(c, J_{K-1}\right) d_{K-2}\left(c, J_{K-1}\right) \tag{3.211}$$

$$a_{K-2}\left(d, J_{K-1}\right) b_{K-2}\left(d, J_{K-1}\right) c_{K-2}\left(d, J_{K-1}\right) d_{K-2}\left(d, J_{K-1}\right) \tag{3.212}$$

relatively the small tetrahedron $A_{K-2}B_{K-2}C_{K-2}D_{K-2}$ with the serial index $J_{K-2} \in \left[1, N_{K-2}^3\right]$ for all of the four vertices A_{K-1} (Eq. 3.209), B_{K-1} (Eq. 3.210), C_{K-1} (Eq. 3.211), and D_{K-1} (Eq. 3.212) of each of the tetrahedrons $A_{K-1}B_{K-1}C_{K-1}D_{K-1}$ with the serial indices $J_{K-1} \in \left[1, N_{K-1}^3\right]$ independent on $J_1, J_2, \ldots, J_{K-2}$ are known and represented into the tables of the third, fifth, and sixth types in Chap. 4.

Thus, one can find the set of the serial indices $I_{K-1} \in \left[1, \frac{(N_{K-1}+1)(N_{K-1}+2)(N_{K-1}+3)}{6}\right]$ for all of the four vertices A_{K-1} (Eq. 3.213), B_{K-1} (Eq. 3.214), C_{K-1} (Eq. 3.215), and D_{K-1} (Eq. 3.216) of each of the tetrahedrons $A_{K-1}B_{K-1}C_{K-1}D_{K-1}$ with the serial indices $J_{K-1} \in \left[1, N_{K-1}^3\right]$ using Eq. 3.148 for N_{K-1}.

$$
\begin{aligned}
I_{K-1}\left(a, J_{K-1}\right) = {} & I_{K-1}\left(a_{K-2}\left(a, J_{K-1}\right), b_{K-2}\left(a, J_{K-1}\right),\right. \\
& \left. c_{K-2}\left(a, J_{K-1}\right), d_{K-2}\left(a, J_{K-1}\right)\right)
\end{aligned} \tag{3.213}
$$

$$
\begin{aligned}
I_{K-1}\left(b, J_{K-1}\right) = {} & I_{K-1}\left(a_{K-2}\left(b, J_{K-1}\right), b_{K-2}\left(b, J_{K-1}\right),\right. \\
& \left. c_{K-2}\left(b, J_{K-1}\right), d_{K-2}\left(b, J_{K-1}\right)\right)
\end{aligned} \tag{3.214}
$$

$$I_{K-1}(c, J_{K-1}) = I_{K-1}(a_{K-2}(c, J_{K-1}), b_{K-2}(c, J_{K-1}), c_{K-2}(c, J_{K-1}), d_{K-2}(c, J_{K-1})) \tag{3.215}$$

$$I_{K-1}(c, J_{K-1}) = I_{K-1}(a_{K-2}(d, J_{K-1}), b_{K-2}(d, J_{K-1}), c_{K-2}(d, J_{K-1}), d_{K-2}(d, J_{K-1})) \tag{3.216}$$

Therefore, using calculations carries out during Step $K-1.1$ the set of the four-digital indices:

$$a(a, J_1, \ldots, J_{K-1})\; b(a, J_1, \ldots, J_{K-1})\; c(a, J_1, \ldots, J_{K-1})\; d(a, J_1, \ldots, J_{K-1}) \tag{3.217}$$

$$a(b, J_1, \ldots, J_{K-1})\; b(b, J_1, \ldots, J_{K-1})\; c(b, J_1, \ldots, J_{K-1})\; d(b, J_1, \ldots, J_{K-1}) \tag{3.218}$$

$$a(c, J_1, \ldots, J_{K-1})\; b(c, J_1, \ldots, J_{K-1})\; c(c, J_1, \ldots, J_{K-1})\; d(c, J_1, \ldots, J_{K-1}) \tag{3.219}$$

$$a(d, J_1, \ldots, J_{K-1})\; b(d, J_1, \ldots, J_{K-1})\; c(d, J_1, \ldots, J_{K-1})\; d(d, J_1, \ldots, J_{K-1}) \tag{3.220}$$

relatively the initial tetrahedron *ABCD* for all of the four vertices A_{K-1} (Eq. 3.217), B_{K-1} (Eq. 3.218), C_{K-1} (Eq. 3.219), and D_{K-1} (Eq. 3.220) of each of the tetrahedrons $A_{K-1}B_{K-1}C_{K-1}D_{K-1}$ with the serial indices $J_{K-1} \in \left[1, N_{K-1}^3\right]$ are known. For example, using Eq. 3.205 one can find:

$$b(a, J_1, J_2, \ldots, J_{K-1}) = b(I_{K-1}(a, J_{K-1}), J_1, J_2 \ldots, J_{K-2}). \tag{3.221}$$

Step K.0 All Steps K are carried out for each of the small tetrahedrons $A_{K-1}B_{K-1}C_{K-1}D_{K-1}$ with the serial indices $J_{K-1} \in \left[1, N_{K-1}^3\right]$.

Step K.1 For each of the small tetrahedrons $A_{K-1}B_{K-1}C_{K-1}D_{K-1}$ with the serial indices $J_{K-1} \in \left[1, N_{K-1}^3\right]$, the local indices of $\frac{(N_K+1)(N_K+2)(N_K+3)}{6}$ reference points into the small tetrahedron $A_{K-1}B_{K-1}C_{K-1}D_{K-1}$ are known. For each of these $\frac{(N_K+1)(N_K+2)(N_K+3)}{6}$ reference points with the four-digital indices $a_{K-1}b_{K-1}c_{K-1}d_{K-1}$ relative the small tetrahedron $A_{K-1}B_{K-1}C_{K-1}D_{K-1}$ with the serial index J_{K-1}, one can calculate new four-digital indices relatively the initial tetrahedron *ABCD* using Eqs. 3.222–3.225.

$$\begin{aligned} a(a_{K-1}b_{K-1}c_{K-1}, J_1, \ldots, J_{K-1}) = {} & a_{K-1}a(a, J_1, \ldots, J_{K-1}) \\ & + b_{K-1}a(b, J_1, \ldots, J_{K-1}) \\ & + c_{K-1}a(c, J_1, \ldots, J_{K-1}) \\ & + d_{K-1}a(d, J_1, \ldots, J_{K-1}) \end{aligned} \tag{3.222}$$

$$
\begin{aligned}
b\left(a_{K-1}b_{K-1}c_{K-1}, J_1, \ldots, J_{K-1}\right) = a_{K-1}b\left(a, J_1, \ldots, J_{K-1}\right) \\
+ b_{K-1}b\left(b, J_1, \ldots, J_{K-1}\right) \\
+ c_{K-1}b\left(c, J_1, \ldots, J_{K-1}\right) \\
+ d_{K-1}b\left(d, J_1, \ldots, J_{K-1}\right)
\end{aligned} \tag{3.223}
$$

$$
\begin{aligned}
c\left(a_{K-1}b_{K-1}c_{K-1}, J_1, \ldots, J_{K-1}\right) = a_{K-1}c\left(a, J_1, \ldots, J_{K-1}\right) \\
+ b_{K-1}c\left(b, J_1, \ldots, J_{K-1}\right) \\
+ c_{K-1}c\left(c, J_1, \ldots, J_{K-1}\right) \\
+ d_{K-1}c\left(d, J_1, \ldots, J_{K-1}\right)
\end{aligned} \tag{3.224}
$$

$$
\begin{aligned}
d\left(a_{K-1}b_{K-1}c_{K-1}, J_1, \ldots, J_{K-1}\right) = a_{K-1}d\left(a, J_1, \ldots, J_{K-1}\right) \\
+ b_{K-1}d\left(b, J_1, \ldots, J_{K-1}\right) \\
+ c_{K-1}d\left(c, J_1, \ldots, J_{K-1}\right) \\
+ d_{K-1}d\left(d, J_1, \ldots, J_{K-1}\right)
\end{aligned} \tag{3.225}
$$

Note that the algorithms on Steps 1 and K differ from the algorithm on Steps $2, 3, \ldots, K-1$.

Therefore, if the local four-digital index $a_{K-1}b_{K-1}c_{K-1}d_{K-1}$ relative the tetrahedron $A_{K-1}B_{K-1}C_{K-1}D_{K-1}$ with the serial index J_{K-1} of the reference point under consideration and the set of the indices of the small tetrahedron $J_1, J_2, \ldots, J_{K-1}$, $J_k \in N_k^3$, in which this reference point under consideration lies are known, one can find the global four-digital index relative the initial tetrahedron *ABCD* using Eqs. 3.222–3.225.

Consider an example: $K = 2$, $N_1 = N_2 = 2$. The indices of the vertices of the tetrahedron with the index $J_1 = 2$ are 1100, 0200, 0110, and 0101 relative the initial tetrahedron. Thus, the global index of the reference point with the local index 1100 relative the small tetrahedron with the index $J_1 = 2$ will be 1300 and can be calculated using Eqs. 3.226–3.229.

$$a = 1 \cdot 1 + 1 \cdot 0 + 0 \cdot 0 + 0 \cdot 0 = 1 \tag{3.226}$$

$$b = 1 \cdot 1 + 1 \cdot 2 + 0 \cdot 1 + 0 \cdot 1 = 3 \tag{3.227}$$

$$c = 1 \cdot 0 + 1 \cdot 0 + 0 \cdot 1 + 0 \cdot 0 = 0 \tag{3.228}$$

$$d = 1 \cdot 0 + 1 \cdot 0 + 0 \cdot 0 + 0 \cdot 1 = 0 \tag{3.229}$$

3.3.3 Hybrid Parabolic Linear Interpolation

In this Section, a hybrid parabolic linear interpolation method is proposed. This interpolation method combines the monotonicity of a piecewise-linear function in some small tetrahedrons with the second order of the quadratic function in other small tetrahedrons, where this does not lead to the formation of additional extrema that do not coincide with the reference points. The resulting function ought to remain continuous, when passing from one tetrahedron to another one. Such hybrid interpolation provides the numerical method with the property of monotonicity and avoids the non-physical oscillations of the solution.

The reference points 2000, 0200, 0020, 0002, 1100, 0110, 0011, 1001, 1010, and 0101 for the polynomial interpolation of the second order are used. The algorithm for restoring a monotone function in a grid tetrahedron is as follows:

Step 1 Define the trial function using the normal polynomial interpolation of the second order.

Step 2 Determine the values in the centers of the edges as follows. If the trial function has not extremum on this edge, then take the value of the test function; otherwise use a linear interpolation on the given edge. Therefore, if there are values u_{A} and u_{B} into the vertices of the edge of the grid tetrahedron and the value u_{AB} in the middle of this edge, then one can use the following inequalities. If $u_{\mathrm{AB}} \in [u_{\mathrm{A}}, u_{\mathrm{B}}]$ or $u_{\mathrm{AB}} \in [u_{\mathrm{B}}, u_{\mathrm{A}}]$, then $u_{\mathrm{AB}}^{\mathrm{NEW}} = u_{\mathrm{AB}}$. In another case, $u_{\mathrm{AB}}^{\mathrm{NEW}} = (u_{\mathrm{A}} + u_{\mathrm{B}})/2$.
This procedure should be performed for all the central-edge reference points 1100, 0110, 0011, 1001, 1010, and 0101. If the central-edge reference point is 1100, then the vertices of the edge are 2000 and 0200. If the central-edge reference point is 0110, then the vertices of the edge are 0200 and 0020. If the central-edge reference point is 0011, then the vertices of the edge are 0020 and 0002. If the central-edge reference point is 1001, then the vertices of the edge are 2000 and 0002. If the central-edge reference point is 1010, then the vertices of the edge are 2000 and 0020. If the central-edge reference point is 0101, then the vertices of the edge are 0200 and 0002.

Step 3 Determine the final value of the interpolant using the normal polynomial interpolation of the second order with new values in the centers of the edges determined on Step 2.

3.3.4 Interpolation Using Min—Max Limiter

Algorithm for constructing an interpolant with a limiter on tetrahedral grids based on interpolation by a polynomial of order N involves the following steps:

Step 1 The value of the test function to the given point $\vec{r}$ using polynomial interpolation of order N is determined. Let it be equal to $u_N\left(\vec{r}\right)$.

Step 2 The small tetrahedron or octahedron, in which the point $\vec{r}$ falls, should be determined. If the point falls into the octahedron, then one of the three possible variants can be chosen. The axis is dividing this octahedron into four tetrahedral, each of them has a volume being equal to the volumes of other small tetrahedra, but not being similar to them. Next, one of these four tetrahedral, in which of them the point falls into, should be determined.

Step 3 Then $u_N\left(\vec{r}\right)$ should be compared with the minimum m and maximum M values at the vertices of this small tetrahedron, in which the point falls.

Step 3.1 If $m \le u_N\left(\vec{r}\right) \le M$, then the value of the interpolant at the point $\vec{r}$ is taking equal to $u_N\left(\vec{r}\right)$.

Step 3.2 If $u_N\left(\vec{r}\right) < m$, then the value of the interpolant at the point $\vec{r}$ is equal to m.

Step 3.3 If $u_N\left(\vec{r}\right) > M$, then the value of the interpolant at the point $\vec{r}$ is equal to M.

The use of interpolation with a Min-Max limiter makes it possible to eliminate the non-physical oscillations of polynomials arising in the presence of discontinuities in the interpolated functions.

3.3.5 Parabolic Interpolation on Reference Points for Interpolation of Fourth Order

Consider a tetrahedron *ABCD* with the volume V and 35 reference points 4000, 0400, 0040, 0004, 3100, 0310, 0031, 1003, 1300, 0130, 0013, 3001, 3010, 0301, 1030, 0103, 2200, 0220, 0022, 2002, 2020, 0202, 2110, 1210, 1120, 0211, 0121, 0112, 2101, 1201, 1102, 2011, 1021, 1012, and 1111 into this tetrahedron for $N = 4$.

One can divide this tetrahedron *ABCD* into 8 small tetrahedrons with volumes $V/8$ using a set of 10 reference points in the Table 3.1 for $N = 2$.

One can use the algorithm discussed in Sect. 3.3.1 to understand, in which of the 8 small tetrahedrons the point under consideration lies. Then one can use the formulae for the parabolic interpolation discussed in Sect. 3.2.3 and use the Table 3.2 containing the congruence of four-digital indices for N = 4 and four-digital indices for N = 2 for each of the 8 small tetrahedrons.

Table 3.1 Congruence of four-digital indices of reference points for $N = 2$ in order to divide the initial tetrahedron into 8 small tetrahedrons and initial four-digital indices for $N = 4$

Index to divide, $N = 2$	2000	0200	0020	0002	1100	0110	0011	1001	1010	0101
Initial index, $N = 4$	4000	0400	0040	0004	2200	0220	0022	2002	2020	0202

Table 3.2 Congruence of four-digital indices of reference points for $N = 2$ and initial four-digital indices for $N = 4$ for each of the 8 small tetrahedrons and for each of the axes

Axis No	No	2000	0200	0020	0002	1100	0110	0011	1001	1010	0101
–	1	4000	2200	2020	2002	3100	2110	2011	3001	3010	2101
	2	2200	0400	0220	0202	1300	0310	0211	1201	1210	0301
	3	2020	0220	0040	0022	1120	0130	0031	1021	1030	0121
	4	2002	0202	0022	0004	1102	0112	0013	1003	1012	0103
1	5	0220	2002	2200	2020	1111	2101	2110	1120	1210	2011
	6	0220	2002	2200	0202	1111	2101	1201	0211	1210	1102
	7	0220	2002	0022	2020	1111	1012	1021	1120	0121	2011
	8	0220	2002	0022	0202	1111	1012	0112	0211	0121	1102
2	5	2020	0202	2200	0220	1111	1201	1210	1120	2110	0211
	6	2020	0202	2200	2002	1111	1201	2101	2011	2110	1102
	7	2020	0202	0022	0220	1111	0112	0121	1120	1021	0211
	8	2020	0202	0022	2002	1111	0112	1012	2011	1021	1102
3	5	2200	0022	0220	2020	1111	0121	1120	2110	1210	1021
	6	2200	0022	0220	0202	1111	0121	0211	1201	1210	0112
	7	2200	0022	2002	2020	1111	1012	2011	2110	2101	1021
	8	2200	0022	2002	0202	1111	1012	1102	1201	2101	0112

3.4 Conclusions

The analytical formulae for high-order interpolation on unstructured tetrahedral grids, such as the polynomial interpolation, hybrid parabolic linear interpolation on the unstructured tetrahedral grids, interpolation on the unstructured tetrahedral grids using Min-Max limiter, and parabolic interpolation on the reference points for interpolation of fourth order are suggested in this Chapter. The cases of order from 1 to 5 inclusively are considered. These interpolation techniques can be used during creation new unstructured tetrahedral or regular gird instead of previous one as an element of numerical method for finding 3D solutions on the unstructured tetrahedral grids and during visualization of some space field and images' creation or transformation. Also, the hierarchical nested unstructured tetrahedral grids and formulae for recalculation from local to global indices are discussed in this Chapter. This type of grids can be used also as an element of numerical method on the unstructured tetrahedral grids and for the visualization, creation, and transformation of images. These formulae for recalculation indices are used to decrease the amount of calculation and software operation time.

Acknowledgements This work has been performed at Non-state Educational Institution "Educational Scientific and Experimental Center of Moscow Institute of Physics and Technology" and supported by the Russian Science Foundation, grant no. 17-71-20088. This work has been carried out using computing resources of the federal collective usage center Complex for Simulation and Data Processing for Mega-science Facilities at NRC "Kurchatov Institute", http://ckp.nrcki.ru/.

References

1. Favorskaya AV, Muratov MV, Petrov IB, Sannikov IV (2014) Grid-characteristic method on unstructured tetrahedral meshes. Comput Math Math Phys 54(5):837–847
2. Favorskaya AV, Petrov IB (2016) A study of high-order grid-characteristic methods on unstructured grids. Numerical Anal Appl 9(2):171–178
3. Favorskaya AV, Petrov IB (2016) Wave responses from oil reservoirs in the Arctic shelf zone. Dokl Earth Sci 466(2):214–217
4. Favorskaya A, Petrov I, Golubev V, Khokhlov N (2017) Numerical simulation of earthquakes impact on facilities by grid-characteristic method. Procedia Comput Sci 112:1206–1215
5. Favorskaya AV, Petrov IB (2017) Numerical modeling of dynamic wave effects in rock masses. Dokl Math 95(3):287–290
6. Vassilevski YV, Beklemysheva KA, Grigoriev GK, Kazakov AO, Kulberg NS, Petrov IB, Salamatova VY, Vasyukov AV (2016) Transcranial ultrasound of cerebral vessels in silico: proof of concept. Russ J Numerical Anal Math Model 31(5):317–328
7. Petrov IB, Favorskaya AV, Khokhlov NI, Miryakha VA, Sannikov AV, Golubev VI (2015) Monitoring the state of the moving train by use of high performance systems and modern computation methods. Math Model Comput Simulations 7(1):51–61
8. Favorskaya A, Petrov I, Grinevskiy A (2017) Numerical simulation of fracturing in geological medium. Procedia Comput Sci 112:1216–1224
9. Petrov I, Vasyukov A, Beklemysheva K, Ermakov A, Favorskaya A (2016) Numerical modeling of non-destructive testing of composites. Procedia Comput Sci 96:930–938

10. Kotelnikov SA, Favorskaya AV, Petrov IB, Khokhlov NI, Miryakha VA (2016) Numerical simulation of non-destructive ultrasonic railway control. In: Civil-Comp Proceedings 110
11. Ball JA, Gohberg I (2013) Interpolation of rational matrix functions, vol 45. Birkhauser
12. Lama RK, Kwon GR (2015) New interpolation method based on combination of discrete cosine transform and wavelet transform. In: Information networking (ICOIN), 2015 international conference. IEEE, pp 363–366
13. Li J, Heap AD (2014) Spatial interpolation methods applied in the environmental sciences: a review. Environ Model Softw 53:173–189
14. Agarwal S, Khade S, Dandawate Y, Khandekar P (2015) Three dimensional image reconstruction using interpolation of distance and image registration. In: Computer, communication and control (IC4), 2015 international conference. IEEE, pp 1–5
15. Dong W, Zhang L, Lukac R, Shi G (2013) Sparse representation based image interpolation with nonlocal autoregressive modeling. IEEE Trans Image Process 22(4):1382–1394
16. Greco L, Cuomo M (2014) An implicit G1 multi patch B-spline interpolation for Kirchhoff-Love space rod. Comput Methods Appl Mech Eng 269:173–197
17. Hu J, Zhang S (2015) A family of symmetric mixed finite elements for linear elasticity on tetrahedral grids. Sci China Math 58(2):297–307
18. Nguyen MN, Bui TQ, Truong TT, Trinh NA, Singh IV, Yu T, Doan DH (2016) Enhanced nodal gradient 3D consecutive-interpolation tetrahedral element (CTH4) for heat transfer analysis. Int J Heat Mass Transf 103:14–27
19. Alauzet F (2016) A parallel matrix-free conservative solution interpolation on unstructured tetrahedral meshes. Comput Methods Appl Mech Eng 299:116–142
20. Paille GP, Ray N, Poulin P, Sheffer A, Levy B (2015) Dihedral angle-based maps of tetrahedral meshes. ACM Transactions on Graphics (TOG) 34(4):54
21. Burago NG, Nikitin IS, Yakushev VL (2016) Hybrid numerical method with adaptive overlapping meshes for solving nonstationary problems in continuum mechanics. Comput Math Math Phys 56(6):1065–1074
22. Lee DT, Schachter BJ (1980) Two algorithms for constructing a Delaunay triangulation. Int J Comput Inform Sci 9(3):219–242

Chapter 4
Piecewise Linear Interpolation on Unstructured Tetrahedral Grids

Alena V. Favorskaya

Abstract This chapter develops analytical expressions for piecewise linear interpolation on unstructured tetrahedral meshes. These expressions are crucial because most of the arithmetic operations are performed analytically and these operations no longer need to be repeated during using the software. So, these expressions allow to perform the supercomputer calculations with less amount of resources as dynamic computer memory. The interpolation on unstructured grids plays a key role for numerical solving of last amount of problems in seismic exploration of oil and gas, non-destructive testing of different up-to-date complex materials, investigation of seismic stability of different complex objects like nuclear power plants, and especially for summation of injuries of human bodies for medicine.

Keywords Interpolation · High-order interpolation · Unstructured grids Tetrahedral grids · Polynomial interpolation · Piecewise interpolation · Hybrid interpolation

4.1 Introduction

In this chapter, analytical expressions for piecewise linear interpolation on unstructured tetrahedral grids is discussed. Other analytical expressions for the

A. V. Favorskaya (✉)
Non-State Educational Institution "Educational Scientific and Experimental Center of Moscow Institute of Physics and Technology", 9, Institutsky Pereulok st., Dolgoprudny, Moscow Region 141700, Russian Federation
e-mail: aleanera@yandex.ru

A. V. Favorskaya
Department of Computer Science and Numerical Mathematics, Moscow Institute of Physics and Technology, 9, Institutsky Pereulok st., Dolgoprudny, Moscow Region 141700, Russian Federation

A. V. Favorskaya
Scientific Research Institute for System Studies of the Russian Academy of Sciences, 36(1), Nahimovskij ave., Moscow 117218, Russian Federation

A. V. Favorskaya and I. B. Petrov (eds.), *Innovations in Wave Processes Modelling and Decision Making*, Smart Innovation, Systems and Technologies 90,
https://doi.org/10.1007/978-3-319-76201-2_4

interpolation on unstructured tetrahedral grids are considered in Chap. 3. The crucial issues of reasons for using an interpolation on the unstructured meshes are discussed in Chap. 2. For example, a piecewise linear interpolation can be used as an element of grid-characteristic method [1–9]. The issues of interpolation require very specific mathematical approaches, so there are a lot of recent works in this area [10–19]. More detailed review of the works [10–15] is given in Chap. 2. More detailed review of the works [16–19] is done in Chap. 3.

In this chapter, only the single case is considered, when the point under consideration lays in the big tetrahedron *ABCD* and the relative volumes discussed in Sect. 3.2 for this point under consideration exist. Therefore, the following inequalities are always satisfied:

$$(v_A \in [0, 1]) \wedge (v_B \in [0, 1]) \wedge (v_C \in [0, 1]) \wedge (v_D \in [0, 1]). \tag{4.1}$$

In order to diminish the amount of mathematical expressions, six types of tables are determined in Sect. 4.3 and adduce these tables for different N in Sects. 4.4–4.7, respectively.

The chapter is structured as follows. In Sect. 4.2, a case of octahedron is discussed. In Sect. 4.3, the types of the tables are introduced. In Sects. 4.4–4.7, the tables for different amount of reference points and appropriate algorithm are presented. The conclusions are given in Sect. 4.8.

4.2 Case of Octahedron

The planes discussed in Sect. 3.2.1, divide the big tetrahedron *ABCD* into $\frac{N(N^2+2)}{3}$ small tetrahedrons and $\frac{N(N-1)(N+1)}{6}$ octahedrons. The volume of the big tetrahedron *ABCD* is equal to V. Then the volume of every small tetrahedrons is equal to V/N^3. All of these small tetrahedrons are similar to the big tetrahedron *ABCD*. The volume of every octahedron is equal to $4V/N^3$. One can draw an axis in every octahedron by one of three ways. This axis will divide the octahedron into 4 different small tetrahedrons. The volume of every of these different tetrahedrons is equal to V/N^3 as well. However, they are not similar to the big tetrahedron *ABCD*. In order to simplify our discussion let us consider degree $N = 2$. In this case, 4 smaller tetrahedrons and 1 octahedron are drawn in Fig. 4.1. The vertices of these 4 small tetrahedrons are written in Table 4.1 from Sect. 4.4 and the vertices of this octahedron are written in Table 4.2 from Sect. 4.4. Note that if the axis is determined for each of $\frac{N(N-1)(N+1)}{6}$ octahedrons, the total amount of small tetrahedrons into the initical tetrahedron will be equal to N^3.

These different tetrahedrons for all 3 ways of drawing the axis into the octahedron for $N = 2$ are shown in Figs. 4.2, 4.3 and 4.4 respectively.

Thus, N^3 small tetrahedrons of two types are. The small tetrahedrons of the first type are similar for the big one. The small tetrahedrons of the second type are not

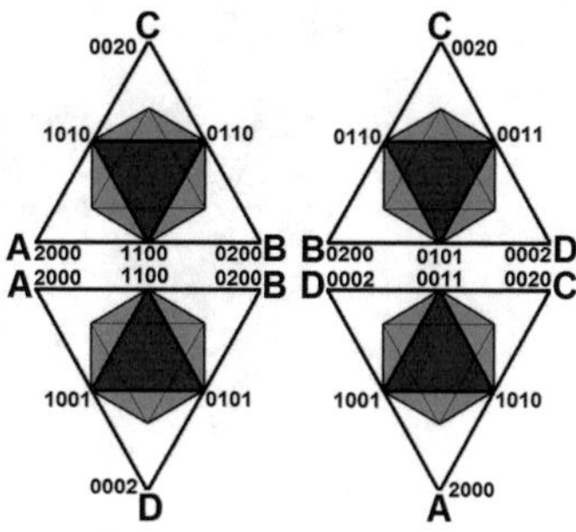

Fig. 4.1 Octahedron based on reference points for polynomial interpolation with order 2

Table 4.1 The vertices of the small tetrahedrons for $N = 2$, *table of the third type*

No	A'	B'	C'	D'
1	2000	1100	1010	1001
2	1100	0200	0110	0101
3	1010	0110	0020	0011
4	1001	0101	0011	0002

Table 4.2 The vertices of the octahedron for $N = 2$, *table of the third type*

No	A'	B'	C'	D'	E'	F'
5	1100	0110	0011	1001	1010	0101

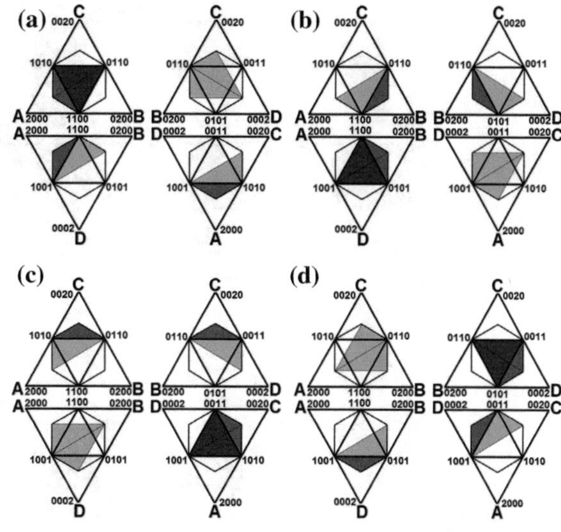

Fig. 4.2 Small different tetrahedrons in the case of using the first way of drawing axis in the octahedron based on reference points for polynomial interpolation with order 2: **a** the first small different tetrahedron, **b** the second small different tetrahedron, **c** the third small different tetrahedron, **d** the fourth small different tetrahedron

similar to the big one. But they have the same volume as small tetrahedrons of the first type. Also each octahedron contains from four small tetrahedrons of the second type. And there are three different ways to delete the octahedron by four small tetrahedrons of the second type.

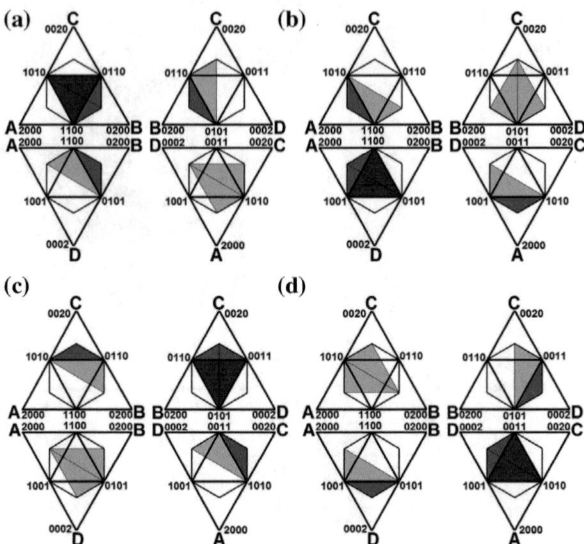

Fig. 4.3 Small different tetrahedrons in the case of using the second way of drawing axis in the octahedron based on reference points for polynomial interpolation with order 2: **a** the first small different tetrahedron, **b** the second small different tetrahedron, **c** the third small different tetrahedron, **d** the fourth small different tetrahedron

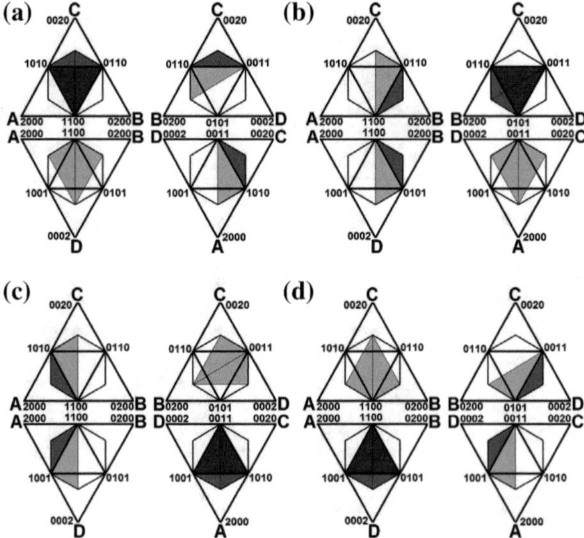

Fig. 4.4 Small different tetrahedrons in the case of using the third way of drawing axis in the octahedron based on reference points for polynomial interpolation with order 2: **a** the first small different tetrahedron, **b** the second small different tetrahedron, **c** the third small different tetrahedron, **d** the fourth small different tetrahedron

Table 4.3 Inequalities for finding the small tetrahedron or the octahedron, in which the point under consideration lies for $N = 2$, *table of the first type*

No	$2v_A$	$2v_B$	$2v_C$	$2v_D$	No	$2v_A$	$2v_B$	$2v_C$	$2v_D$
1	>1	–	–	–	4	–	–	–	>1
2	–	>1	–	–	5	–	–	–	–
3	–	–	>1	–					

Table 4.4 Logical expressions being true if the point under consideration lies into the small tetrahedron or into the octahedron for $N = 2$, *table of the second type*

No	$2v_A \in$	$2v_B \in$	$2v_C \in$	$2v_D \in$	No	$2v_A \in$	$2v_B \in$	$2v_C \in$	$2v_D \in$
1	$(1, 2]$	$[0, 1]$	$[0, 1]$	$[0, 1]$	4	$[0, 1]$	$[0, 1]$	$[0, 1]$	$(1, 2]$
2	$[0, 1]$	$(1, 2]$	$[0, 1]$	$[0, 1]$	5	$[0, 1]$	$[0, 1]$	$[0, 1]$	$[0, 1]$
3	$[0, 1]$	$[0, 1]$	$(1, 2]$	$[0, 1]$					

Table 4.5 Formulae for relative volumes corresponding the appropriate vertices of the small tetrahedrons for $N = 2$, *table of the fourth type*

No	$v_{A'}$	$v_{B'}$	$v_{C'}$	$v_{D'}$
1	$2v_A - 1$	$2v_B$	$2v_C$	$2v_D$
2	$2v_A$	$2v_B - 1$	$2v_C$	$2v_D$
3	$2v_A$	$2v_B$	$2v_C - 1$	$2v_D$
4	$2v_A$	$2v_B$	$2v_C$	$2v_D - 1$

Table 4.6 Axis points for each of 3 ways to draw them for the octahedron and corresponding expressions for v_1 and v_2 for $N = 2$, *table of the fifth type*

No	Axis no	A'	B'	v_1	v_2
5	1	0110	1001	$(2v_A + 2v_B - 2v_C - 2v_D)/2$	$(2v_A + 2v_C - 2v_B - 2v_D)/2$
	2	1010	0101	$(2v_A + 2v_B - 2v_C - 2v_D)/2$	$(2v_B + 2v_C - 2v_D - 2v_A)/2$
	3	1100	0011	$(2v_B + 2v_C - 2v_D - 2v_A)/2$	$(2v_A + 2v_C - 2v_B - 2v_D)/2$

4.3 Types of Tables and Their Description

In order to diminish the amount of mathematical expressions, six types of tables are defined. The lines in the tables help to catch the topological similarity of the formulae and due to this it is suitable to search for errors and misprints in the program code.

Tables of the first type contain the inequalities for finding the small tetrahedron or the octahedron. Tables 4.3 and 4.7 in Sect. 4.4, Table 4.11 in Sect. 4.5, Table 4.18 in Sect. 4.6, and Table 4.25 in Sect. 4.7 are the tables of the first type.

In order to obtain an inequality based on the filled cell in row I and in column J from the table of the first type, one can take the relative volume in the zero row and column J and put it before the condition in the cell IJ. For example, the inequality based on the cell 3, 3 in Table 4.3 could be written as follows:

Table 4.7 Inequalities for finding the different small tetrahedron, in which the point under consideration lies, *table of the first type*

No	v_1	v_2
1	>0	>0
2	>0	–
3	–	>0
4	–	–

Table 4.8 Logical expressions being true if the point under consideration lies into the different small tetrahedron, *table of the second type*

No	$v_1 \in$	$v_2 \in$
1	$(0, 1]$	$(0, 1]$
2	$(0, 1]$	$[-1, 0]$
3	$[-1, 0]$	$(0, 1]$
4	$[-1, 0]$	$[-1, 0]$

Table 4.9 The vertices C', D' of the different small tetrahedrons and the formulae for relative volumes corresponding the appropriate vertices A', B' of the different small tetrahedrons for $N = 2$, *table of the sixth type*

No	Axis no	Small tetrahedron no	C'	D'	$v_{A'}$	$v_{B'}$
5	1	1	1100	1010	$1 - 2v_A$	$2v_D$
		2	1100	0101	$2v_C$	$1 - 2v_B$
		3	0011	1010	$2v_B$	$1 - 2v_C$
		4	0011	0101	$1 - 2v_D$	$2v_A$
	2	1	1100	0110	$1 - 2v_B$	$2v_D$
		2	1100	1001	$2v_C$	$1 - 2v_A$
		3	0011	0110	$2v_A$	$1 - 2v_C$
		4	0011	1001	$1 - 2v_D$	$2v_B$
	3	1	0110	1010	$1 - 2v_C$	$2v_D$
		2	0110	0101	$2v_A$	$1 - 2v_B$
		3	1001	1010	$2v_B$	$1 - 2v_A$
		4	1001	0101	$1 - 2v_D$	$2v_C$

Table 4.10 Formulae for relative volumes corresponding the appropriate vertices C', D' of the different small tetrahedrons, *table of the fourth type*

Small tetrahedron no	$v_{C'}$	$v_{D'}$
1	v_1	v_2
2	v_1	$-v_2$
3	$-v_1$	v_2
4	$-v_1$	$-v_2$

Table 4.11 Inequalities for finding the small tetrahedron or the octahedron, in which the point under consideration lies for $N = 3$, *table of the first type*

No	$3v_A$	$3v_B$	$3v_C$	$3v_D$	No	$3v_A$	$3v_B$	$3v_C$	$3v_D$	No	$3v_A$	$3v_B$	$3v_C$	$3v_D$
1	>2	–	–	–	6	–	>1	>1	–	11	>1	–	–	–
2	–	>2	–	–	7	–	–	>1	>1	12	–	>1	–	–
3	–	–	>2	–	8	>1	–	–	>1	13	–	–	>1	–
4	–	–	–	>2	9	>1	–	>1	–	14	–	–	–	>1
5	>1	>1	–	–	10	–	>1	–	>1	15	–	–	–	–

Table 4.12 Logical expressions being true if the point under consideration lies into the small tetrahedron or into the octahedron for $N = 3$, *table of the second type*

No	$3v_A \in$	$3v_B \in$	$3v_C \in$	$3v_D \in$	No	$3v_A \in$	$3v_B \in$	$3v_C \in$	$3v_D \in$	No	$3v_A \in$	$3v_B \in$	$3v_C \in$	$3v_D \in$
1	(2, 3]	[0, 1]	[0, 1]	[0, 1]	6	[0, 1]	(1, 2]	(1, 2]	[0, 1]	11	(1, 2]	[0, 1]	[0, 1]	[0, 1]
2	[0, 1]	(2, 3]	[0, 1]	[0, 1]	7	[0, 1]	[0, 1]	(1, 2]	(1, 2]	12	[0, 1]	(1, 2]	[0, 1]	[0, 1]
3	[0, 1]	[0, 1]	(2, 3]	[0, 1]	8	(1, 2]	[0, 1]	[0, 1]	(1, 2]	13	[0, 1]	[0, 1]	(1, 2]	[0, 1]
4	[0, 1]	[0, 1]	[0, 1]	(2, 3]	9	(1, 2]	[0, 1]	(1, 2]	[0, 1]	14	[0, 1]	[0, 1]	[0, 1]	(1, 2]
5	(1, 2]	(1, 2]	[0, 1]	[0, 1]	10	[0, 1]	(1, 2]	[0, 1]	(1, 2]	15	[0, 1]	[0, 1]	[0, 1]	[0, 1]

Table 4.13 The vertices of the small tetrahedrons for $N = 3$, *table of the third type*

No	A'	B'	C'	D'	No	A'	B'	C'	D'
1	3000	2100	2010	2001	7	1011	0111	0021	0012
2	1200	0300	0210	0201	8	2001	1101	1011	1002
3	1020	0120	0030	0021	9	2010	1110	1020	1011
4	1002	0102	0012	0003	10	1101	0201	0111	0102
5	2100	1200	1110	1101	15	0111	1011	1101	1110
6	1110	0210	0120	0111					

$$2v_C > 1. \tag{4.2}$$

In order to obtain a logical expression based on the row I in the table of first type, one can write the conjunction of all inequalities based on the cells in this row. For example, the logical expression based on the row 9 from Table 4.25 can be written as follows:

$$(5v_A > 1) \wedge (5v_B > 3). \tag{4.3}$$

Tables of the second type contain the logical expressions being true if the point under consideration is into the small tetrahedron or into the octahedron with number in the zero column. Tables 4.4 and 4.8 in Sect. 4.4, Table 4.12 in Sect. 4.5, Table 4.19 in Sect. 4.6, and Table 4.26 in Sect. 4.7 are the tables of the second type.

In order to obtain a logical expression based on the cell in row I and in column J from the table of the second type, one can take the relative volume in the zero row and column J and put it before the segment in the cell IJ. For example, the logical expression based on the cell 2, 4 in Table 4.4 can be written as follows:

$$2v_D \in [0, 1]. \tag{4.4}$$

In order to obtain a logical expression based on the row I in the table of the second type, one can write the conjunction of all logical expression based on the cells in this row I. For example, the logical expression based on the row 3 in Table 4.4 can be written as follows:

$$(2v_A \in [0, 1]) \wedge (2v_B \in [0, 1]) \wedge (2v_C \in (1, 2]) \wedge (2v_D \in [0, 1]). \tag{4.5}$$

Tables of the third type contain the reference points corresponding to the vertices of the small tetrahedrons or the octahedrons. Tables 4.1 and 4.2 in Sect. 4.4, Tables 4.13 and 4.14 in Sect. 4.5, Tables 4.20 and 4.21 in Sect. 4.6, and Tables 4.27 and 4.28 in Sect. 4.7 are the tables of the third type.

Tables of the fourth type contain the formulae for relative volumes corresponding to the appropriate vertices of the small tetrahedrons. Tables 4.5 and 4.10 in Sect. 4.4,

Table 4.14 The vertices of the octahedrons for $N = 3$, *table of the third type*

No	A'	B'	C'	D'	E'	F'	No	A'	B'	C'	D'	E'	F'
11	2100	1110	1011	2001	2010	1101	13	1110	0120	0021	1011	1020	0111
12	1200	0210	0111	1101	1110	0201	14	1101	0111	0012	1002	1011	0102

Table 4.15 Formulae for relative volumes corresponding the appropriate vertices of the small tetrahedrons for $N = 3$, *table of the fourth type*

No	$v_{A'}$	$v_{B'}$	$v_{C'}$	$v_{D'}$	No	$v_{A'}$	$v_{B'}$	$v_{C'}$	$v_{D'}$
1	$3v_A - 2$	$3v_B$	$3v_C$	$3v_D$	7	$3v_A$	$3v_B$	$3v_C - 1$	$3v_D - 1$
2	$3v_A$	$3v_B - 2$	$3v_C$	$3v_D$	8	$3v_A - 1$	$3v_B$	$3v_C$	$3v_D - 1$
3	$3v_A$	$3v_B$	$3v_C - 2$	$3v_D$	9	$3v_A - 1$	$3v_B$	$3v_C - 1$	$3v_D$
4	$3v_A$	$3v_B$	$3v_C$	$3v_D - 2$	10	$3v_A$	$3v_B - 1$	$3v_C$	$3v_D - 1$
5	$3v_A - 1$	$3v_B - 1$	$3v_C$	$3v_D$	15	$1 - 3v_A$	$1 - 3v_B$	$1 - 3v_C$	$1 - 3v_D$
6	$3v_A$	$3v_B - 1$	$3v_C - 1$	$3v_D$					

Table 4.16 Axis points for each of 3 ways to draw them for all of the octahedrons and corresponding expressions for v_1 and v_2 for $N = 3$, *table of the fifth type*

No	Axis no	A'	B'	v_1	v_2
11	1	1110	2001	$(3v_A + 3v_B - 3v_C - 3v_D - 1)/2$	$(3v_A + 3v_C - 3v_B - 3v_D - 1)/2$
	2	2010	1101	$(3v_A + 3v_B - 3v_C - 3v_D - 1)/2$	$(3v_B + 3v_C - 3v_D - 3v_A + 1)/2$
	3	2100	1011	$(3v_B + 3v_C - 3v_D - 3v_A + 1)/2$	$(3v_A + 3v_C - 3v_B - 3v_D - 1)/2$
12	1	0210	1101	$(3v_A + 3v_B - 3v_C - 3v_D - 1)/2$	$(3v_A + 3v_C - 3v_B - 3v_D + 1)/2$
	2	1110	0201	$(3v_A + 3v_B - 3v_C - 3v_D - 1)/2$	$(3v_B + 3v_C - 3v_D - 3v_A - 1)/2$
	3	1200	0111	$(3v_B + 3v_C - 3v_D - 3v_A - 1)/2$	$(3v_A + 3v_C - 3v_B - 3v_D + 1)/2$
13	1	0120	1011	$(3v_A + 3v_B - 3v_C - 3v_D + 1)/2$	$(3v_A + 3v_C - 3v_B - 3v_D - 1)/2$
	2	1020	0111	$(3v_A + 3v_B - 3v_C - 3v_D + 1)/2$	$(3v_B + 3v_C - 3v_D - 3v_A - 1)/2$
	3	1110	0021	$(3v_B + 3v_C - 3v_D - 3v_A - 1)/2$	$(3v_A + 3v_C - 3v_B - 3v_D - 1)/2$
14	1	0111	1002	$(3v_A + 3v_B - 3v_C - 3v_D + 1)/2$	$(3v_A + 3v_C - 3v_B - 3v_D + 1)/2$
	2	1011	0102	$(3v_A + 3v_B - 3v_C - 3v_D + 1)/2$	$(3v_B + 3v_C - 3v_D - 3v_A + 1)/2$
	3	1101	0012	$(3v_B + 3v_C - 3v_D - 3v_A + 1)/2$	$(3v_A + 3v_C - 3v_B - 3v_D + 1)/2$

Table 4.15 in Sect. 4.5, Table 4.22 in Sect. 4.6, and Table 4.29 in Sect. 4.7 are the tables of the fourth type.

Tables of the fifth type contain the axis points for each of 3 ways to draw them for all of the octahedrons and corresponding to expressions for v_1 and v_2 defined to diminish the amount of calculation. Table 4.6 in Sect. 4.4, Table 4.16 in Sect. 4.5, Table 4.23 in Sect. 4.6, and Table 4.30 in Sect. 4.7 are the tables of the fifth type. Note that a number of the ways for drawing the axis is called the axis number.

The sixth type is composed of the third and the fourth ones. Table 4.9 in Sect. 4.4, Table 4.17 in Sect. 4.5, Table 4.24 in Sect. 4.6, and Table 4.31 in Sect. 4.7 are the tables of the sixth type.

Note that the 4 different small tetrahedrons lied into one octahedron always have the vertices A' and B' from the corresponding table of the fifth type and coincides with the axis points. Other two vertices C' and D' can be found in appropriate line in the corresponding table of sixth type. For example, the vertices of third different

Table 4.17 The vertices C', D' of the different small tetrahedrons and the formulae for relative volumes corresponding the appropriate vertices A', B' of the different small tetrahedrons for $N = 3$, *table of the sixth type*

No	Axis no	Small tetrahedron no	C'	D'	$v_{A'}$	$v_{B'}$
11	1	1	2100	2010	$2 - 3v_A$	$3v_D$
		2	2100	1101	$3v_C$	$1 - 3v_B$
		3	1011	2010	$3v_B$	$1 - 3v_C$
		4	1011	1101	$1 - 3v_D$	$3v_A - 1$
	2	1	2100	1110	$1 - 3v_B$	$3v_D$
		2	2100	2001	$3v_C$	$2 - 3v_A$
		3	1011	1110	$3v_A - 1$	$1 - 3v_C$
		4	1011	2001	$1 - 3v_D$	$3v_B$
	3	1	1110	2010	$1 - 3v_C$	$3v_D$
		2	1110	1101	$3v_A - 1$	$1 - 3v_B$
		3	2001	2010	$3v_B$	$2 - 3v_A$
		4	2001	1101	$1 - 3v_D$	$3v_C$
12	1	1	1200	1110	$1 - 3v_A$	$3v_D$
		2	1200	0201	$3v_C$	$2 - 3v_B$
		3	0111	1110	$3v_B - 1$	$1 - 3v_C$
		4	0111	0201	$1 - 3v_D$	$3v_A$
	2	1	1200	0210	$2 - 3v_B$	$3v_D$
		2	1200	1101	$3v_C$	$1 - 3v_A$
		3	0111	0210	$3v_A$	$1 - 3v_C$
		4	0111	1101	$1 - 3v_D$	$3v_B - 1$
	3	1	0210	1110	$1 - 3v_C$	$3v_D$
		2	0210	0201	$3v_A$	$2 - 3v_B$
		3	1101	1110	$3v_B - 1$	$1 - 3v_A$
		4	1101	0201	$1 - 3v_D$	$3v_C$
13	1	1	1110	1020	$1 - 3v_A$	$3v_D$
		2	1110	0111	$3v_C - 1$	$1 - 3v_B$
		3	0021	1020	$3v_B$	$2 - 3v_C$
		4	0021	0111	$1 - 3v_D$	$3v_A$
	2	1	1110	0120	$1 - 3v_B$	$3v_D$
		2	1110	1011	$3v_C - 1$	$1 - 3v_A$
		3	0021	0120	$3v_A$	$2 - 3v_C$
		4	0021	1011	$1 - 3v_D$	$3v_B$
	3	1	0120	1020	$2 - 3v_C$	$3v_D$
		2	0120	0111	$3v_A$	$1 - 3v_B$
		3	1011	1020	$3v_B$	$1 - 3v_A$
		4	1011	0111	$1 - 3v_D$	$3v_C - 1$

(continued)

Table 4.17 (continued)

No	Axis no	Small tetrahedron no	C'	D'	$v_{A'}$	$v_{B'}$
14	1	1	1101	1011	$1-3v_A$	$3v_D-1$
		2	1101	0102	$3v_C$	$1-3v_B$
		3	0012	1011	$3v_B$	$1-3v_C$
		4	0012	0102	$2-3v_D$	$3v_A$
	2	1	1101	0111	$1-3v_B$	$3v_D-1$
		2	1101	1002	$3v_C$	$1-3v_A$
		3	0012	0111	$3v_A$	$1-3v_C$
		4	0012	1002	$2-3v_D$	$3v_B$
	3	1	0111	1011	$1-3v_C$	$3v_D-1$
		2	0111	0102	$3v_A$	$1-3v_B$
		3	1002	1011	$3v_B$	$1-3v_A$
		4	1002	0102	$2-3v_D$	$3v_C$

small tetrahedron into the octahedron with number 18 for $N = 5$ and the second way for drawing the axis are 1310, 0401, 0311, and 0410.

Note that the formulae for the relative volumes for the vertices C' and D' of all of the 4 different small tetrahedrons are always given in the Table 4.10 in Sect. 4.4. The relative volumes for the vertices A' and B' are given in the appropriate table of the sixth type. For example, the relative volumes of the third different small tetrahedron into the octahedron with the number 18 for $N = 5$ and the second way for drawing the axis are given by Eqs. 4.6–4.9.

$$v_{A'} = 5v_A \tag{4.6}$$

$$v_{B'} = 1 - 5v_C \tag{4.7}$$

$$v_{C'} = -v_1 \tag{4.8}$$

$$v_{D'} = v_2 \tag{4.9}$$

Note that only the values Nv_A, Nv_B, Nv_C, and Nv_D are used in all types of tables.

4.4 Tables and Algorithms for N = 2

If all of the logical expressions corresponding to the rows 1, 2, …, $I-1$ in Table 4.3 are false and the logical expression corresponding to the row I in Table 4.3 is true, then the logical expression corresponding to the row I in Table 4.4 is true and, therefore, the point under consideration lies into the small tetrahedron I with vertices given in Table 4.1 or into the octahedron I with the vertices given in Table 4.2. If it is the

Table 4.18 Inequalities for finding the small tetrahedron or the octahedron, in which the point under consideration lies for $N = 4$, *table of the first type*

No	$4v_A$	$4v_B$	$4v_C$	$4v_D$	No	$4v_A$	$4v_B$	$4v_C$	$4v_D$	No	$4v_A$	$4v_B$	$4v_C$	$4v_D$
1	>3	–	–	–	13	>2	–	>1	–	25	–	–	<1	<1
2	–	>3	–	–	14	–	>2	–	>1	26	<1	–	–	<1
3	–	–	>3	–	15	>1	–	>2	–	27	<1	<1	–	–
4	–	–	–	>3	16	–	>1	–	>2	28	–	<1	<1	–
5	>2	>1	–	–	17	>2	–	–	–	29	–	<1	–	<1
6	–	>2	>1	–	18	–	>2	–	–	30	<1	–	<1	–
7	–	–	>2	>1	19	–	–	>2	–	31	<1	–	–	–
8	>1	–	–	>2	20	–	–	–	>2	32	–	<1	–	–
9	>1	>2	–	–	21	–	<1	<1	<1	33	–	–	<1	–
10	–	>1	>2	–	22	<1	–	<1	<1	34	–	–	–	–
11	–	–	>1	>2	23	<1	<1	–	<1					
12	>2	–	–	>1	24	<1	<1	<1	–					

Table 4.19 Logical expressions being true if the point under consideration lies into the small tetrahedron or into the octahedron for $N = 4$, *table of the second type*

No	$4v_A \in$	$4v_B \in$	$4v_C \in$	$4v_D \in$	No	$4v_A \in$	$4v_B \in$	$4v_C \in$	$4v_D \in$	No	$4v_A \in$	$4v_B \in$	$4v_C \in$	$4v_D \in$
1	(3, 4]	[0, 1]	[0, 1]	[0, 1]	13	(2, 3]	[0, 1]	(1, 2]	[0, 1]	25	[1, 2]	[1, 2]	[0, 1)	[0, 1)
2	[0, 1]	(3, 4]	[0, 1]	[0, 1]	14	[0, 1]	(2, 3]	[0, 1]	(1, 2]	26	[0, 1)	[1, 2]	[1, 2]	[0, 1)
3	[0, 1]	[0, 1]	(3, 4]	[0, 1]	15	(1, 2]	[0, 1]	(2, 3]	[0, 1]	27	[0, 1)	[0, 1)	[1, 2]	[1, 2]
4	[0, 1]	[0, 1]	[0, 1]	(3, 4]	16	[0, 1]	(1, 2]	[0, 1]	(2, 3]	28	[1, 2]	[0, 1)	[0, 1)	[1, 2]
5	(2, 3]	(1, 2]	[0, 1]	[0, 1]	17	(2, 3]	[0, 1]	[0, 1]	[0, 1]	29	[1, 2]	[0, 1)	[1, 2]	[0, 1)
6	[0, 1]	(2, 3]	(1, 2]	[0, 1]	18	[0, 1]	(2, 3]	[0, 1]	[0, 1]	30	[0, 1)	[1, 2]	[0, 1)	[1, 2]
7	[0, 1]	[0, 1]	(2, 3]	(1, 2]	19	[0, 1]	[0, 1]	(2, 3]	[0, 1]	31	[0, 1)	[1, 2]	[1, 2]	[1, 2]
8	(1, 2]	[0, 1]	[0, 1]	(2, 3]	20	[0, 1]	[0, 1]	[0, 1]	(2, 3]	32	[1, 2]	[0, 1)	[1, 2]	[1, 2]
9	(1, 2]	(2, 3]	[0, 1]	[0, 1]	21	[1, 2]	[0, 1)	[0, 1)	[0, 1)	33	[1, 2]	[1, 2]	[0, 1)	[1, 2]
10	[0, 1]	(1, 2]	(2, 3]	[0, 1]	22	[0, 1)	[1, 2]	[0, 1)	[0, 1)	34	[1, 2]	[1, 2]	[1, 2]	[0, 1]
11	[0, 1]	[0, 1]	(1, 2]	(2, 3]	23	[0, 1)	[0, 1)	[1, 2]	[0, 1)					
12	(2, 3]	[0, 1]	[0, 1]	(1, 2]	24	[0, 1)	[0, 1)	[0, 1)	[1, 2]					

Table 4.20 The vertices of the small tetrahedrons for $N = 4$, *table of the third type*

No	A'	B'	C'	D'	No	A'	B'	C'	D'
1	4000	3100	3010	3001	13	3010	2110	2020	2011
2	1300	0400	0310	0301	14	1201	0301	0211	0202
3	1030	0130	0040	0031	15	2020	1120	1030	1021
4	1003	0103	0013	0004	16	1102	0202	0112	0103
5	3100	2200	2110	2101	21	1111	2011	2101	2110
6	1210	0310	0220	0211	22	0211	1111	1201	1210
7	1021	0121	0031	0022	23	0121	1021	1111	1120
8	2002	1102	1012	1003	24	0112	1012	1102	1111
9	2200	1300	1210	1201	31	1111	0211	0121	0112
10	1120	0220	0130	0121	32	2011	1111	1021	1012
11	1012	0112	0022	0013	33	2101	1201	1111	1102
12	3001	2101	2011	2002	34	2110	1210	1120	1111

Table 4.21 The vertices of the octahedrons for $N = 4$, *table of the third type*

No	A'	B'	C'	D'	E'	F'	No	A'	B'	C'	D'	E'	F'
17	3100	2110	2011	3001	3010	2101	26	1210	0220	0121	1111	1120	0211
18	1300	0310	0211	1201	1210	0301	27	1111	0121	0022	1012	1021	0112
19	1120	0130	0031	1021	1030	0121	28	2101	1111	1012	2002	2011	1102
20	1102	0112	0013	1003	1012	0103	29	2110	1120	1021	2011	2020	1111
25	2200	1210	1111	2101	2110	1201	30	1201	0211	0112	1102	1111	0202

Table 4.22 Formulae for relative volumes corresponding the appropriate vertices of the small tetrahedrons for $N = 4$, *table of the fourth type*

No	$v_{A'}$	$v_{B'}$	$v_{C'}$	$v_{D'}$	No	$v_{A'}$	$v_{B'}$	$v_{C'}$	$v_{D'}$
1	$4v_A - 3$	$4v_B$	$4v_C$	$4v_D$	13	$4v_A - 2$	$4v_B$	$4v_C - 1$	$4v_D$
2	$4v_A$	$4v_B - 3$	$4v_C$	$4v_D$	14	$4v_A$	$4v_B - 2$	$4v_C$	$4v_D - 1$
3	$4v_A$	$4v_B$	$4v_C - 3$	$4v_D$	15	$4v_A - 1$	$4v_B$	$4v_C - 2$	$4v_D$
4	$4v_A$	$4v_B$	$4v_C$	$4v_D - 3$	16	$4v_A$	$4v_B - 1$	$4v_C$	$4v_D - 2$
5	$4v_A - 2$	$4v_B - 1$	$4v_C$	$4v_D$	21	$2 - 4v_A$	$1 - 4v_B$	$1 - 4v_C$	$1 - 4v_D$
6	$4v_A$	$4v_B - 2$	$4v_C - 1$	$4v_D$	22	$1 - 4v_A$	$2 - 4v_B$	$1 - 4v_C$	$1 - 4v_D$
7	$4v_A$	$4v_B$	$4v_C - 2$	$4v_D - 1$	23	$1 - 4v_A$	$1 - 4v_B$	$2 - 4v_C$	$1 - 4v_D$
8	$4v_A - 1$	$4v_B$	$4v_C$	$4v_D - 2$	24	$1 - 4v_A$	$1 - 4v_B$	$1 - 4v_C$	$2 - 4v_D$
9	$4v_A - 1$	$4v_B - 2$	$4v_C$	$4v_D$	31	$4v_A$	$4v_B - 1$	$4v_C - 1$	$4v_D - 1$
10	$4v_A$	$4v_B - 1$	$4v_C - 2$	$4v_D$	32	$4v_A - 1$	$4v_B$	$4v_C - 1$	$4v_D - 1$
11	$4v_A$	$4v_B$	$4v_C - 1$	$4v_D - 2$	33	$4v_A - 1$	$4v_B - 1$	$4v_C$	$4v_D - 1$
12	$4v_A - 2$	$4v_B$	$4v_C$	$4v_D - 1$	34	$4v_A - 1$	$4v_B - 1$	$4v_C - 1$	$4v_D$

Table 4.23 Axis points for each of 3 ways to draw it for all of the octahedrons and corresponding expressions for v_1 and v_2 for $N = 4$, *table of the fifth type*

No	Axis no	A'	B'	v_1	v_2
17	1	2110	3001	$(4v_A + 4v_B - 4v_C - 4v_D - 2)/2$	$(4v_A + 4v_C - 4v_B - 4v_D - 2)/2$
	2	3010	2101	$(4v_A + 4v_B - 4v_C - 4v_D - 2)/2$	$(4v_B + 4v_C - 4v_D - 4v_A + 2)/2$
	3	3100	2011	$(4v_B + 4v_C - 4v_D - 4v_A + 2)/2$	$(4v_A + 4v_C - 4v_B - 4v_D - 2)/2$
18	1	0310	1201	$(4v_A + 4v_B - 4v_C - 4v_D - 2)/2$	$(4v_A + 4v_C - 4v_B - 4v_D + 2)/2$
	2	1210	0301	$(4v_A + 4v_B - 4v_C - 4v_D - 2)/2$	$(4v_B + 4v_C - 4v_D - 4v_A - 2)/2$
	3	1300	0211	$(4v_B + 4v_C - 4v_D - 4v_A - 2)/2$	$(4v_A + 4v_C - 4v_B - 4v_D + 2)/2$
19	1	0130	1021	$(4v_A + 4v_B - 4v_C - 4v_D + 2)/2$	$(4v_A + 4v_C - 4v_B - 4v_D - 2)/2$
	2	1030	0121	$(4v_A + 4v_B - 4v_C - 4v_D + 2)/2$	$(4v_B + 4v_C - 4v_D - 4v_A - 2)/2$
	3	1120	0031	$(4v_B + 4v_C - 4v_D - 4v_A - 2)/2$	$(4v_A + 4v_C - 4v_B - 4v_D - 2)/2$
20	1	0112	1003	$(4v_A + 4v_B - 4v_C - 4v_D + 2)/2$	$(4v_A + 4v_C - 4v_B - 4v_D + 2)/2$
	2	1012	0103	$(4v_A + 4v_B - 4v_C - 4v_D + 2)/2$	$(4v_B + 4v_C - 4v_D - 4v_A + 2)/2$
	3	1102	0013	$(4v_B + 4v_C - 4v_D - 4v_A + 2)/2$	$(4v_A + 4v_C - 4v_B - 4v_D + 2)/2$
25	1	1210	2101	$(4v_A + 4v_B - 4v_C - 4v_D - 2)/2$	$(4v_A + 4v_C - 4v_B - 4v_D)/2$
	2	2110	1201	$(4v_A + 4v_B - 4v_C - 4v_D - 2)/2$	$(4v_B + 4v_C - 4v_D - 4v_A)/2$
	3	2200	1111	$(4v_B + 4v_C - 4v_D - 4v_A)/2$	$(4v_A + 4v_C - 4v_B - 4v_D)/2$
26	1	0220	1111	$(4v_A + 4v_B - 4v_C - 4v_D)/2$	$(4v_A + 4v_C - 4v_B - 4v_D)/2$
	2	1120	0211	$(4v_A + 4v_B - 4v_C - 4v_D)/2$	$(4v_B + 4v_C - 4v_D - 4v_A - 2)/2$
	3	1210	0121	$(4v_B + 4v_C - 4v_D - 4v_A - 2)/2$	$(4v_A + 4v_C - 4v_B - 4v_D)/2$
27	1	0121	1012	$(4v_A + 4v_B - 4v_C - 4v_D + 2)/2$	$(4v_A + 4v_C - 4v_B - 4v_D)/2$
	2	1021	0112	$(4v_A + 4v_B - 4v_C - 4v_D + 2)/2$	$(4v_B + 4v_C - 4v_D - 4v_A)/2$
	3	1111	0022	$(4v_B + 4v_C - 4v_D - 4v_A)/2$	$(4v_A + 4v_C - 4v_B - 4v_D)/2$
28	1	1111	2002	$(4v_A + 4v_B - 4v_C - 4v_D)/2$	$(4v_A + 4v_C - 4v_B - 4v_D)/2$
	2	2011	1102	$(4v_A + 4v_B - 4v_C - 4v_D)/2$	$(4v_B + 4v_C - 4v_D - 4v_A + 2)/2$
	3	2101	1012	$(4v_B + 4v_C - 4v_D - 4v_A + 2)/2$	$(4v_A + 4v_C - 4v_B - 4v_D)/2$
29	1	1120	2011	$(4v_A + 4v_B - 4v_C - 4v_D)/2$	$(4v_A + 4v_C - 4v_B - 4v_D - 2)/2$
	2	2020	1111	$(4v_A + 4v_B - 4v_C - 4v_D)/2$	$(4v_B + 4v_C - 4v_D - 4v_A)/2$
	3	2110	1021	$(4v_B + 4v_C - 4v_D - 4v_A)/2$	$(4v_A + 4v_C - 4v_B - 4v_D - 2)/2$
30	1	0211	1102	$(4v_A + 4v_B - 4v_C - 4v_D)/2$	$(4v_A + 4v_C - 4v_B - 4v_D + 2)/2$
	2	1111	0202	$(4v_A + 4v_B - 4v_C - 4v_D)/2$	$(4v_B + 4v_C - 4v_D - 4v_A)/2$
	3	1201	0112	$(4v_B + 4v_C - 4v_D - 4v_A)/2$	$(4v_A + 4v_C - 4v_B - 4v_D + 2)/2$

tetrahedron I, one can find the relative volumes for its vertices in Table 4.5. If it is the octahedron I, one of 3 ways for drawing the axis into it should be chosen. Then one can calculate the appropriate v_1 and v_2 using the suitable formulae given in Table 4.6. One can use the Table 4.7 to find a number of the different small tetrahedron into this octahedron I. If all of the logical expressions corresponding to the rows 1, …, $J - 1$ in Table 4.7 are false and the logical expression corresponding the row J in Table 4.7

Table 4.24 The vertices C', D' of the different small tetrahedrons and the formulae for relative volumes corresponding the appropriate vertices A', B' of the different small tetrahedrons for $N = 4$, *table of the sixth type*

No	Axis no	Small tetrahedron no	C'	D'	$v_{A'}$	$v_{B'}$
17	1	1	3100	3010	$3 - 4v_A$	$4v_D$
		2	3100	2101	$4v_C$	$1 - 4v_B$
		3	2011	3010	$4v_B$	$1 - 4v_C$
		4	2011	2101	$1 - 4v_D$	$4v_A - 2$
	2	1	3100	2110	$1 - 4v_B$	$4v_D$
		2	3100	3001	$4v_C$	$3 - 4v_A$
		3	2011	2110	$4v_A - 2$	$1 - 4v_C$
		4	2011	3001	$1 - 4v_D$	$4v_B$
	3	1	2110	3010	$1 - 4v_C$	$4v_D$
		2	2110	2101	$4v_A - 2$	$1 - 4v_B$
		3	3001	3010	$4v_B$	$3 - 4v_A$
		4	3001	2101	$1 - 4v_D$	$4v_C$
18	1	1	1300	1210	$1 - 4v_A$	$4v_D$
		2	1300	0301	$4v_C$	$3 - 4v_B$
		3	0211	1210	$4v_B - 2$	$1 - 4v_C$
		4	0211	0301	$1 - 4v_D$	$4v_A$
	2	1	1300	0310	$3 - 4v_B$	$4v_D$
		2	1300	1201	$4v_C$	$1 - 4v_A$
		3	0211	0310	$4v_A$	$1 - 4v_C$
		4	0211	1201	$1 - 4v_D$	$4v_B - 2$
	3	1	0310	1210	$1 - 4v_C$	$4v_D$
		2	0310	0301	$4v_A$	$3 - 4v_B$
		3	1201	1210	$4v_B - 2$	$1 - 4v_A$
		4	1201	0301	$1 - 4v_D$	$4v_C$
19	1	1	1120	1030	$1 - 4v_A$	$4v_D$
		2	1120	0121	$4v_C - 2$	$1 - 4v_B$
		3	0031	1030	$4v_B$	$3 - 4v_C$
		4	0031	0121	$1 - 4v_D$	$4v_A$
	2	1	1120	0130	$1 - 4v_B$	$4v_D$
		2	1120	1021	$4v_C - 2$	$1 - 4v_A$
		3	0031	0130	$4v_A$	$3 - 4v_C$
		4	0031	1021	$1 - 4v_D$	$4v_B$
	3	1	0130	1030	$3 - 4v_C$	$4v_D$

(continued)

Table 4.24 (continued)

No	Axis no	Small tetrahedron no	C'	D'	$v_{A'}$	$v_{B'}$
		2	0130	0121	$4v_A$	$1 - 4v_B$
		3	1021	1030	$4v_B$	$1 - 4v_A$
		4	1021	0121	$1 - 4v_D$	$4v_C - 2$
20	1	1	1102	1012	$1 - 4v_A$	$4v_D - 2$
		2	1102	0103	$4v_C$	$1 - 4v_B$
		3	0013	1012	$4v_B$	$1 - 4v_C$
		4	0013	0103	$3 - 4v_D$	$4v_A$
	2	1	1102	0112	$1 - 4v_B$	$4v_D - 2$
		2	1102	1003	$4v_C$	$1 - 4v_A$
		3	0013	0112	$4v_A$	$1 - 4v_C$
		4	0013	1003	$3 - 4v_D$	$4v_B$
	3	1	0112	1012	$1 - 4v_C$	$4v_D - 2$
		2	0112	0103	$4v_A$	$1 - 4v_B$
		3	1003	1012	$4v_B$	$1 - 4v_A$
		4	1003	0103	$3 - 4v_D$	$4v_C$
25	1	1	2200	2110	$2 - 4v_A$	$4v_D$
		2	2200	1201	$4v_C$	$2 - 4v_B$
		3	1111	2110	$4v_B - 1$	$1 - 4v_C$
		4	1111	1201	$1 - 4v_D$	$4v_A - 1$
	2	1	2200	1210	$2 - 4v_B$	$4v_D$
		2	2200	2101	$4v_C$	$2 - 4v_A$
		3	1111	1210	$4v_A - 1$	$1 - 4v_C$
		4	1111	2101	$1 - 4v_D$	$4v_B - 1$
	3	1	1210	2110	$1 - 4v_C$	$4v_D$
		2	1210	1201	$4v_A - 1$	$2 - 4v_B$
		3	2101	2110	$4v_B - 1$	$2 - 4v_A$
		4	2101	1201	$1 - 4v_D$	$4v_C$
26	1	1	1210	1120	$1 - 4v_A$	$4v_D$
		2	1210	0211	$4v_C - 1$	$2 - 4v_B$
		3	0121	1120	$4v_B - 1$	$2 - 4v_C$
		4	0121	0211	$1 - 4v_D$	$4v_A$
	2	1	1210	0220	$2 - 4v_B$	$4v_D$
		2	1210	1111	$4v_C - 1$	$1 - 4v_A$
		3	0121	0220	$4v_A$	$2 - 4v_C$
		4	0121	1111	$1 - 4v_D$	$4v_B - 1$

(continued)

Table 4.24 (continued)

No	Axis no	Small tetrahedron no	C'	D'	$v_{A'}$	$v_{B'}$
	3	1	0220	1120	$2-4v_C$	$4v_D$
		2	0220	0211	$4v_A$	$2-4v_B$
		3	1111	1120	$4v_B-1$	$1-4v_A$
		4	1111	0211	$1-4v_D$	$4v_C-1$
27	1	1	1111	1021	$1-4v_A$	$4v_D-1$
		2	1111	0112	$4v_C-1$	$1-4v_B$
		3	0022	1021	$4v_B$	$2-4v_C$
		4	0022	0112	$2-4v_D$	$4v_A$
	2	1	1111	0121	$1-4v_B$	$4v_D-1$
		2	1111	1012	$4v_C-1$	$1-4v_A$
		3	0022	0121	$4v_A$	$2-4v_C$
		4	0022	1012	$2-4v_D$	$4v_B$
	3	1	0121	1021	$2-4v_C$	$4v_D-1$
		2	0121	0112	$4v_A$	$1-4v_B$
		3	1012	1021	$4v_B$	$1-4v_A$
		4	1012	0112	$2-4v_D$	$4v_C-1$
28	1	1	2101	2011	$2-4v_A$	$4v_D-1$
		2	2101	1102	$4v_C$	$1-4v_B$
		3	1012	2011	$4v_B$	$1-4v_C$
		4	1012	1102	$2-4v_D$	$4v_A-1$
	2	1	2101	1111	$1-4v_B$	$4v_D-1$
		2	2101	2002	$4v_C$	$2-4v_A$
		3	1012	1111	$4v_A-1$	$1-4v_C$
		4	1012	2002	$2-\mathrm{v}$	$4v_B$
	3	1	1111	2011	$1-4v_C$	$4v_D-1$
		2	1111	1102	$4v_A-1$	$1-4v_B$
		3	2002	2011	$4v_B$	$2-4v_A$
		4	2002	1102	$2-4v_D$	$4v_C$
29	1	1	2200	2110	$2-4v_A$	$4v_D$
		2	2200	1201	$4v_C-1$	$1-4v_B$
		3	1111	2110	$4v_B$	$2-4v_C$
		4	1111	1201	$1-4v_D$	$4v_A-1$
	2	1	2110	1120	$1-4v_B$	$4v_D$
		2	2110	2011	$4v_C-1$	$2-4v_A$
		3	1021	1120	$4v_A-1$	$2-4v_C$

(continued)

Table 4.24 (continued)

No	Axis no	Small tetrahedron no	C'	D'	$v_{A'}$	$v_{B'}$
		4	1021	2011	$1-4v_D$	$4v_B$
	3	1	1120	2020	$2-4v_C$	$4v_D$
		2	1120	1111	$4v_A-1$	$1-4v_B$
		3	2011	2020	$4v_B$	$2-4v_A$
		4	2011	1111	$1-4v_D$	$4v_C-1$
30	1	1	1201	1111	$1-4v_A$	$4v_D-1$
		2	1201	0202	$4v_C$	$2-4v_B$
		3	0112	1111	$4v_B-1$	$1-4v_C$
		4	0112	0202	$2-4v_D$	$4v_A$
	2	1	1201	0211	$2-4v_B$	$4v_D-1$
		2	1201	1102	$4v_C$	$1-4v_A$
		3	0112	0211	$4v_A$	$1-4v_C$
		4	0112	1102	$2-4v_D$	$4v_B-1$
	3	1	0211	1111	$1-4v_C$	$4v_D-1$
		2	0211	0202	$4v_A$	$2-4v_B$
		3	1102	1111	$4v_B-1$	$1-4v_A$
		4	1102	0202	$2-4v_D$	$4v_C$

is true, then the logical expression corresponding the row J in Table 4.8 is true and, therefore, the point under consideration lies into the different small tetrahedron with the number J. One can find its vertices in Tables 4.6 and 4.9 and corresponding relative volumes in Tables 4.9 and 4.10.

For example, $I = 2$. If

$$(2v_A \le 1) \wedge (2v_B > 1), \tag{4.10}$$

then

$$(2v_A \in [0,1]) \wedge (2v_B \in (1,2]) \wedge (2v_C \in [0,1]) \wedge (2v_D \in [0,1]) \tag{4.11}$$

and, therefore, the point under consideration lies into the small tetrahedron with the number 2 and vertices 1100, 0200, 0110, and 0101. The result of piecewise linear interpolation can be determined using Eq. 4.12.

$$u(\vec{r}) = 2v_A(\vec{r})u_{1100} + \left(2v_B(\vec{r}) - 1\right)u_{0200} + 2v_C(\vec{r})u_{0110} + 2v_D(\vec{r})u_{0101} \tag{4.12}$$

Table 4.25 Inequalities for finding the small tetrahedron or the octahedron, in which the point under consideration lies for $N = 5$, *table of the first type*

No	$5v_A$	$5v_B$	$5v_C$	$5v_D$	No	$5v_A$	$5v_B$	$5v_C$	$5v_D$
1	>4	–	–	–	34	>1	>2	–	>1
2	–	>4	–	–	35	>1	>1	–	>2
3	–	–	>4	–	36	>2	–	>1	>1
4	–	–	–	>4	37	>1	–	>2	>1
5	>3	>1	–	–	38	>1	–	>1	>2
6	–	>3	>1	–	39	>2	>1	–	–
7	–	–	>3	>1	40	–	>2	>1	–
8	>1	–	–	>3	41	–	–	>2	>1
9	>1	>3	–	–	42	>1	–	–	>2
10	–	>1	>3	–	43	>1	>2	–	–
11	–	–	>1	>3	44	–	>1	>2	–
12	>3	–	–	>1	45	–	–	>1	>2
13	>3	–	>1	–	46	>2	–	–	>1
14	–	>3	–	>1	47	>2	–	>1	–
15	>1	–	>3	–	48	–	>2	–	>1
16	–	>1	–	>3	49	>1	–	>2	–
17	>3	–	–	–	50	–	>1	–	>2
18	–	>3	–	–	51	>2	–	–	–
19	–	–	>3	–	52	–	>2	–	–
20	–	–	–	>3	53	–	–	>2	–
21	>2	>2	–	–	54	–	–	–	>2
22	–	>2	>2	–	55	–	–	<1	<1
23	–	–	>2	>2	56	<1	–	–	<1
24	>2	–	–	>2	57	<1	<1	–	–
25	>2	–	>2	–	58	–	<1	<1	–
26	–	>2	–	>2	59	–	<1	–	<1
27	>2	>1	>1	–	60	<1	–	<1	–
28	>1	>2	>1	–	61	<1	–	–	–
29	>1	>1	>2	–	62	–	<1	–	–
30	–	>2	>1	>1	63	–	–	<1	–
31	–	>1	>2	>1	64	–	–	–	<1
32	–	>1	>1	>2	65	–	–	–	–
33	>2	>1	–	>1					

4.5 Tables and Algorithms for N = 3

Consider degree $N = 3$. 11 smaller tetrahedrons and 4 octahedrons are. If all of the logical expressions corresponding to the rows 1, 2, …, $I - 1$ in Table 4.11 are false and the logical expression corresponding to the row I in Table 4.11 is true, then the

Table 4.26 Logical expressions being true if the point under consideration lies into the small tetrahedron or into the octahedron for $N = 5$, *table of the second type*

No	$5v_A \in$	$5v_B \in$	$5v_C \in$	$5v_D \in$	No	$5v_A \in$	$5v_B \in$	$5v_C \in$	$5v_D \in$
1	(4, 5]	[0, 1]	[0, 1]	[0, 1]	34	(1, 2]	(2, 3]	[0, 1]	(1, 2]
2	[0, 1]	(4, 5]	[0, 1]	[0, 1]	35	(1, 2]	(1, 2]	[0, 1]	(2, 3]
3	[0, 1]	[0, 1]	(4, 5]	[0, 1]	36	(2, 3]	[0, 1]	(1, 2]	(1, 2]
4	[0, 1]	[0, 1]	[0, 1]	(4, 5]	37	(1, 2]	[0, 1]	(2, 3]	(1, 2]
5	(3, 4]	(1, 2]	[0, 1]	[0, 1]	38	(1, 2]	[0, 1]	(1, 2]	(2, 3]
6	[0, 1]	(3, 4]	(1, 2]	[0, 1]	39	(2, 3]	(1, 2]	[0, 1]	[0, 1]
7	[0, 1]	[0, 1]	(3, 4]	(1, 2]	40	[0, 1]	(2, 3]	(1, 2]	[0, 1]
8	(1, 2]	[0, 1]	[0, 1]	(3, 4]	41	[0, 1]	[0, 1]	(2, 3]	(1, 2]
9	(1, 2]	(3, 4]	[0, 1]	[0, 1]	42	(1, 2]	[0, 1]	[0, 1]	(2, 3]
10	[0, 1]	(1, 2]	(3, 4]	[0, 1]	43	(1, 2]	(2, 3]	[0, 1]	[0, 1]
11	[0, 1]	[0, 1]	(1, 2]	(3, 4]	44	[0, 1]	(1, 2]	(2, 3]	[0, 1]
12	(3, 4]	[0, 1]	[0, 1]	(1, 2]	45	[0, 1]	[0, 1]	(1, 2]	(2, 3]
13	(3, 4]	[0, 1]	(1, 2]	[0, 1]	46	(2, 3]	[0, 1]	[0, 1]	(1, 2]
14	[0, 1]	(3, 4]	[0, 1]	(1, 2]	47	(2, 3]	[0, 1]	(1, 2]	[0, 1]
15	(1, 2]	[0, 1]	(3, 4]	[0, 1]	48	[0, 1]	(2, 3]	[0, 1]	(1, 2]
16	[0, 1]	(1, 2]	[0, 1]	(3, 4]	49	(1, 2]	[0, 1]	(2, 3]	[0, 1]
17	(3, 4]	[0, 1]	[0, 1]	[0, 1]	50	[0, 1]	(1, 2]	[0, 1]	(2, 3]
18	[0, 1]	(3, 4]	[0, 1]	[0, 1]	51	(2, 3]	[0, 1]	[0, 1]	[0, 1]

(continued)

Table 4.26 (continued)

No	$5v_A \in$	$5v_B \in$	$5v_C \in$	$5v_D \in$	No	$5v_A \in$	$5v_B \in$	$5v_C \in$	$5v_D \in$
19	[0, 1]	[0, 1]	(3, 4]	[0, 1]	52	[0, 1]	(2, 3]	[0, 1]	[0, 1]
20	[0, 1]	[0, 1]	[0, 1]	(3, 4]	53	[0, 1]	[0, 1]	(2, 3]	[0, 1]
21	(2, 3]	(2, 3]	[0, 1]	[0, 1]	54	[0, 1]	[0, 1]	[0, 1]	(2, 3]
22	[0, 1]	(2, 3]	(2, 3]	[0, 1]	55	[1, 2]	[1, 2]	[0, 1)	[0, 1)
23	[0, 1]	[0, 1]	(2, 3]	(2, 3]	56	[0, 1)	[1, 2]	[1, 2]	[0, 1)
24	(2, 3]	[0, 1]	[0, 1]	(2, 3]	57	[0, 1)	[0, 1)	[1, 2]	[1, 2]
25	(2, 3]	[0, 1]	(2, 3]	[0, 1]	58	[1, 2]	[0, 1)	[0, 1)	[1, 2]
26	[0, 1]	(2, 3]	[0, 1]	(2, 3]	59	[1, 2]	[0, 1)	[1, 2]	[0, 1)
27	(2, 3]	(1, 2]	(1, 2]	[0, 1]	60	[0, 1)	[1, 2]	[0, 1)	[1, 2]
28	(1, 2]	(2, 3]	(1, 2]	[0, 1]	61	[0, 1)	[1, 2]	[1, 2]	[1, 2]
29	(1, 2]	(1, 2]	(2, 3]	[0, 1]	62	[1, 2]	[0, 1)	[1, 2]	[1, 2]
30	[0, 1]	(2, 3]	(1, 2]	(1, 2]	63	[1, 2]	[1, 2]	[0, 1)	[1, 2]
31	[0, 1]	(1, 2]	(2, 3]	(1, 2]	64	[1, 2]	[1, 2]	[1, 2]	[0, 1)
32	[0, 1]	(1, 2]	(1, 2]	(2, 3]	65	[1, 2]	[1, 2]	[1, 2]	[1, 2]
33	(2, 3]	(1, 2]	[0, 1]	(1, 2]					

Table 4.27 The vertices of the small tetrahedrons for $N = 5$, *table of the third type*

No	A'	B'	C'	D'	No	A'	B'	C'	D'
1	5000	4100	4010	4001	28	2210	1310	1220	1211
2	1400	0500	0410	0401	29	2120	1220	1130	1121
3	1040	0140	0050	0041	30	1211	0311	0221	0212
4	1004	0104	0014	0005	31	1121	0221	0131	0122
5	4100	3200	3110	3101	32	1112	0212	0122	0113
6	1310	0410	0320	0311	33	3101	2201	2111	2102
7	1031	0131	0041	0032	34	2201	1301	1211	1202
8	2003	1103	1013	1004	35	2102	1202	1112	1103
9	2300	1400	1310	1301	36	3011	2111	2021	2012
10	1130	0230	0140	0131	37	2021	1121	1031	1022
11	1013	0113	0023	0014	38	2012	1112	1022	1013
12	4001	3101	3011	3002	51	2111	3011	3101	3110
13	4010	3110	3020	3011	52	0311	1211	1301	1310
14	1301	0401	0311	0302	53	0131	1031	1121	1130
15	2030	1130	1040	1031	54	0113	1013	1103	1112
16	1103	0203	0113	0104	55	1211	2111	2201	2210
21	3200	2300	2210	2201	56	0221	1121	1211	1220
22	1220	0320	0230	0221	57	0122	1022	1112	1121
23	1022	0122	0032	0023	58	1112	2012	2102	2111
24	3002	2102	2012	2003	59	1121	2021	2111	2120
25	3020	2120	2030	2021	60	0212	1112	1202	1211
26	1202	0302	0212	0203	65	2111	1211	1121	1112
27	3110	2210	2120	2111					

Table 4.28 The vertices of the octahedrons for $N = 5$, *table of the third type*

No	A'	B'	C'	D'	E'	F'
17	4100	3110	3011	4001	4010	3101
18	1400	0410	0311	1301	1310	0401
19	1130	0140	0041	1031	1040	0131
20	1103	0113	0014	1004	1013	0104
39	3200	2210	2111	3101	3110	2201
40	1310	0320	0221	1211	1220	0311
41	1121	0131	0032	1022	1031	0122
42	2102	1112	1013	2003	2012	1103
43	2300	1310	1211	2201	2210	1301
44	1220	0230	0131	1121	1130	0221
45	1112	0122	0023	1013	1022	0113
46	3101	2111	2012	3002	3011	2102
47	3110	2120	2021	3011	3020	2111
48	1301	0311	0212	1202	1211	0302
49	2120	1130	1031	2021	2030	1121
50	1202	0212	0113	1103	1112	0203
61	1211	0221	0122	1112	1121	0212
62	2111	1121	1022	2012	2021	1112
63	2201	1211	1112	2102	2111	1202
64	2210	1220	1121	2111	2120	1211

logical expression corresponding to the row I in Table 4.12 is true and, therefore, the point under consideration lies into the small tetrahedron I with vertices given in Table 4.13 or into the octahedron I with the vertices given in Table 4.14. If it is the tetrahedron I, one can find the relative volumes for its vertices in Table 4.15. If it is the octahedron I, one of 3 ways for drawing the axis into it should be chosen. Then one can calculate the appropriate v_1 and v_2 using the suitable formulae given in Table 4.16. One can use the Table 4.7 in Sect. 4.4 to find a number of different small tetrahedron into this octahedron I. If all of the logical expressions corresponding to the rows 1, …, $J - 1$ in Table 4.7 are false and the logical expression corresponding to the row J in Table 4.7 is true, then the logical expression corresponding to the row J in Table 4.8 in Sect. 4.4 is true and, therefore, the point under consideration lies into different small tetrahedron with the number J. One can find its vertices in Tables 4.16 and 4.17 and corresponding relative volumes in Table 4.10 in Sect. 4.4 and in Table 4.17.

For example, $I = 11$. If all of the logical expressions corresponding to the rows 1, 2, 3, 4, 5, 6, 7, 8, 9, and 10 in Table 4.11 are false and

$$3v_A > 1, \tag{4.13}$$

then

Table 4.29 Formulae for relative volumes corresponding the appropriate vertices of the small tetrahedrons for $N = 5$, *table of the fourth type*

No	$v_{A'}$	$v_{B'}$	$v_{C'}$	$v_{D'}$	No	$v_{A'}$	$v_{B'}$	$v_{C'}$	$v_{D'}$
1	$5v_A - 4$	$5v_B$	$5v_C$	$5v_D$	28	$5v_A - 1$	$5v_B - 2$	$5v_C - 1$	$5v_D$
2	$5v_A$	$5v_B - 4$	$5v_C$	$5v_D$	29	$5v_A - 1$	$5v_B - 1$	$5v_C - 2$	$5v_D$
3	$5v_A$	$5v_B$	$5v_C - 4$	$5v_D$	30	$5v_A$	$5v_B - 2$	$5v_C - 1$	$5v_D - 1$
4	$5v_A$	$5v_B$	$5v_C$	$5v_D - 4$	31	$5v_A$	$5v_B - 1$	$5v_C - 2$	$5v_D - 1$
5	$5v_A - 3$	$5v_B - 1$	$5v_C$	$5v_D$	32	$5v_A$	$5v_B - 1$	$5v_C - 1$	$5v_D - 2$
6	$5v_A$	$5v_B - 3$	$5v_C - 1$	$5v_D$	33	$5v_A - 2$	$5v_B - 1$	$5v_C$	$5v_D - 1$
7	$5v_A$	$5v_B$	$5v_C - 3$	$5v_D - 1$	34	$5v_A - 1$	$5v_B - 2$	$5v_C$	$5v_D - 1$
8	$5v_A - 1$	$5v_B$	$5v_C$	$5v_D - 3$	35	$5v_A - 1$	$5v_B - 1$	$5v_C$	$5v_D - 2$
9	$5v_A - 1$	$5v_B - 3$	$5v_C$	$5v_D$	36	$5v_A - 2$	$5v_B$	$5v_C - 1$	$5v_D - 1$
10	$5v_A$	$5v_B - 1$	$5v_C - 3$	$5v_D$	37	$5v_A - 1$	$5v_B$	$5v_C - 2$	$5v_D - 1$
11	$5v_A$	$5v_B$	$5v_C - 1$	$5v_D - 3$	38	$5v_A - 1$	$5v_B$	$5v_C - 1$	$5v_D - 2$
12	$5v_A - 3$	$5v_B$	$5v_C$	$5v_D - 1$	51	$3 - 5v_A$	$1 - 5v_B$	$1 - 5v_C$	$1 - 5v_D$
13	$5v_A - 3$	$5v_B$	$5v_C - 1$	$5v_D$	52	$1 - 5v_A$	$3 - 5v_B$	$1 - 5v_C$	$1 - 5v_D$
14	$5v_A$	$5v_B - 3$	$5v_C$	$5v_D - 1$	53	$1 - 5v_A$	$1 - 5v_B$	$3 - 5v_C$	$1 - 5v_D$
15	$5v_A - 1$	$5v_B$	$5v_C - 3$	$5v_D$	54	$1 - 5v_A$	$1 - 5v_B$	$1 - 5v_C$	$3 - 5v_D$
16	$5v_A$	$5v_B - 1$	$5v_C$	$5v_D - 3$	55	$2 - 5v_A$	$2 - 5v_B$	$1 - 5v_C$	$1 - 5v_D$
21	$5v_A - 2$	$5v_B - 2$	$5v_C$	$5v_D$	56	$1 - 5v_A$	$2 - 5v_B$	$2 - 5v_C$	$1 - 5v_D$
22	$5v_A$	$5v_B - 2$	$5v_C - 2$	$5v_D$	57	$1 - 5v_A$	$1 - 5v_B$	$2 - 5v_C$	$2 - 5v_D$
23	$5v_A$	$5v_B$	$5v_C - 2$	$5v_D - 2$	58	$2 - 5v_A$	$1 - 5v_B$	$1 - 5v_C$	$2 - 5v_D$
24	$5v_A - 2$	$5v_B$	$5v_C$	$5v_D - 2$	59	$2 - 5v_A$	$1 - 5v_B$	$2 - 5v_C$	$1 - 5v_D$
25	$5v_A - 2$	$5v_B$	$5v_C - 2$	$5v_D$	60	$1 - 5v_A$	$2 - 5v_B$	$1 - 5v_C$	$2 - 5v_D$
26	$5v_A$	$5v_B - 2$	$5v_C$	$5v_D - 2$	65	$5v_A - 1$	$5v_B - 1$	$5v_C - 1$	$5v_D - 1$
27	$5v_A - 2$	$5v_B - 1$	$5v_C - 1$	$5v_D$					

Table 4.30 Axis points for each of 3 ways to draw them for all of the octahedrons and corresponding expressions for v_1 and v_2 for $N = 5$, *table of the fifth type*

No	Axis no	A'	B'	v_1	v_2
17	1	3110	4001	$(5v_A + 5v_B - 5v_C - 5v_D - 3)/2$	$(5v_A + 5v_C - 5v_B - 5v_D - 3)/2$
	2	4010	3101	$(5v_A + 5v_B - 5v_C - 5v_D - 3)/2$	$(5v_B + 5v_C - 5v_D - 5v_A + 3)/2$
	3	4100	3011	$(5v_B + 5v_C - 5v_D - 5v_A + 3)/2$	$(5v_A + 5v_C - 5v_B - 5v_D - 3)/2$
18	1	0410	1301	$(5v_A + 5v_B - 5v_C - 5v_D - 3)/2$	$(5v_A + 5v_C - 5v_B - 5v_D + 3)/2$
	2	1310	0401	$(5v_A + 5v_B - 5v_C - 5v_D - 3)/2$	$(5v_B + 5v_C - 5v_D - 5v_A - 3)/2$
	3	1400	0311	$(5v_B + 5v_C - 5v_D - 5v_A - 3)/2$	$(5v_A + 5v_C - 5v_B - 5v_D + 3)/2$
19	1	0140	1031	$(5v_A + 5v_B - 5v_C - 5v_D + 3)/2$	$(5v_A + 5v_C - 5v_B - 5v_D - 3)/2$
	2	1040	0131	$(5v_A + 5v_B - 5v_C - 5v_D + 3)/2$	$(5v_B + 5v_C - 5v_D - 5v_A - 3)/2$
	3	1130	0041	$(5v_B + 5v_C - 5v_D - 5v_A - 3)/2$	$(5v_A + 5v_C - 5v_B - 5v_D - 3)/2$
20	1	0113	1004	$(5v_A + 5v_B - 5v_C - 5v_D + 3)/2$	$(5v_A + 5v_C - 5v_B - 5v_D + 3)/2$
	2	1013	0104	$(5v_A + 5v_B - 5v_C - 5v_D + 3)/2$	$(5v_B + 5v_C - 5v_D - 5v_A + 3)/2$
	3	1103	0014	$(5v_B + 5v_C - 5v_D - 5v_A + 3)/2$	$(5v_A + 5v_C - 5v_B - 5v_D + 3)/2$
39	1	2210	3101	$(5v_A + 5v_B - 5v_C - 5v_D - 3)/2$	$(5v_A + 5v_C - 5v_B - 5v_D - 1)/2$
	2	3110	2201	$(5v_A + 5v_B - 5v_C - 5v_D - 3)/2$	$(5v_B + 5v_C - 5v_D - 5v_A + 1)/2$
	3	3200	2111	$(5v_B + 5v_C - 5v_D - 5v_A + 1)/2$	$(5v_A + 5v_C - 5v_B - 5v_D - 1)/2$
40	1	0320	1211	$(5v_A + 5v_B - 5v_C - 5v_D - 1)/2$	$(5v_A + 5v_C - 5v_B - 5v_D + 1)/2$
	2	1220	0311	$(5v_A + 5v_B - 5v_C - 5v_D - 1)/2$	$(5v_B + 5v_C - 5v_D - 5v_A - 3)/2$
	3	1310	0221	$(5v_B + 5v_C - 5v_D - 5v_A - 3)/2$	$(5v_A + 5v_C - 5v_B - 5v_D + 1)/2$
41	1	0131	1022	$(5v_A + 5v_B - 5v_C - 5v_D + 3)/2$	$(5v_A + 5v_C - 5v_B - 5v_D - 1)/2$
	2	1031	0122	$(5v_A + 5v_B - 5v_C - 5v_D + 3)/2$	$(5v_B + 5v_C - 5v_D - 5v_A - 1)/2$
	3	1121	0032	$(5v_B + 5v_C - 5v_D - 5v_A - 1)/2$	$(5v_A + 5v_C - 5v_B - 5v_D - 1)/2$
42	1	1112	2003	$(5v_A + 5v_B - 5v_C - 5v_D + 1)/2$	$(5v_A + 5v_C - 5v_B - 5v_D + 1)/2$
	2	2012	1103	$(5v_A + 5v_B - 5v_C - 5v_D + 1)/2$	$(5v_B + 5v_C - 5v_D - 5v_A + 3)/2$
	3	2102	1013	$(5v_B + 5v_C - 5v_D - 5v_A + 3)/2$	$(5v_A + 5v_C - 5v_B - 5v_D + 1)/2$
43	1	1310	2201	$(5v_A + 5v_B - 5v_C - 5v_D - 3)/2$	$(5v_A + 5v_C - 5v_B - 5v_D + 1)/2$
	2	2210	1301	$(5v_A + 5v_B - 5v_C - 5v_D - 3)/2$	$(5v_B + 5v_C - 5v_D - 5v_A - 1)/2$
	3	2300	1211	$(5v_B + 5v_C - 5v_D - 5v_A - 1)/2$	$(5v_A + 5v_C - 5v_B - 5v_D + 1)/2$
44	1	0230	1121	$(5v_A + 5v_B - 5v_C - 5v_D + 1)/2$	$(5v_A + 5v_C - 5v_B - 5v_D - 1)/2$
	2	1130	0221	$(5v_A + 5v_B - 5v_C - 5v_D + 1)/2$	$(5v_B + 5v_C - 5v_D - 5v_A - 3)/2$
	3	1220	0131	$(5v_B + 5v_C - 5v_D - 5v_A - 3)/2$	$(5v_A + 5v_C - 5v_B - 5v_D - 1)/2$
45	1	0122	1013	$(5v_A + 5v_B - 5v_C - 5v_D + 3)/2$	$(5v_A + 5v_C - 5v_B - 5v_D + 1)/2$
	2	1022	0113	$(5v_A + 5v_B - 5v_C - 5v_D + 3)/2$	$(5v_B + 5v_C - 5v_D - 5v_A + 1)/2$
	3	1112	0023	$(5v_B + 5v_C - 5v_D - 5v_A + 1)/2$	$(5v_A + 5v_C - 5v_B - 5v_D + 1)/2$

(continued)

Table 4.30 (continued)

No	Axis no	A'	B'	v_1	v_2
46	1	2111	3002	$(5v_A + 5v_B - 5v_C - 5v_D - 1)/2$	$(5v_A + 5v_C - 5v_B - 5v_D - 1)/2$
	2	3011	2102	$(5v_A + 5v_B - 5v_C - 5v_D - 1)/2$	$(5v_B + 5v_C - 5v_D - 5v_A + 3)/2$
	3	3101	2012	$(5v_B + 5v_C - 5v_D - 5v_A + 3)/2$	$(5v_A + 5v_C - 5v_B - 5v_D - 1)/2$
47	1	2120	3011	$(5v_A + 5v_B - 5v_C - 5v_D - 1)/2$	$(5v_A + 5v_C - 5v_B - 5v_D - 3)/2$
	2	3020	2111	$(5v_A + 5v_B - 5v_C - 5v_D - 1)/2$	$(5v_B + 5v_C - 5v_D - 5v_A + 1)/2$
	3	3110	2021	$(5v_B + 5v_C - 5v_D - 5v_A + 1)/2$	$(5v_A + 5v_C - 5v_B - 5v_D - 3)/2$
48	1	0311	1202	$(5v_A + 5v_B - 5v_C - 5v_D - 1)/2$	$(5v_A + 5v_C - 5v_B - 5v_D + 3)/2$
	2	1211	0302	$(5v_A + 5v_B - 5v_C - 5v_D - 1)/2$	$(5v_B + 5v_C - 5v_D - 5v_A - 1)/2$
	3	1301	0212	$(5v_B + 5v_C - 5v_D - 5v_A - 1)/2$	$(5v_A + 5v_C - 5v_B - 5v_D + 3)/2$
49	1	1130	2021	$(5v_A + 5v_B - 5v_C - 5v_D + 1)/2$	$(5v_A + 5v_C - 5v_B - 5v_D - 3)/2$
	2	2030	1121	$(5v_A + 5v_B - 5v_C - 5v_D + 1)/2$	$(5v_B + 5v_C - 5v_D - 5v_A - 1)/2$
	3	2120	1031	$(5v_B + 5v_C - 5v_D - 5v_A - 1)/2$	$(5v_A + 5v_C - 5v_B - 5v_D - 3)/2$
50	1	0212	1103	$(5v_A + 5v_B - 5v_C - 5v_D + 1)/2$	$(5v_A + 5v_C - 5v_B - 5v_D + 3)/2$
	2	1112	0203	$(5v_A + 5v_B - 5v_C - 5v_D + 1)/2$	$(5v_B + 5v_C - 5v_D - 5v_A + 1)/2$
	3	1202	0113	$(5v_B + 5v_C - 5v_D - 5v_A + 1)/2$	$(5v_A + 5v_C - 5v_B - 5v_D + 3)/2$
61	1	0221	1112	$(5v_A + 5v_B - 5v_C - 5v_D + 1)/2$	$(5v_A + 5v_C - 5v_B - 5v_D + 1)/2$
	2	1121	0212	$(5v_A + 5v_B - 5v_C - 5v_D + 1)/2$	$(5v_B + 5v_C - 5v_D - 5v_A - 1)/2$
	3	1211	0122	$(5v_B + 5v_C - 5v_D - 5v_A - 1)/2$	$(5v_A + 5v_C - 5v_B - 5v_D + 1)/2$
62	1	1121	2012	$(5v_A + 5v_B - 5v_C - 5v_D + 1)/2$	$(5v_A + 5v_C - 5v_B - 5v_D - 1)/2$
	2	2021	1112	$(5v_A + 5v_B - 5v_C - 5v_D + 1)/2$	$(5v_B + 5v_C - 5v_D - 5v_A + 1)/2$
	3	2111	1022	$(5v_B + 5v_C - 5v_D - 5v_A + 1)/2$	$(5v_A + 5v_C - 5v_B - 5v_D - 1)/2$
63	1	1211	2102	$(5v_A + 5v_B - 5v_C - 5v_D - 1)/2$	$(5v_A + 5v_C - 5v_B - 5v_D + 1)/2$
	2	2111	1202	$(5v_A + 5v_B - 5v_C - 5v_D - 1)/2$	$(5v_B + 5v_C - 5v_D - 5v_A + 1)/2$
	3	2201	1112	$(5v_B + 5v_C - 5v_D - 5v_A + 1)/2$	$(5v_A + 5v_C - 5v_B - 5v_D + 1)/2$
64	1	1220	2111	$(5v_A + 5v_B - 5v_C - 5v_D - 1)/2$	$(5v_A + 5v_C - 5v_B - 5v_D - 1)/2$
	2	2120	1211	$(5v_A + 5v_B - 5v_C - 5v_D - 1)/2$	$(5v_B + 5v_C - 5v_D - 5v_A - 1)/2$
	3	2210	1121	$(5v_B + 5v_C - 5v_D - 5v_A - 1)/2$	$(5v_A + 5v_C - 5v_B - 5v_D - 1)/2$

$$(3v_A \in (1, 2]) \wedge (3v_B \in [0, 1]) \wedge (3v_C \in [0, 1]) \wedge (3v_D \in [0, 1]) \tag{4.14}$$

and, therefore, the point under consideration lies into the octahedron with the number 11 and vertices 2100, 1110, 1011, 2001, 2010, and 1101. If the third way for drawing the axis is chosen, one can calculate v_1 and v_2 using Eqs. 4.15–4.16.

$$v_1 = \frac{3v_B + 3v_C - 3v_D - 3v_A + 1}{2} \tag{4.15}$$

$$v_2 = \frac{3v_A + 3v_C - 3v_B - 3v_D - 1}{2} \tag{4.16}$$

Then if, for example,

$$(v_1 > 0) \wedge (v_2 > 0), \tag{4.17}$$

then the following logical expression is true

$$(v_1 \in (0, 1]) \wedge (v_2 \in (0, 1]), \tag{4.18}$$

and, therefore, the point under consideration lies into the first different small tetrahedron 2100, 1011, 1110, 2010 and one can find the result of piecewise linear interpolation using Eq. 4.19.

$$u\left(\vec{r}\right) = \left(1 - 3v_C\left(\vec{r}\right)\right)u_{2100} + 3v_D\left(\vec{r}\right)u_{1011} + v_1\left(\vec{r}\right)u_{1110} + v_2\left(\vec{r}\right)u_{2010} \tag{4.19}$$

4.6 Tables and Algorithms for N = 4

Consider degree $N = 4$. 24 smaller tetrahedrons and 10 octahedrons are. If all of the logical expressions corresponding to the rows 1, 2, …, $I - 1$ in Table 4.18 are false and the logical expression corresponding to the row I in Table 4.18 is true, then the logical expression corresponding to the row I in Table 4.19 is true and, therefore, the point under consideration lies into the small tetrahedron I with vertices given in Table 4.20 or into the octahedron I with the vertices given in Table 4.21. If it is the tetrahedron I, one can find the relative volumes for its vertices in Table 4.22. If it is the octahedron I, one of 3 ways for drawing the axis into it should be chosen. Then one can calculate the appropriate v_1 and v_2 using the suitable formulae given in Table 4.23. One can use the Table 4.7 in Sect. 4.4 to find a number of different small tetrahedron into this octahedron I. If all of the logical expressions corresponding to the rows 1, …, $J - 1$ in Table 4.7 are false and the logical expression corresponding to the row J in Table 4.7 is true, then the logical expression corresponding to the row J in Table 4.8 in Sect. 4.4 is true and, therefore, the point under consideration lies into different small tetrahedron with the number J. One can find its vertices in Tables 4.23 and 4.24 and corresponding relative volumes in Table 4.10 in Sect. 4.4 and in Table 4.24.

4.7 Tables and Algorithms for N = 5

Consider degree $N = 5$. 45 smaller tetrahedrons and 20 octahedrons are. If all of the logical expressions corresponding to the rows $1, 2, \ldots, I - 1$ in Table 4.25 are false and the logical expression corresponding to the row I in Table 4.25 is true, then the logical

Table 4.31 The vertices C', D' of the different small tetrahedrons and the formulae for relative volumes corresponding the appropriate vertices A', B' of the different small tetrahedrons for $N = 5$, *table of the sixth type*

No	Axis no	Small tetrahedron no	C'	D'	$v_{A'}$	$v_{B'}$
17	1	1	4100	4010	$4 - 5v_A$	$5v_D$
		2	4100	3101	$5v_C$	$1 - 5v_B$
		3	3011	4010	$5v_B$	$1 - 5v_C$
		4	3011	3101	$1 - 5v_D$	$5v_A - 3$
	2	1	4100	3110	$1 - 5v_B$	$5v_D$
		2	4100	4001	$5v_C$	$4 - 5v_A$
		3	3011	3110	$5v_A - 3$	$1 - 5v_C$
		4	3011	4001	$1 - 5v_D$	$5v_B$
	3	1	3110	4010	$1 - 5v_C$	$5v_D$
		2	3110	3101	$5v_A - 3$	$1 - 5v_B$
		3	4001	4010	$5v_B$	$4 - 5v_A$
		4	4001	3101	$1 - 5v_D$	$5v_C$
18	1	1	1400	1310	$1 - 5v_A$	$5v_D$
		2	1400	0401	$5v_C$	$4 - 5v_B$
		3	0311	1310	$5v_B - 3$	$1 - 5v_C$
		4	0311	0401	$1 - 5v_D$	$5v_A$
	2	1	1400	0410	$4 - 5v_B$	$5v_D$
		2	1400	1301	$5v_C$	$1 - 5v_A$
		3	0311	0410	$5v_A$	$1 - 5v_C$
		4	0311	1301	$1 - 5v_D$	$5v_B - 3$
	3	1	0410	1310	$1 - 5v_C$	$5v_D$
		2	0410	0401	$5v_A$	$4 - 5v_B$
		3	1301	1310	$5v_B - 3$	$1 - 5v_A$
		4	1301	0401	$1 - 5v_D$	$5v_C$
19	1	1	1130	1040	$1 - 5v_A$	$5v_D$
		2	1130	0131	$5v_C - 3$	$1 - 5v_B$
		3	0041	1040	$5v_B$	$4 - 5v_C$
		4	0041	0131	$1 - 5v_D$	$5v_A$
	2	1	1130	0140	$1 - 5v_B$	$5v_D$
		2	1130	1031	$5v_C - 3$	$1 - 5v_A$
		3	0041	0140	$5v_A$	$4 - 5v_C$
		4	0041	1031	$1 - 5v_D$	$5v_B$
	3	1	0140	1040	$4 - 5v_C$	$5v_D$

(continued)

Table 4.31 (continued)

No	Axis no	Small tetrahedron no	C'	D'	$v_{A'}$	$v_{B'}$
		2	0140	0131	$5v_A$	$1-5v_B$
		3	1031	1040	$5v_B$	$1-5v_A$
		4	1031	0131	$1-5v_D$	$5v_C-3$
20	1	1	1103	1013	$1-5v_A$	$5v_D-3$
		2	1103	0104	$5v_C$	$1-5v_B$
		3	0014	1013	$5v_B$	$1-5v_C$
		4	0014	0104	$4-5v_D$	$5v_A$
	2	1	1103	0113	$1-5v_B$	$5v_D-3$
		2	1103	1004	$5v_C$	$1-5v_A$
		3	0014	0113	$5v_A$	$1-5v_C$
		4	0014	1004	$4-5v_D$	$5v_B$
	3	1	0113	1013	$1-5v_C$	$5v_D-3$
		2	0113	0104	$5v_A$	$1-5v_B$
		3	1004	1013	$5v_B$	$1-5v_A$
		4	1004	0104	$4-5v_D$	$5v_C$
39	1	1	3200	3110	$3-5v_A$	$5v_D$
		2	3200	2201	$5v_C$	$2-5v_B$
		3	2111	3110	$5v_B-1$	$1-5v_C$
		4	2111	2201	$1-5v_D$	$5v_A-2$
	2	1	3200	2210	$2-5v_B$	$5v_D$
		2	3200	3101	$5v_C$	$3-5v_A$
		3	2111	2210	$5v_A-2$	$1-5v_C$
		4	2111	3101	$1-5v_D$	$5v_B-1$
	3	1	2210	3110	$1-5v_C$	$5v_D$
		2	2210	2201	$5v_A-2$	$2-5v_B$
		3	3101	3110	$5v_B-1$	$3-5v_A$
		4	3101	2201	$1-5v_D$	$5v_C$
40	1	1	1310	1220	$1-5v_A$	$5v_D$
		2	1310	0311	$5v_C-1$	$3-5v_B$
		3	0221	1220	$5v_B-2$	$2-5v_C$
		4	0221	0311	$1-5v_D$	$5v_A$
	2	1	1310	0320	$3-5v_B$	$5v_D$
		2	1310	1211	$5v_C-1$	$1-5v_A$
		3	0221	0320	$5v_A$	$2-5v_C$
		4	0221	1211	$1-5v_D$	$5v_B-2$

(continued)

Table 4.31 (continued)

No	Axis no	Small tetrahedron no	C'	D'	$v_{A'}$	$v_{B'}$
	3	1	0320	1220	$2 - 5v_C$	$5v_D$
		2	0320	0311	$5v_A$	$3 - 5v_B$
		3	1211	1220	$5v_B - 2$	$1 - 5v_A$
		4	1211	0311	$1 - 5v_D$	$5v_C - 1$
41	1	1	1121	1031	$1 - 5v_A$	$5v_D - 1$
		2	1121	0122	$5v_C - 2$	$1 - 5v_B$
		3	0032	1031	$5v_B$	$3 - 5v_C$
		4	0032	0122	$2 - 5v_D$	$5v_A$
	2	1	1121	0131	$1 - 5v_B$	$5v_D - 1$
		2	1121	1022	$5v_C - 2$	$1 - 5v_A$
		3	0032	0131	$5v_A$	$3 - 5v_C$
		4	0032	1022	$2 - 5v_D$	$5v_B$
	3	1	0131	1031	$3 - 5v_C$	$5v_D - 1$
		2	0131	0122	$5v_A$	$1 - 5v_B$
		3	1022	1031	$5v_B$	$1 - 5v_A$
		4	1022	0122	$2 - 5v_D$	$5v_C - 2$
42	1	1	2102	2012	$2 - 5v_A$	$5v_D - 2$
		2	2102	1103	$5v_C$	$1 - 5v_B$
		3	1013	2012	$5v_B$	$1 - 5v_C$
		4	1013	1103	$3 - 5v_D$	$5v_A - 1$
	2	1	2102	1112	$1 - 5v_B$	$5v_D - 2$
		2	2102	2003	$5v_C$	$2 - 5v_A$
		3	1013	1112	$5v_A - 1$	$1 - 5v_C$
		4	1013	2003	$3 - 5v_D$	$5v_B$
	3	1	1112	2012	$1 - 5v_C$	$5v_D - 2$
		2	1112	1103	$5v_A - 1$	$1 - 5v_B$
		3	2003	2012	$5v_B$	$2 - 5v_A$
		4	2003	1103	$3 - 5v_D$	$5v_C$
43	1	1	2300	2210	$2 - 5v_A$	$5v_D$
		2	2300	1301	$5v_C$	$3 - 5v_B$
		3	1211	2210	$5v_B - 2$	$1 - 5v_C$
		4	1211	1301	$1 - 5v_D$	$5v_A - 1$
	2	1	2300	1310	$3 - 5v_B$	$5v_D$
		2	2300	2201	$5v_C$	$2 - 5v_A$
		3	1211	1310	$5v_A - 1$	$1 - 5v_C$

(continued)

Table 4.31 (continued)

No	Axis no	Small tetrahedron no	C'	D'	$v_{A'}$	$v_{B'}$
		4	1211	2201	$1-5v_D$	$5v_B-2$
	3	1	1310	2210	$1-5v_C$	$5v_D$
		2	1310	1301	$5v_A-1$	$3-5v_B$
		3	2201	2210	$5v_B-2$	$2-5v_A$
		4	2201	1301	$1-5v_D$	$5v_C$
44	1	1	1220	1130	$1-5v_A$	$5v_D$
		2	1220	0221	$5v_C-2$	$2-5v_B$
		3	0131	1130	$5v_B-1$	$3-5v_C$
		4	0131	0221	$1-5v_D$	$5v_A$
	2	1	1220	0230	$2-5v_B$	$5v_D$
		2	1220	1121	$5v_C-2$	$1-5v_A$
		3	0131	0230	$5v_A$	$3-5v_C$
		4	0131	1121	$1-5v_D$	$5v_B-1$
	3	1	0230	1130	$3-5v_C$	$5v_D$
		2	0230	0221	$5v_A$	$2-5v_B$
		3	1121	1130	$5v_B-1$	$1-5v_A$
		4	1121	0221	$1-5v_D$	$5v_C-2$
45	1	1	1112	1022	$1-5v_A$	$5v_D-2$
		2	1112	0113	$5v_C-1$	$1-5v_B$
		3	0023	1022	$5v_B$	$2-5v_C$
		4	0023	0113	$3-5v_D$	$5v_A$
	2	1	1112	0122	$1-5v_B$	$5v_D-2$
		2	1112	1013	$5v_C-1$	$1-5v_A$
		3	0023	0122	$5v_A$	$2-5v_C$
		4	0023	1013	$3-5v_D$	$5v_B$
	3	1	0122	1022	$2-5v_C$	$5v_D-2$
		2	0122	0113	$5v_A$	$1-5v_B$
		3	1013	1022	$5v_B$	$1-5v_A$
		4	1013	0113	$3-5v_D$	$5v_C-1$
46	1	1	3101	3011	$3-5v_A$	$5v_D-1$
		2	3101	2102	$5v_C$	$1-5v_B$
		3	2012	3011	$5v_B$	$1-5v_C$
		4	2012	2102	$2-5v_D$	$5v_A-2$
	2	1	3101	2111	$1-5v_B$	$5v_D-1$
		2	3101	3002	$5v_C$	$3-5v_A$

(continued)

Table 4.31 (continued)

No	Axis no	Small tetrahedron no	C'	D'	$v_{A'}$	$v_{B'}$
		3	2012	2111	$5v_A - 2$	$1 - 5v_C$
		4	2012	3002	$2 - 5v_D$	$5v_B$
	3	1	2111	3011	$1 - 5v_C$	$5v_D - 1$
		2	2111	2102	$5v_A - 2$	$1 - 5v_B$
		3	3002	3011	$5v_B$	$3 - 5v_A$
		4	3002	2102	$2 - 5v_D$	$5v_C$
47	1	1	3110	3020	$3 - 5v_A$	$5v_D$
		2	3110	2111	$5v_C - 1$	$1 - 5v_B$
		3	2021	3020	$5v_B$	$2 - 5v_C$
		4	2021	2111	$1 - 5v_D$	$5v_A - 2$
	2	1	3110	2120	$1 - 5v_B$	$5v_D$
		2	3110	3011	$5v_C - 1$	$3 - 5v_A$
		3	2021	2120	$5v_A - 2$	$2 - 5v_C$
		4	2021	3011	$1 - 5v_D$	$5v_B$
	3	1	2120	3020	$2 - 5v_C$	$5v_D$
		2	2120	2111	$5v_A - 2$	$1 - 5v_B$
		3	3011	3020	$5v_B$	$3 - 5v_A$
		4	3011	2111	$1 - 5v_D$	$5v_C - 1$
48	1	1	1301	1211	$1 - 5v_A$	$5v_D - 1$
		2	1301	0302	$5v_C$	$3 - 5v_B$
		3	0212	1211	$5v_B - 2$	$1 - 5v_C$
		4	0212	0302	$2 - 5v_D$	$5v_A$
	2	1	1301	0311	$3 - 5v_B$	$5v_D - 1$
		2	1301	1202	$5v_C$	$1 - 5v_A$
		3	0212	0311	$5v_A$	$1 - 5v_C$
		4	0212	1202	$2 - 5v_D$	$5v_B - 2$
	3	1	0311	1211	$1 - 5v_C$	$5v_D - 1$
		2	0311	0302	$5v_A$	$3 - 5v_B$
		3	1202	1211	$5v_B - 2$	$1 - 5v_A$
		4	1202	0302	$2 - 5v_D$	$5v_C$
49	1	1	2120	2030	$2 - 5v_A$	$5v_D$
		2	2120	1121	$5v_C - 2$	$1 - 5v_B$
		3	1031	2030	$5v_B$	$3 - 5v_C$
		4	1031	1121	$1 - 5v_D$	$5v_A - 1$
	2	1	2120	1130	$1 - 5v_B$	$5v_D$

(continued)

Table 4.31 (continued)

No	Axis no	Small tetrahedron no	C'	D'	$v_{A'}$	$v_{B'}$
		2	2120	2021	$5v_C - 2$	$2 - 5v_A$
		3	1031	1130	$5v_A - 1$	$3 - 5v_C$
		4	1031	2021	$1 - 5v_D$	$5v_B$
	3	1	1130	2030	$3 - 5v_C$	$5v_D$
		2	1130	1121	$5v_A - 1$	$1 - 5v_B$
		3	2021	2030	$5v_B$	$2 - 5v_A$
		4	2021	1121	$1 - 5v_D$	$5v_C - 2$
50	1	1	1202	1112	$1 - 5v_A$	$5v_D - 2$
		2	1202	0203	$5v_C$	$2 - 5v_B$
		3	0113	1112	$5v_B - 1$	$1 - 5v_C$
		4	0113	0203	$3 - 5v_D$	$5v_A$
	2	1	1202	0212	$2 - 5v_B$	$5v_D - 2$
		2	1202	1103	$5v_C$	$1 - 5v_A$
		3	0113	0212	$5v_A$	$1 - 5v_C$
		4	0113	1103	$3 - 5v_D$	$5v_B - 1$
	3	1	0212	1112	$1 - 5v_C$	$5v_D - 2$
		2	0212	0203	$5v_A$	$2 - 5v_B$
		3	1103	1112	$5v_B - 1$	$1 - 5v_A$
		4	1103	0203	$3 - 5v_D$	$5v_C$
61	1	1	1211	1121	$1 - 5v_A$	$5v_D - 1$
		2	1211	0212	$5v_C - 1$	$2 - 5v_B$
		3	0122	1121	$5v_B - 1$	$2 - 5v_C$
		4	0122	0212	$2 - 5v_D$	$5v_A$
	2	1	1211	0221	$2 - 5v_B$	$5v_D - 1$
		2	1211	1112	$5v_C - 1$	$1 - 5v_A$
		3	0122	0221	$5v_A$	$2 - 5v_C$
		4	0122	1112	$2 - 5v_D$	$5v_B - 1$
	3	1	0221	1121	$2 - 5v_C$	$5v_D - 1$
		2	0221	0212	$5v_A$	$2 - 5v_B$
		3	1112	1121	$5v_B - 1$	$1 - 5v_A$
		4	1112	0212	$2 - 5v_D$	$5v_C - 1$
62	1	1	2111	2021	$2 - 5v_A$	$5v_D - 1$
		2	2111	1112	$5v_C - 1$	$1 - 5v_B$
		3	1022	2021	$5v_B$	$2 - 5v_C$
		4	1022	1112	$2 - 5v_D$	$5v_A - 1$

(continued)

Table 4.31 (continued)

No	Axis no	Small tetrahedron no	C'	D'	$v_{A'}$	$v_{B'}$
	2	1	2111	1121	$1-5v_B$	$5v_D-1$
		2	2111	2012	$5v_C-1$	$2-5v_A$
		3	1022	1121	$5v_A-1$	$2-5v_C$
		4	1022	2012	$2-5v_D$	$5v_B$
	3	1	1121	2021	$2-5v_C$	$5v_D-1$
		2	1121	1112	$5v_A-1$	$1-5v_B$
		3	2012	2021	$5v_B$	$2-5v_A$
		4	2012	1112	$2-5v_D$	$5v_C-1$
63	1	1	2201	2111	$2-5v_A$	$5v_D-1$
		2	2201	1202	$5v_C$	$2-5v_B$
		3	1112	2111	$5v_B-1$	$1-5v_C$
		4	1112	1202	$2-5v_D$	$5v_A-1$
	2	1	2201	1211	$2-5v_B$	$5v_D-1$
		2	2201	2102	$5v_C$	$2-5v_A$
		3	1112	1211	$5v_A-1$	$1-5v_C$
		4	1112	2102	$2-5v_D$	$5v_B-1$
	3	1	1211	2111	$1-5v_C$	$5v_D-1$
		2	1211	1202	$5v_A-1$	$2-5v_B$
		3	2102	2111	$5v_B-1$	$2-5v_A$
		4	2102	1202	$2-5v_D$	$5v_C$
64	1	1	2210	2120	$2-5v_A$	$5v_D$
		2	2210	1211	$5v_C-1$	$2-5v_B$
		3	1121	2120	$5v_B-1$	$2-5v_C$
		4	1121	1211	$1-5v_D$	$5v_A-1$
	2	1	2210	1220	$2-5v_B$	$5v_D$
		2	2210	2111	$5v_C-1$	$2-5v_A$
		3	1121	1220	$5v_A-1$	$2-5v_C$
		4	1121	2111	$1-5v_D$	$5v_B-1$
	3	1	1220	2120	$2-5v_C$	$5v_D$
		2	1220	1211	$5v_A-1$	$2-5v_B$
		3	2111	2120	$5v_B-1$	$2-5v_A$
		4	2111	1211	$1-5v_D$	$5v_C-1$

expression corresponding to the row I in Table 4.26 is true and, therefore, the point under consideration lies into the small tetrahedron I with vertices given in Table 4.27 or into the octahedron I with the vertices given in Table 4.28. If it is the tetrahedron I, one can find the relative volumes for its vertices in Table 4.29. If it is the octahedron I, one of 3 ways for drawing the axis into it should be chosen. Then one can calculate the appropriate v_1 and v_2 using the suitable formulae given in Table 4.30. One can use the Table 4.7 in Sect. 4.4 to find a number of the different small tetrahedron into this octahedron I. If all of the logical expressions corresponding to the rows 1, ..., $J - 1$ in Table 4.7 are false and the logical expression corresponding to the row J in Table 4.7 is true, then the logical expression corresponding to the row J in Table 4.8 in Sect. 4.4 is true and, therefore, the point under consideration lies into different small tetrahedron with the number J. One can find its vertices in Tables 4.30 and 4.31 and corresponding relative volumes in Table 4.10 in Sect. 4.4 and in Table 4.31.

4.8 Conclusions

The analytical formulae for piecewise linear interpolation on the unstructured tetrahedral grids are suggested in this chapter. The cases of reference points used for the polynomial interpolation discussed in Chap. 3 for order from 1 to 5 inclusively are considered. These interpolation techniques can be used during the creation of a new unstructured triangular or regular gird instead of previous one (as an element of numerical method for finding 3D solutions on unstructured tetrahedral grids), visualization of some space field, and pictures' creation or converting. Also an element for applying the hierarchical nested unstructured tetrahedral grids and formulae for recalculation from local to global indices are discussed in Chap. 3. Analytical formulae reduce the computational recourses, i.e. the software operation time and amount of dynamic computer memory. In this chapter, one can find a vast amount of analytical expressions joint into convenient tables ready for using. These analytical expressions and tables help to achieve the great numerical modelling results in a case of the deficiency of hardware resources.

Acknowledgements This work has been performed at Non-state Educational Institution "Educational Scientific and Experimental Center of Moscow Institute of Physics and Technology" and supported by the Russian Science Foundation, grant no. 17-71-20088. This work has been carried out using computing resources of the federal collective usage center Complex for Simulation and Data Processing for Mega-science Facilities at NRC "Kurchatov Institute", http://ckp.nrcki.ru/.

References

1. Favorskaya AV, Muratov MV, Petrov IB, Sannikov IV (2014) Grid-characteristic method on unstructured tetrahedral meshes. Comput Math Math Phys 54(5):837–847
2. Favorskaya AV, Petrov IB (2016) Wave responses from oil reservoirs in the Arctic shelf zone. Dokl Earth Sci 466(2):214–217
3. Khokhlov N, Yavich N, Malovichko M, Petrov I (2015) Solution of large-scale seismic modeling problems. Procedia Comput Sci 66:191–199
4. Favorskaya AV, Petrov IB (2017) Numerical modeling of dynamic wave effects in rock masses. Dokl Math 95(3):287–290
5. Vassilevski YV, Beklemysheva KA, Grigoriev GK, Kazakov AO, Kulberg NS, Petrov IB, Salamatova VY, Vasyukov AV (2016) Transcranial ultrasound of cerebral vessels in silico: proof of concept. Russ J Numer Anal Math Model 31(5):317–328
6. Beklemysheva KA, Vasyukov AV, Petrov IB (2015) Numerical simulation of dynamic processes in biomechanics using the grid-characteristic method. Comput Math Math Phys 55(8):1346–1355
7. Beklemysheva KA, Danilov AA, Petrov IB, Salamatova VYu, Vassilevski YV, Vasyukov AV (2015) Virtual blunt injury of human thorax: age-dependent response of vascular system. Russ J Numer Anal Math Model 30(5):259–268
8. Petrov I, Vasyukov A, Beklemysheva K, Ermakov A, Favorskaya A (2016) Numerical modeling of non-destructive testing of composites. Procedia Comput Sci 96:930–938
9. Beklemysheva KA, Vasyukov AV, Ermakov AS, Petrov IB (2016) Numerical simulation of the failure of composite materials by using the grid-characteristic method. Math Models Comput Simul 8(5):557–567
10. Ball JA, Gohberg I (2013) Interpolation of rational matrix functions, vol 45. Birkhauser, Basel
11. Lama RK, Kwon GR (2015) New interpolation method based on combination of discrete cosine transform and wavelet transform. In: Information Networking (ICOIN), 2015 International Conference, IEEE, 363–366
12. Li J, Heap AD (2014) Spatial interpolation methods applied in the environmental sciences: a review. Environ Model Softw 53:173–189
13. Agarwal S, Khade S, Dandawate Y, Khandekar P (2015) Three dimensional image reconstruction using interpolation of distance and image registration. In: Computer, Communication and Control (IC4), 2015 International Conference, IEEE, 1–5
14. Dong W, Zhang L, Lukac R, Shi G (2013) Sparse representation based image interpolation with nonlocal autoregressive modeling. IEEE Trans Image Process 22(4):1382–1394
15. Greco L, Cuomo M (2014) An implicit G1 multi patch B-spline interpolation for Kirchhoff-Love space rod. Comput Methods Appl Mech Eng 269:173–197
16. Hu J, Zhang S (2015) A family of symmetric mixed finite elements for linear elasticity on tetrahedral grids. Sci China Math 58(2):297–307
17. Nguyen MN, Bui TQ, Truong TT, Trinh NA, Singh IV, Yu T, Doan DH (2016) Enhanced nodal gradient 3D consecutive-interpolation tetrahedral element (CTH4) for heat transfer analysis. Int J Heat Mass Transf 103:14–27
18. Alauzet F (2016) A parallel matrix-free conservative solution interpolation on unstructured tetrahedral meshes. Comput Methods Appl Mech Eng 299:116–142
19. Paille GP, Ray N, Poulin P, Sheffer A, Levy B (2015) Dihedral angle-based maps of tetrahedral meshes. ACM Trans Graph (TOG) 34(4):54

Chapter 5
Grid-Characteristic Method

Alena V. Favorskaya and Igor B. Petrov

Abstract In this chapter, a family of grid-characteristic methods for numerical simulation is considered. These methods are developed and used to solve a wide range of applied problems: traumatology, ultrasound studies of the human body, ultrasonic operations, seismic exploration of oil and gas, seismic resistance of residential and industrial facilities, non-destructive testing of railways and innovative materials including composites, development territories with complex natural conditions, shock effects on complex-shaped structures, and global seismic of various planets of the solar system. The methods allow to simulate the wave processes in heterogeneous media of complex topology and dynamic process of destruction of these media. Also these methods help to investigate clearly small heterogeneous features that represent breaks in the integration domain. Grid-characteristic methods are used to solve the hyperbolic systems of equations describing the wave processes. In this chapter, the elastic waves in isotropic and anisotropic cases and acoustic waves are considered. The method is well paralleled and actively implemented in software using the high-performance computing systems.

Keywords Grid-characteristic method · Numerical method
Linear-elastic media · Elastic waves · Acoustic waves · Anisotropy
Waves modelling · Orthorhombic anisotropy · Vertically transversal anisotropy

A. V. Favorskaya (✉)
Non-state Educational Institution "Educational Scientific and Experimental Center of Moscow Institute of Physics and Technology", 9, Institutsky Pereulok st., Dolgoprudny, Moscow Region 141700, Russian Federation
e-mail: aleanera@yandex.ru

A. V. Favorskaya · I. B. Petrov
Department of Computer Science and Numerical Mathematics, Moscow Institute of Physics and Technology, 9, Institutsky Pereulok st., Dolgoprudny, Moscow Region 141700, Russian Federation
e-mail: petrov@mipt.ru

A. V. Favorskaya · I. B. Petrov
Scientific Research Institute for System Studies of the Russian Academy of Sciences, 36(1), Nahimovskij ave, Moscow 117218, Russian Federation

A. V. Favorskaya and I. B. Petrov (eds.), *Innovations in Wave Processes Modelling and Decision Making*, Smart Innovation, Systems and Technologies 90,
https://doi.org/10.1007/978-3-319-76201-2_5

Horizontally transversal anisotropy · Boundary conditions · Interface conditions
Non-reflecting conditions

5.1 Introduction

Grid-characteristic method (GCM) [1–11] is widely used for numerical investigation of dynamic wave processes, called Wave Logica [1–5], in heterogeneous anisotropic [1, 2], isotropic elastic, and acoustic environments using regular [3, 4], tetrahedral [5], triangular [6], and curvilinear [6, 7] meshes, different solutions schemes [8], and direct modelling of the fractures [1] and their growth [3]. This family of numerical methods is based on the grid-characteristic method [11], which is a finite-difference method for the numerical solving of direct methods.

LeVeque provided a comprehensive overview of numerical modelling of hyperbolic systems of equations [12]. Direct methods, integral-equation methods, and asymptotic methods applying for oil and gas seismic exploration problems were discussed by Carcione et al. [13]. In works [14–18], Nikitin et al. obtained averaged effective models of layered and block media with varying degrees of accuracy and various contact conditions on interlayer (interblock) boundaries for elastic, viscous, and viscoplastic media and proposed numerical schemes for their solution. The study of the dispersive behavior of spectral element method was done by Ainsworth and Wajid [19]. Also, Ainsworth et al. studied the discontinuous Galerkin finite element methods for the second-order wave equation [20]. Alford investigated the accuracy of finite-difference modelling of the acoustic wave equation [21]. The accuracy of heterogeneous staggered-grid finite-difference modelling of Rayleigh seismic waves was improved by Bohlen and Saenger [22]. A 27-point scheme for a 3D frequency-domain scalar wave equation based on an average-derivative method was suggested by Chen [23]. Numerical methods for transient acoustic, elastic, and electromagnetic waves were considered by Cohen and Gaunaurd [24]. Numerical dispersion of spectral element methods of arbitrary order for the isotropic elastic wave equation in two and three dimensions was studied by Seriani and Oliveira [25]. Saenger et al. discussed a propagation of the elastic waves using a modified finite-difference grid [26]. The accuracy of the finite-difference and finite-element schemes with respect to P-wave to S-wave speed ratio was investigated by Moczo et al. [27]. A comprehensive introduction to finite-difference technique and its applications to earthquake motion were provided by Moczo et al. [28]. Nonreflecting boundary conditions for the wave equation were proposed by Hagstrom et al. [29–31] and independently by Appelö and Kreiss for elastic waves [32].

The conventional approach of numerical modelling used in the reviewed above papers includes the following items.

- First, the system of equations is chosen to solve.
- Second, the numerical method is chosen to solve this system.
- Third, the software is developed using the chosen method.

There is the alternative approach for numerical experiments and investigation, called grid-characteristic approach, which contains the following steps:

- First, the main physical effects are chosen to study in the numerical investigation.
- Second, the numerical method is chosen and applied to describe the main physical effects.
- Third, the software is developed using this method that describes the main physical effects.

The main difference between the grid-characteristic and conventional approaches is in different targeting of a numerical method. In a case of conventional approach, a numerical method is chosen to solve the system of equation. In a case of grid-characteristic approach, a numerical method is chosen to describe the physical effects having meaning in the investigation under consideration.

The grid-characteristic approach was used by Prof. Petrov and his scientific group at Moscow Institute of Physics and Technology (MIPT) during last 30 years. Petrov was a student of Prof. Belotserkovskii [33]. Also Petrov used a grid-characteristic method for solving the practical problems as a student of Kholodov [11]. Thus, the elaboration of this grid-characteristic method was directed by the needs of practical investigations in different areas [1–10]. The main principles of the family of grid-characteristic methods are discussed in this chapter. The comparison of the grid-characteristic method and other forward modelling techniques was discussed by Khokhlov et al. [6].

In Sect. 5.2, the widely used systems of equations are considered, i.e. the system of equations describing the state of an infinitesimal volume of a continuous linear-elastic medium in isotropic [3, 4] and anisotropic [1, 2, 34–37] cases and a system describing the acoustic field [38]. Numerical modelling of the solutions of these systems makes it possible to solve problems of a wide class mentioned in Chap. 2 and in this Chapter. Also these systems of equations are used during solving the problems of seismic prospecting and exploration of oil and gas [1, 4] including the shelf zones [4]. The solution of the system of equations describing a linear-elastic media is calculated for describing the wave processes in rocks, soils, ice, and in some cases in the oil-containing reservoirs. The system describing an acoustic field is solved for modelling wave processes in water using approximation of an ideal fluid [38] and in some cases in the oil-containing reservoirs. Also, a numerical solution of the system of equations describing a linear-elastic medium in an isotropic case enables to simulate seismic stability of ground and underground structures [3] and ultrasonic non-destructive testing of railway track elements [39]. The solution of the system of equations describing a linear-elastic medium in isotropic and anisotropic cases makes it possible to investigate composite materials as shown in Chap. 6.

In Sect. 5.3, a family of grid-characteristic numerical methods for the solution of hyperbolic systems of equations is considered. This family of numerical methods can be applied on different structured [3, 4] and unstructured (triangular [6] and tetrahedral [5]) meshes. Also, this family of numerical methods might be used to solve different system of equations, including the systems describing the state continuous linear-elastic medium [1–10, 13], including an anisotropic case [1, 34–37], and acoustic field [38]. It should be noted that adding a heterogeneous structure and taking into account dynamic delamination allows to describe quite accurately a wide class of natural phenomena with a linear-elastic case [1–10, 39]. One can see some examples in Chap. 6. The study on approximation of this family of grid-characteristic methods on the unstructured triangular and tetrahedral grids is given in [8]. The study on stability of solution schemes, into which this family of methods is transformed in the a one-dimensional case, is done in [8] as well.

In Sect. 5.4, the boundary and interface conditions preserving the order of the difference scheme that is used for calculating interior points are considered. Boundary conditions of given traction, given velocity, mixed and non-reflecting boundary conditions, the boundary conditions using imaginary points [3], interface conditions of continuity of the velocity and traction, free sliding interface conditions, interface conditions of dynamic friction [9] might be used during solving the system of equations that describes elastic wave including an anisotropic case.

Boundary conditions with a given normal projection of the boundary velocity and given pressure are considered for a case of the acoustic field. The interface conditions are also considered. The contact conditions on the interface between the elastic and acoustic layers are discussed in [4]. The contact conditions at the boundaries of the transition between hierarchical grids with different fineness might be used as well [5]. The contact conditions at the interface of the sub-domain, in which the grid-characteristic method is used, and the sub-domain, in which a Smoothed Particles Hydrodynamics (SPH) method is used are discussed in [10]. They are called as a contact condition for the combined GCM-SPH method [10].

Section 5.5 concludes of the Chapter.

5.2 Systems of Equations Describing Wave Processes

In this Section, the systems of equations describing the wave processes used to solve the problems mentioned above are considered. The general case of anisotropic elastic waves, orthorhombic anisotropy, vertical-transversal anisotropy, and horizontal-transversal anisotropy are discussed in Sects. 5.2.1–5.2.4, respectively. Two last types of anisotropy are widely used for modelling the seismic waves into geological media and composite materials. In Sect. 5.2.5, a system of equations describing elastic waves is considered, while the acoustic waves are discussed in Sect. 5.2.6.

5.2.1 System of Equations Describing Anisotropic Elastic Waves

The condition of an infinitesimal volume of the continuous linear-elastic anisotropic medium is described by the Eqs. 5.1, and 5.2 [1, 34–37].

$$\rho(x, y, z)\, \partial_t v_i(x, y, z, t) = \sum_j \partial_j \sigma_{ij}(x, y, z, t) \tag{5.1}$$

$$\partial_t \sigma_{ij}(x, y, z, t) = \sum_k \sum_l C_{ij,kl}(x, y, z)\left(\partial_k v_l(x, y, z, t) + \partial_l v_k(x, y, z, t)\right) \tag{5.2}$$

In Eqs. 5.1, and 5.2 and further $\partial_t a(x, y, z, t) \equiv \frac{\partial a(x,y,z,t)}{\partial t}$ is the partial derivative of the field $a(x, y, z, t)$ with respect to t. In Eqs. 5.1, and 5.2, the indices near components of vectors and tensors are varied from 1 to 3. In Eqs. 5.1, 5.2 and further, t is the time, x, y, z are the coordinates, $\rho(x, y, z)$ is the density of the material, $\vec{v}(x, y, z, t)$ is the velocity of motion, $\boldsymbol{\sigma}(x, y, z, t)$ is the symmetric Cauchy stress tensor, $C_{ij,kl}(x, y, z)$ is the tensor of elastic constants of the fourth rank given by Eq. 5.3.

$$\begin{aligned} C_{ij,kl} = {} & c_{i,k}\delta_{ij}\delta_{kl} + \sum_{m=1}^{3} c_{i,m+3}\delta_{ij}\left|\varepsilon_{mkl}\right| + \sum_{m=1}^{3} c_{m+3,k}\left|\varepsilon_{mij}\right|\delta_{kl} \\ & + \sum_{m=1}^{3}\sum_{n=1}^{3} c_{m+3,n+3}\left|\varepsilon_{mij}\right|\left|\varepsilon_{nkl}\right| \end{aligned} \tag{5.3}$$

In Eq. 5.3 and further, δ_{ij} are the components of a unit tensor of the second rank, ε_{mij} are the components of an absolutely antisymmetric unit tensor of the third rank.

The coefficients of the fourth-rank elastic constants tensor are often written in a view of matrix 6×6 provided by Eq. 5.4.

$$\begin{pmatrix} c_{11} & c_{12} & c_{13} & c_{14} & c_{15} & c_{16} \\ c_{12} & c_{22} & c_{23} & c_{24} & c_{25} & c_{26} \\ c_{13} & c_{23} & c_{33} & c_{34} & c_{35} & c_{36} \\ c_{14} & c_{24} & c_{34} & c_{44} & c_{45} & c_{46} \\ c_{15} & c_{25} & c_{35} & c_{45} & c_{55} & c_{56} \\ c_{16} & c_{26} & c_{36} & c_{46} & c_{56} & c_{66} \end{pmatrix} \tag{5.4}$$

The correspondence between the coefficients of the structure given by Eq. 5.4 and the tensor in Eq. 5.3 can be written by Eqs. 5.5–5.13, where the numbers from the structure (Eq. 5.4) are written on the left side and the pairs of indices of the tensor (Eq. 5.3) are written on the right side.

$$1 \leftrightarrow 11 \tag{5.5}$$

$$2 \leftrightarrow 22 \tag{5.6}$$

$$3 \leftrightarrow 33 \tag{5.7}$$

$$4 \leftrightarrow 23 \tag{5.8}$$

$$4 \leftrightarrow 32 \tag{5.9}$$

$$5 \leftrightarrow 13 \tag{5.10}$$

$$5 \leftrightarrow 31 \tag{5.11}$$

$$6 \leftrightarrow 12 \tag{5.12}$$

$$6 \leftrightarrow 21. \tag{5.13}$$

5.2.2 Case of Orthorhombic Anisotropy

Let *OXY* be a plane of symmetry. Then the components of the tensor of elastic constants $c_{14}, c_{15}, c_{24}, c_{25}, c_{34}, c_{35}, c_{46}, c_{56}$ given in this Cartesian coordinate system *OXYZ* are equal to zero as written in Eq. 5.14.

$$c_{14} = c_{15} = c_{24} = c_{25} = c_{34} = c_{35} = c_{46} = c_{56} = 0 \tag{5.14}$$

A structure given by Eq. 5.4 in accordance with Eq. 5.14 is written in a view of Eq. 5.15.

$$\begin{pmatrix} c_{11} & c_{12} & c_{13} & 0 & 0 & c_{16} \\ c_{12} & c_{22} & c_{23} & 0 & 0 & c_{26} \\ c_{13} & c_{23} & c_{33} & 0 & 0 & c_{36} \\ 0 & 0 & 0 & c_{44} & c_{45} & 0 \\ 0 & 0 & 0 & c_{45} & c_{55} & 0 \\ c_{16} & c_{26} & c_{36} & 0 & 0 & c_{66} \end{pmatrix} \tag{5.15}$$

In a case of orthorhombic anisotropy, three mutually perpendicular axes of symmetry can be distinguished. If the Cartesian coordinate system, in which the tensor of elastic constants is written, coincides with these axes, then the components of the elastic constant tensor $c_{14}, c_{15}, c_{24}, c_{25}, c_{34}, c_{35}, c_{46}, c_{56}, c_{16}, c_{26}, c_{36}, c_{45}$ will be equal to zero in accordance with Eq. 5.16.

$$c_{14} = c_{15} = c_{24} = c_{25} = c_{34} = c_{35} = c_{46} = c_{56} = c_{16} = c_{26} = c_{36} = c_{45} = 0 \tag{5.16}$$

The structure given by Eq. 5.4 in accordance with Eq. 5.16 in this Cartesian coordinate system will be strongly sparse and is written by Eq. 5.17.

$$\begin{pmatrix} c_{11} & c_{12} & c_{13} & 0 & 0 & 0 \\ c_{12} & c_{22} & c_{23} & 0 & 0 & 0 \\ c_{13} & c_{23} & c_{33} & 0 & 0 & 0 \\ 0 & 0 & 0 & c_{44} & 0 & 0 \\ 0 & 0 & 0 & 0 & c_{55} & 0 \\ 0 & 0 & 0 & 0 & 0 & c_{66} \end{pmatrix} \tag{5.17}$$

5.2.3 *Case of Vertical-Transversal Anisotropy*

A particular case of orthorhombic anisotropy is the vertical-transversal anisotropy. In a case of vertically transversal anisotropy, the components of the tensor of elastic constants c_{14}, c_{15}, c_{24}, c_{25}, c_{34}, c_{35}, c_{46}, c_{56},c_{16}, c_{26}, c_{36}, c_{45} are being equal to zero in accordance with Eq. 5.18.

$$c_{14} = c_{15} = c_{24} = c_{25} = c_{34} = c_{35} = c_{46} = c_{56} = c_{16} = c_{26} = c_{36} = c_{45} = 0 \tag{5.18}$$

The components of the elastic tensor c_{11}, c_{22}, c_{44}, c_{55}, c_{66}, c_{12}, c_{13}, c_{23} are related by Eqs. 5.19–5.22.

$$c_{11} = c_{22} \tag{5.19}$$

$$c_{13} = c_{23} \tag{5.20}$$

$$c_{44} = c_{55} \tag{5.21}$$

$$c_{12} = c_{11} - 2c_{66} \tag{5.22}$$

The structure given by Eq. 5.4 in accordance with Eqs. 5.18–5.22 is written by Eq. 5.23.

$$\begin{pmatrix} c_{11} & c_{11} - 2c_{66} & c_{13} & 0 & 0 & 0 \\ c_{11} - 2c_{66} & c_{11} & c_{13} & 0 & 0 & 0 \\ c_{13} & c_{13} & c_{33} & 0 & 0 & 0 \\ 0 & 0 & 0 & c_{44} & 0 & 0 \\ 0 & 0 & 0 & 0 & c_{44} & 0 \\ 0 & 0 & 0 & 0 & 0 & c_{66} \end{pmatrix} \tag{5.23}$$

5.2.4 Case of Horizontal-Transversal Anisotropy

Another particular case of orthorhombic anisotropy is the horizontal-transversal anisotropy. In a case of horizontally transversal anisotropy, the components of the tensor of elastic constants $c_{14}, c_{15}, c_{25}, c_{34}, c_{35}, c_{46}, c_{56}, c_{16}, c_{26}, c_{36}, c_{45}$ are equal to zero in accordance with Eq. 5.24.

$$c_{14} = c_{15} = c_{24} = c_{25} = c_{34} = c_{35} = c_{46} = c_{56} = c_{16} = c_{26} = c_{36} = c_{45} = 0 \tag{5.24}$$

The components of the elastic tensor $c_{22}, c_{33}, c_{44}, c_{55}, c_{66}, c_{12}, c_{13}, c_{23}$ are related by Eqs. 5.25–5.28.

$$c_{22} = c_{33} \tag{5.25}$$

$$c_{12} = c_{13} \tag{5.26}$$

$$c_{55} = c_{66} \tag{5.27}$$

$$c_{23} = c_{22} - 2c_{44} \tag{5.28}$$

The structure given by Eq. 5.4 in accordance with Eqs. 5.24–5.28 is written by Eq. 5.29.

$$\begin{pmatrix} c_{11} & c_{12} & c_{12} & 0 & 0 & 0 \\ c_{12} & c_{22} & c_{22} - 2c_{44} & 0 & 0 & 0 \\ c_{12} & c_{22} - 2c_{44} & c_{22} & 0 & 0 & 0 \\ 0 & 0 & 0 & c_{44} & 0 & 0 \\ 0 & 0 & 0 & 0 & c_{55} & 0 \\ 0 & 0 & 0 & 0 & 0 & c_{55} \end{pmatrix} \tag{5.29}$$

5.2.5 System of Equations Describing Isotropic Elastic Waves

The system of equations describing the state of an infinitesimal element of the linearly elastic medium is considered and given by Eqs. 5.30–5.31.

$$\rho(x, y, z)\, \partial_t \vec{v}(x, y, z, t) = (\nabla \cdot \boldsymbol{\sigma}(x, y, z, t))^{\mathrm{T}} \tag{5.30}$$

$$\partial_t \boldsymbol{\sigma}(x, y, z, t) = \lambda(x, y, z) \left(\nabla \cdot \vec{v}(x, y, z, t)\right) \mathbf{I} + \mu(x, y, z) \left(\left(\nabla : \vec{v}(x, y, z, t)\right) + \left(\nabla : \vec{v}(x, y, z, t)\right)^{\mathrm{T}}\right) \tag{5.31}$$

Equation 5.30 is a local equation of motion. In Eqs. 5.30–5.31 and further, ∇ is the vector-gradient given by Eq. 5.32.

$$\nabla = \begin{bmatrix} \partial_x \\ \partial_y \\ \partial_z \end{bmatrix} \tag{5.32}$$

Equation 5.31 is obtained by differentiating the Hooke law with respect to time. In Eq. 5.31, λ and μ are the Lame parameters that define the elastic properties of the material, I is the unit tensor of the second rank. Equations 5.30–5.31 in a 2D case take the same form but one should use the dependence from x and y instead of dependence from x, y, and z. All 3D vectors and tensors of rank 2 should be changed to 2D vectors and tensors of rank 2 as well.

It should be noted that the physical characteristics of a linearly elastic medium are always described by two parameters, between which there is a one-to-one correspondence. One can use Lame parameters. One can use the Poisson's ratio and the Young's modulus. The speeds of longitudinal waves (P-waves) and transverse waves (S-waves) can be used. These speeds are most convenient for describing geological environments in the interests of seismic exploration of oil and gas.

In Eq. 5.31 and further, $\vec{a} : \vec{b}$ is the tensor product of vectors $\vec{a}$ and $\vec{b}$. The components of this tensor product can be calculated using Eq. 5.33.

$$\left(\vec{a} : \vec{b}\right)^{ij} = a^i b^j \tag{5.33}$$

The speed of P-waves in a linear-elastic medium can be found from Eq. 5.34.

$$c_{\text{P}} = \sqrt{\frac{\lambda + 2\mu}{\rho}} \tag{5.34}$$

The S-waves' speed is calculated in accordance with Eq. 5.35.

$$c_{\text{S}} = \sqrt{\frac{\mu}{\rho}} \tag{5.35}$$

Notice that Eqs. 5.30, and 5.31 one can obtain from Eqs. 5.1, and 5.2 using the structure from Eq. 5.4 in the form given by Eq. 5.36.

$$\begin{pmatrix} \lambda+2\mu & \lambda & \lambda & 0 & 0 & 0 \\ \lambda & \lambda+2\mu & \lambda & 0 & 0 & 0 \\ \lambda & \lambda & \lambda+2\mu & 0 & 0 & 0 \\ 0 & 0 & 0 & \mu & 0 & 0 \\ 0 & 0 & 0 & 0 & \mu & 0 \\ 0 & 0 & 0 & 0 & 0 & \mu \end{pmatrix} \tag{5.36}$$

Also, Eq. 5.31 may be represented in the form given by Eq. 5.37 using Eqs. 5.34, and 5.35. This form is more convenient for wave processes modelling.

$$\partial_t \boldsymbol{\sigma} = \left(\rho c_{\mathrm{P}}^2 - 2\rho c_{\mathrm{S}}^2\right)\left(\nabla \cdot \vec{v}\right)\mathbf{I} + \rho c_{\mathrm{S}}^2 \left(\left(\nabla : \vec{v}\right) + \left(\nabla : \vec{v}\right)^{\mathrm{T}}\right) \tag{5.37}$$

5.2.6 System of Equations Describing Isotropic Acoustic Waves

A system of equations describing the acoustic field is given by Eqs. 5.38–5.39. These formulae describe the acoustic pressure field $p\left(x, y, z, t\right)$ and the velocity vector field $\vec{v}\left(x, y, z, t\right)$. This system of equations might be used for the numerical simulation of the liquid in the approximation of an ideal fluid [38].

$$\rho\left(x, y, z\right) \partial_t \vec{v}\left(x, y, z, t\right) = -\nabla p\left(x, y, z, t\right) \tag{5.38}$$

$$\partial_t p\left(x, y, z, t\right) = -\rho\left(x, y, z\right) c\left(x, y, z\right)^2 \left(\nabla \cdot \vec{v}\left(x, y, z, t\right)\right) \tag{5.39}$$

In Eq. 5.39, the speed of sound in an ideal fluid is denoted by $c\left(x, y, z\right)$. Equations 5.38–5.39 in a 2D case take the same form but one should use the dependence from x and y only instead of dependence from x, y, and z. All 3D vectors and tensors of rank 2 should be changed to 2D vectors and tensors of rank 2 as well.

Notice that Eqs. 5.38–5.39 can be obtained from Eqs. 5.1 to 5.2 using the structure from Eq. 5.4 in the form given by Eq. 5.40.

$$\begin{pmatrix} \lambda & \lambda & \lambda & 0 & 0 & 0 \\ \lambda & \lambda & \lambda & 0 & 0 & 0 \\ \lambda & \lambda & \lambda & 0 & 0 & 0 \\ 0 & 0 & 0 & 0 & 0 & 0 \\ 0 & 0 & 0 & 0 & 0 & 0 \\ 0 & 0 & 0 & 0 & 0 & 0 \end{pmatrix} \tag{5.40}$$

Also notice that Eqs. 5.38, and 5.39 are derived from Eqs. 5.30, and 5.31 using the formulae given in Eqs. 5.41–5.43 for a 3D case and Eqs. 5.41, 5.42, and 5.44 in a 2D case.

$$c_P = c \tag{5.41}$$

$$c_S = 0 \tag{5.42}$$

$$\sigma = \begin{pmatrix} -p/3 & 0 & 0 \\ 0 & -p/3 & 0 \\ 0 & 0 & -p/3 \end{pmatrix} \tag{5.43}$$

$$\sigma = \begin{pmatrix} -p/2 & 0 \\ 0 & -p/2 \end{pmatrix} \tag{5.44}$$

5.3 Grid-Characteristic Method Describing

The main aspects of the software algorithm for grid-characteristic method are discussed at Sect. 5.3.1. The general statements of grid-characteristic methods are presented in Sect. 5.3.2. In the general case, the grid-characteristic methods for anisotropic linear-elastic media are described in Sect. 5.3.3. Grid-characteristic method for a case of orthorhombic anisotropic elastic waves is discussed in Sect. 5.3.4. The method for the isotropic case of linear-elastic media is considered in Sect. 5.3.5. The method for a case of acoustic waves one can find in Sect. 5.3.6.

5.3.1 Scheme of the Algorithm for Grid-Characteristic Method Using

In order to diminish the amount of Random Access Memory (RAM) needed for calculations, two layers of coordinate mesh are used. Denote them as *Array0* and *Array1*. *Array0* is also called n time layer. *Array1* is also called $n + 1$ time layer. One can use the following algorithm to perform the calculations. One step of this algorithm is called “time-step”. It is characterized by a time moment t varying from 0 to T using time step τ.

Time step 0. $t = 0$.

0.0. The initial conditions giving unknown values at the zero time moment in all region of integration are used to fill *Array0*.

0.1. *Array0* might be written to the hard disk as unknown values for $(x, y, z, 0)$.

Time step 1. $t = \tau$.

1.0. The data in the *Array1* are calculated for the X-direction in the inner points using the data in the *Array0*.

1.1. The data in the *Array1* are corrected in the points placed at the boundaries and interfaces for the X-direction using information about boundary conditions, interface conditions, and non-zero external force.

1.2. *Array0* = *Array1*.

1.3. The data in the *Array1* are calculated for the Y-direction in the inner points using the data in the *Array0*.

1.4. The data in the *Array1* are corrected in the points placed at the boundaries and interfaces for the Y-direction using information about boundary conditions, interface conditions, and non-zero external force.

1.5. *Array0* = *Array1*.

1.6. The data in the *Array1* are calculated for the Z-direction in the inner points using the data in the *Array0*.

1.7. The data in the *Array1* are corrected in the points placed at the boundaries and interfaces for the Z-direction using information about boundary conditions, interface conditions, and non-zero external force.

1.8. *Array0* = *Array1*.

1.9. *Array0* might be written to the hard disk as unknown values for (x, y, z, τ).

<u>Time step n</u>. $t = n\tau$.

n.0. The data in the *Array1* are calculated for the X-direction in the inner points using the data in the *Array0*.

n.1. The data in the *Array1* are corrected in the points placed at the boundaries and interfaces for the X-direction using information about boundary conditions, interface conditions, and non-zero external force.

n.2. *Array0* = *Array1*.

n.3. The data in the *Array1* are calculated for the Y-direction in the inner points using the data in the *Array0*.

n.4. The data in the *Array1* are corrected in the points placed at the boundaries and interfaces for the Y-direction using information about boundary conditions, interface conditions, and non-zero external force.

n.5. *Array0* = *Array1*.

n.6. The data in the *Array1* are calculated for the Z-direction in the inner points using the data in the *Array0*.

n.7. The data in the *Array1* are corrected in the points placed at the boundaries and interfaces for the Z-direction using information about boundary conditions, interface conditions, and non-zero external force.

n.8. *Array0* = *Array1*.

n.9. *Array0* might be written to the hard disk as unknown values for $(x, y, z, n\tau)$.

The 2D case algorithm is developed similar to 3D case.

One can use different source mechanisms like boundary conditions, external force, or initial conditions.

5.3.2 General Statements of Grid-Characteristic Methods

At each time step the arbitrary directions (ξ_1, ξ_2, ξ_3) in a 3D case or (ξ_1, ξ_2) in a 2D case, the basis are chosen in a case of unstructured triangular or tetrahedral grids. These new coordinate systems make it possible to ensure the isotropy of the numerical method. Note that these new directions coincide with OX, OY, and OZ axes in a 3D case or with OX and OY axes in a 2D case using the regular or curvilinear hexahedral meshes.

A system of hyperbolic type for a two-dimensional case in this new coordinate system is given by Eq. 5.45.

$$\vec{q}_t + \mathbf{A}_1^{2D}\vec{q}_{\xi_1} + \mathbf{A}_2^{2D}\vec{q}_{\xi_2} = 0 \tag{5.45}$$

A system of hyperbolic type for a three-dimensional case in this new coordinate system is given by Eq. 5.46.

$$\vec{q}_t + \mathbf{A}_1^{3D}\vec{q}_{\xi_1} + \mathbf{A}_2^{3D}\vec{q}_{\xi_2} + \mathbf{A}_3^{3D}\vec{q}_{\xi_3} = 0 \tag{5.46}$$

Further splitting in two or three directions is carried out. One can obtain the system given in Eq. 5.47 for each direction.

$$\vec{q}_t + \mathbf{A}_1\vec{q}_{\xi_1} = 0 \tag{5.47}$$

Equations 5.48, 5.49 are valid for 2D and 3D cases, respectively, for the system of equations given by Eq. 5.47.

$$\vec{q}\,(\xi_1, \xi_2, t + \tau) = \sum_{i=1}^{I} \mathbf{X}_i^{2D,1}\vec{q}\left(\xi_1 - c_i^{2D,1}\tau, \xi_2, t\right) \tag{5.48}$$

$$\vec{q}\,(\xi_1, \xi_2, \xi_3, t + \tau) = \sum_{i=1}^{I} \mathbf{X}_i^{3D,1}\vec{q}\left(\xi_1 - c_i^{3D,1}\tau, \xi_2, \xi_3, t\right) \tag{5.49}$$

In Eq. 5.47 and further,$\mathbf{A}_1$ is the matrix $\mathbf{A}_1^{2D}$ for a 2D case and matrix $\mathbf{A}_1^{3D}$ for a 3D case, respectively. In Eq. 5.48, $\mathbf{X}_i^{2D,1}$ are the matrices expressed in terms of the matrix $\mathbf{A}_1^{2D}$ components and given further for all considered systems of equations, $c_i^{2D,1}$ are the eigenvalues of the matrix $\mathbf{A}_1^{2D}$. In Eq. 5.49, $\mathbf{X}_i^{3D,1}$ are the matrices expressed in terms of the matrix $\mathbf{A}_1^{3D}$ components and given further for all considered systems of equations, $c_i^{3D,1}$ are the eigenvalues of the matrix $\mathbf{A}_1^{3D}$. In Eqs. 5.48, 5.49, I is a number of matrix $\mathbf{A}_1$ eigenvalues. Let the matrix $\mathbf{A}_1$ have a set of I^+ positive eigenvalues, a set of I^- negative eigenvalues, and a set of I^0 zero ones.

For the matrix $\mathbf{X}_i^1$ represented both matrices $\mathbf{X}_i^{2D,1}$ and $\mathbf{X}_i^{3D,1}$, the following relation given by Eq. 5.50 is satisfied.

$$\sum_{i=1}^{I} \mathbf{X}_i^1 = \mathbf{I} \tag{5.50}$$

Thus, one can express the matrices $\mathbf{X}_i^1$ corresponding to zero eigenvalues using the following Eq. 5.51.

$$\sum_{i \in I^0} \mathbf{X}_i^1 = \mathbf{I} - \sum_{i \in I^{0+}} \mathbf{X}_i^1 - \sum_{i \in I^-} \mathbf{X}_i^1 \tag{5.51}$$

Taking into account Eqs. 5.48, 5.49, their equivalents are obtained and given by Eqs. 5.52, 5.53, respectively.

$$\vec{q}\,(\xi_1, \xi_2, t + \tau) = \vec{q}\,(\xi_1, \xi_2, t) + \sum_{i \in I^+ \cup I^-} \mathbf{X}_i^{2D,1} \left(\vec{q} \left(\xi_1 - c_i^{2D,1} \tau, \xi_2, t \right) - \vec{q}\,(\xi_1, \xi_2, t) \right) \tag{5.52}$$

$$\vec{q}\,(\xi_1, \xi_2, \xi_3, t + \tau) = \vec{q}\,(\xi_1, \xi_2, \xi_3, t) + \sum_{i \in I^+ \cup I^-} \mathbf{X}_i^{3D,1} \left(\vec{q} \left(\xi_1 - c_i^{2D,1} \tau, \xi_2, \xi_3, t \right) - \vec{q}\,(\xi_1, \xi_2, \xi_3, t) \right) \tag{5.53}$$

The matrix $\mathbf{A}_1$ has a set of eigenvectors. Thus, it can be represented as Eq. 5.54.

$$\mathbf{A}^1 = \left(\mathbf{\Omega}^1 \right)^{-1} \mathbf{\Lambda}^1 \mathbf{\Omega}^1 \tag{5.54}$$

In Eq. 5.54, $\left(\mathbf{\Omega}^1 \right)^{-1}$ is the matrix composed of eigenvectors of the matrix $\mathbf{A}_1$, $\mathbf{\Lambda}^1$ is the diagonal matrix, whose elements are the eigenvalues of the matrix $\mathbf{A}_1$.

Formulae 5.52 and 5.53 can also be divided into three stages. At the first stage, a multiplication of all unknown values $\vec{q}\,(\xi_1, \xi_2, t)$ or $\vec{q}\,(\xi_1, \xi_2, \xi_3, t)$ stored on the n time layer by the matrix $\mathbf{\Omega}^1$ is performed using Eqs. 5.55, 5.56.

$$\vec{\omega}\,(\xi_1, \xi_2, t) = \mathbf{\Omega}^1 \vec{q}\,(\xi_1, \xi_2, t) \tag{5.55}$$

$$\vec{\omega}\,(\xi_1, \xi_2, \xi_3, t) = \mathbf{\Omega}^1 \vec{q}\,(\xi_1, \xi_2, \xi_3, t) \tag{5.56}$$

At the second stage, the following expressions given by Eqs. 5.57, 5.58 for the 2D and 3D cases, respectively, should be carried out.

$$\vec{\omega}\,(\xi_1, \xi_2, t + \tau) = \vec{\omega}\,(\xi_1, \xi_2, t) + \sum_{i \in I^+ \cup I^-} \vec{\omega} \left(\xi_1 - c_i^{2D,1} \tau, \xi_2, t \right) - \vec{\omega}\,(\xi_1, \xi_2, t) \tag{5.57}$$

$$\vec{\omega}(\xi_1, \xi_2, \xi_3, t+\tau) = \vec{\omega}(\xi_1, \xi_2, \xi_3, t) + \sum_{i \in I^+ \cup I^-} \vec{\omega}\left(\xi_1 - c_i^{3D,1}\tau, \xi_2, \xi_3, t\right) - \vec{\omega}(\xi_1, \xi_2, \xi_3, t) \quad (5.58)$$

At the third stage, a multiplication of all unknown values $\vec{q}(\xi_1, \xi_2, t)$ or $\vec{q}(\xi_1, \xi_2, \xi_3, t)$ stored on the $n+1$ time layer by the matrix $\mathbf{\Omega}^1$ is performed using Eqs. 5.59, 5.60. Thus, $n.0$ stage of the algorithm discussed in Sect. 5.3.1 is done.

$$\vec{q}(\xi_1, \xi_2, \xi_3, t) = \left(\mathbf{\Omega}^1\right)^{-1} \vec{\omega}(\xi_1, \xi_2, \xi_3, t) \quad (5.59)$$

$$\vec{q}(\xi_1, \xi_2, \xi_3, t+\tau) = \left(\mathbf{\Omega}^1\right)^{-1} \vec{\omega}(\xi_1, \xi_2, \xi_3, t+\tau) \quad (5.60)$$

The use of Eqs. 5.59, 5.60 is equivalent for solving the following independent transport equations given by Eq. 5.61.

$$(\omega_i)_t + c_i^1 (\omega_i)_{\xi_1} = 0 \quad (5.61)$$

Using a high order interpolation on the unstructured triangular or tetrahedral grids (Chaps. 2–4) into Eqs. 5.57, 5.58, respectively, one can compete the stage $n.0$ applying the unstructured meshes. Using the regular grids, one-dimensional independent transport Eqs. 5.61 are solved to perform the stage $n.0$. The stages $n.3$ and $n.6$ are performed similarly. For example, in order to solve Eq. 5.61 in a case of regular meshes one can use the numerical scheme given by the Eq. 5.62 for positive eigenvalues c_i^1 and by the Eq. 5.63 for negative eigenvalues c_i^1.

$$\begin{aligned}(\omega_i)_m^{n+1} &= (\omega_i)_m^n + \frac{1}{6}\frac{c_i^1\tau}{h}\left(6(\omega_i)_{m-1}^n - 3(\omega_i)_m^n - 2(\omega_i)_{m+1}^n - (\omega_i)_{m-2}^n\right) \\ &+ \frac{1}{2}\left(\frac{c_i^1\tau}{h}\right)^2\left((\omega_i)_{m-1}^n - 2(\omega_i)_m^n + (\omega_i)_{m+1}^n\right) \\ &+ \frac{1}{6}\left(\frac{c_i^1\tau}{h}\right)^3\left((\omega_i)_{m-2}^n - 3(\omega_i)_{m-1}^n + 3(\omega_i)_m^n - (\omega_i)_{m+1}^n\right) \end{aligned} \quad (5.62)$$

$$\begin{aligned}(\omega_i)_m^{n+1} &= (\omega_i)_m^n + \frac{1}{6}\frac{c_i^1\tau}{h}\left(6(\omega_i)_{m+1}^n - 3(\omega_i)_m^n - 2(\omega_i)_{m-1}^n - (\omega_i)_{m+2}^n\right) \\ &+ \frac{1}{2}\left(\frac{c_i^1\tau}{h}\right)^2\left((\omega_i)_{m+1}^n - 2(\omega_i)_m^n + (\omega_i)_{m-1}^n\right) \\ &+ \frac{1}{6}\left(\frac{c_i^1\tau}{h}\right)^3\left((\omega_i)_{m+2}^n - 3(\omega_i)_{m+1}^n + 3(\omega_i)_m^n - (\omega_i)_{m-1}^n\right) \end{aligned} \quad (5.63)$$

In Eqs. 5.62, 5.63, an index n corresponds to time coordinate t. Index m and step h correspond to spatial coordinate ξ_1.

5.3.3 *Grid-Characteristic Methods for Anisotropic Elastic Waves in the General Case*

The vector of unknowns appearing in Eq. 5.46 is given by Eq. 5.64.

$$\vec{q} = \begin{bmatrix} \vec{v} \\ \boldsymbol{\sigma} \end{bmatrix} = \begin{bmatrix} v_1 \\ v_2 \\ v_3 \\ \sigma_{11} \\ \sigma_{22} \\ \sigma_{33} \\ \sigma_{23} \\ \sigma_{13} \\ \sigma_{12} \end{bmatrix} \tag{5.64}$$

In the Eq. 5.46 the matrices $\mathbf{A}_1$, $\mathbf{A}_2$, $\mathbf{A}_3$, are given by Eqs. 5.65–5.67.

$$\mathbf{A}_1 = -\begin{bmatrix} 0 & 0 & 0 & \frac{1}{\rho} & 0 & 0 & 0 & 0 & 0 \\ 0 & 0 & 0 & 0 & 0 & 0 & 0 & 0 & \frac{1}{\rho} \\ 0 & 0 & 0 & 0 & 0 & 0 & 0 & \frac{1}{\rho} & 0 \\ c_{11} & c_{16} & c_{15} & 0 & 0 & 0 & 0 & 0 & 0 \\ c_{12} & c_{26} & c_{25} & 0 & 0 & 0 & 0 & 0 & 0 \\ c_{13} & c_{36} & c_{35} & 0 & 0 & 0 & 0 & 0 & 0 \\ c_{14} & c_{46} & c_{45} & 0 & 0 & 0 & 0 & 0 & 0 \\ c_{15} & c_{56} & c_{55} & 0 & 0 & 0 & 0 & 0 & 0 \\ c_{16} & c_{66} & c_{56} & 0 & 0 & 0 & 0 & 0 & 0 \end{bmatrix} \tag{5.65}$$

$$\mathbf{A}_2 = -\begin{bmatrix} 0 & 0 & 0 & 0 & 0 & 0 & 0 & 0 & \frac{1}{\rho} \\ 0 & 0 & 0 & 0 & \frac{1}{\rho} & 0 & 0 & 0 & 0 \\ 0 & 0 & 0 & 0 & 0 & 0 & \frac{1}{\rho} & 0 & 0 \\ c_{16} & c_{12} & c_{14} & 0 & 0 & 0 & 0 & 0 & 0 \\ c_{26} & c_{22} & c_{24} & 0 & 0 & 0 & 0 & 0 & 0 \\ c_{36} & c_{23} & c_{34} & 0 & 0 & 0 & 0 & 0 & 0 \\ c_{46} & c_{24} & c_{44} & 0 & 0 & 0 & 0 & 0 & 0 \\ c_{56} & c_{25} & c_{45} & 0 & 0 & 0 & 0 & 0 & 0 \\ c_{66} & c_{26} & c_{46} & 0 & 0 & 0 & 0 & 0 & 0 \end{bmatrix} \tag{5.66}$$

$$\mathbf{A}_3 = -\begin{bmatrix} 0 & 0 & 0 & 0 & 0 & 0 & 0 & \frac{1}{\rho} & 0 \\ 0 & 0 & 0 & 0 & 0 & 0 & \frac{1}{\rho} & 0 & 0 \\ 0 & 0 & 0 & 0 & 0 & \frac{1}{\rho} & 0 & 0 & 0 \\ c_{15} & c_{14} & c_{13} & 0 & 0 & 0 & 0 & 0 & 0 \\ c_{25} & c_{24} & c_{23} & 0 & 0 & 0 & 0 & 0 & 0 \\ c_{35} & c_{34} & c_{33} & 0 & 0 & 0 & 0 & 0 & 0 \\ c_{45} & c_{44} & c_{34} & 0 & 0 & 0 & 0 & 0 & 0 \\ c_{55} & c_{45} & c_{35} & 0 & 0 & 0 & 0 & 0 & 0 \\ c_{56} & c_{46} & c_{36} & 0 & 0 & 0 & 0 & 0 & 0 \end{bmatrix} \tag{5.67}$$

It can be shown that for each pair of selected directions three types of waves can be distinguished. It was proved that in order to find the eigenvalues of the matrix $\mathbf{A}_1$ given by Eq. 5.65 it is necessary to solve the cubic Eq. 5.68.

$$\frac{1}{\rho^3}\left(c_{11}c_{56}^2 + c_{55}c_{16}^2 + c_{66}c_{15}^2 - c_{11}c_{55}c_{66} - 2c_{15}c_{16}c_{56}\right) + \frac{1}{\rho^2}\left(c_{11}c_{55} + c_{11}c_{66} + c_{55}c_{66} - c_{15}^2 - c_{16}^2 - c_{56}^2\right)\beta - \frac{1}{\rho}\left(c_{11} + c_{55} + c_{66}\right)\beta^2 + \beta^3 = 0 \tag{5.68}$$

Also, it was proved that Eq. 5.68 has three real positive roots denoted as β_{11}, β_{12}, and β_{13}. The eigenvalues of the matrix $\mathbf{A}_1$ given by Eq. 5.65 are represented by the following set given in Eq. 5.69.

$$\left\{\sqrt{\beta_{11}}, -\sqrt{\beta_{11}}, \sqrt{\beta_{12}}, -\sqrt{\beta_{12}}, \sqrt{\beta_{13}}, -\sqrt{\beta_{13}}, 0, 0, 0\right\} \tag{5.69}$$

It was shown that for each pair of selected directions it is possible to distinguish three types of waves. It was proved that to find the eigenvalues of the matrix $\mathbf{A}_2$ given by Eq. 5.66 it is necessary to solve the following cubic Eq. 5.70.

$$\frac{1}{\rho^3}\left(c_{22}c_{46}^2 + c_{44}c_{26}^2 + c_{66}c_{24}^2 - c_{22}c_{44}c_{66} - 2c_{24}c_{26}c_{46}\right) + \frac{1}{\rho^2}\left(c_{22}c_{44} + c_{22}c_{66} + c_{44}c_{66} - c_{24}^2 - c_{26}^2 - c_{46}^2\right)\beta - \frac{1}{\rho}\left(c_{22} + c_{44} + c_{66}\right)\beta^2 + \beta^3 = 0 \tag{5.70}$$

It was proved that Eq. 5.70 has three real positive roots. Denote them as β_{21}, β_{22}, and β_{23}. The eigenvalues of the matrix $\mathbf{A}_2$ from Eq. 5.66 are represented by the following set given in Eq. 5.71.

$$\left\{\sqrt{\beta_{21}}, -\sqrt{\beta_{21}}, \sqrt{\beta_{22}}, -\sqrt{\beta_{22}}, \sqrt{\beta_{23}}, -\sqrt{\beta_{23}}, 0, 0, 0\right\} \tag{5.71}$$

It was shown that for each pair of selected directions, three types of waves can be distinguished. It was proved that to find the eigenvalues of the matrix $\mathbf{A}_3$ given by Eq. 5.67 it is necessary to solve the following cubic Eq. 5.72.

$$\frac{1}{\rho^3}\left(c_{33}c_{45}^2 + c_{44}c_{35}^2 + c_{55}c_{34}^2 - c_{33}c_{44}c_{55} - 2c_{34}c_{35}c_{45}\right) + \frac{1}{\rho^2}\left(c_{33}c_{44} + c_{33}c_{55} + c_{44}c_{55} - c_{34}^2 - c_{35}^2 - c_{45}^2\right)\beta - \frac{1}{\rho}(c_{33} + c_{44} + c_{55})\beta^2 + \beta^3 = 0 \tag{5.72}$$

Also it was proved that Eq. 5.72 has three real positive roots. Denote them as β_{31}, β_{32}, and β_{33}. The eigenvalues of the matrix $\mathbf{A}_3$ given by Eq. 5.67 are represented by the following set provided by Eq. 5.73.

$$\left\{\sqrt{\beta_{31}}, -\sqrt{\beta_{31}}, \sqrt{\beta_{32}}, -\sqrt{\beta_{32}}, \sqrt{\beta_{33}}, -\sqrt{\beta_{33}}, 0, 0, 0\right\} \tag{5.73}$$

5.3.4 *Grid-Characteristic Methods for Orthorhombic Anisotropic Elastic Waves*

For the elastic constants tensor given by Eq. 5.15 with the coincidence of the selected axes of the orthorhombic anisotropy with the axes of the coordinate system (ξ_1, ξ_2, ξ_3), the matrices $\mathbf{A}_1$, $\mathbf{A}_2$, $\mathbf{A}_3$ appearing in Eq. 5.46 take the following form given by Eqs. 5.74–5.76. Also, the topology of the eigenvalues of the matrix for this case will be considered in detail. It should be noted that if the axes do not coincide with the axes of symmetry, the structure and the eigenvalues will return to the form of an arbitrary form given in Sect. 5.3.3.

$$\mathbf{A}_1 = -\begin{bmatrix} 0 & 0 & 0 & \frac{1}{\rho} & 0 & 0 & 0 & 0 & 0 \\ 0 & 0 & 0 & 0 & 0 & 0 & 0 & 0 & \frac{1}{\rho} \\ 0 & 0 & 0 & 0 & 0 & 0 & 0 & \frac{1}{\rho} & 0 \\ c_{11} & 0 & 0 & 0 & 0 & 0 & 0 & 0 & 0 \\ c_{12} & 0 & 0 & 0 & 0 & 0 & 0 & 0 & 0 \\ c_{13} & 0 & 0 & 0 & 0 & 0 & 0 & 0 & 0 \\ 0 & 0 & 0 & 0 & 0 & 0 & 0 & 0 & 0 \\ 0 & 0 & c_{55} & 0 & 0 & 0 & 0 & 0 & 0 \\ 0 & c_{66} & 0 & 0 & 0 & 0 & 0 & 0 & 0 \end{bmatrix} \tag{5.74}$$

$$\mathbf{A}_2 = -\begin{bmatrix} 0 & 0 & 0 & 0 & 0 & 0 & 0 & 0 & \frac{1}{\rho} \\ 0 & 0 & 0 & 0 & \frac{1}{\rho} & 0 & 0 & 0 & 0 \\ 0 & 0 & 0 & 0 & 0 & 0 & \frac{1}{\rho} & 0 & 0 \\ 0 & c_{12} & 0 & 0 & 0 & 0 & 0 & 0 & 0 \\ 0 & c_{22} & 0 & 0 & 0 & 0 & 0 & 0 & 0 \\ 0 & c_{23} & 0 & 0 & 0 & 0 & 0 & 0 & 0 \\ 0 & 0 & c_{44} & 0 & 0 & 0 & 0 & 0 & 0 \\ 0 & 0 & 0 & 0 & 0 & 0 & 0 & 0 & 0 \\ c_{66} & 0 & 0 & 0 & 0 & 0 & 0 & 0 & 0 \end{bmatrix} \tag{5.75}$$

$$\mathbf{A}_3 = -\begin{bmatrix} 0 & 0 & 0 & 0 & 0 & 0 & 0 & \frac{1}{\rho} & 0 \\ 0 & 0 & 0 & 0 & 0 & 0 & \frac{1}{\rho} & 0 & 0 \\ 0 & 0 & 0 & 0 & 0 & \frac{1}{\rho} & 0 & 0 & 0 \\ 0 & 0 & c_{13} & 0 & 0 & 0 & 0 & 0 & 0 \\ 0 & 0 & c_{23} & 0 & 0 & 0 & 0 & 0 & 0 \\ 0 & 0 & c_{33} & 0 & 0 & 0 & 0 & 0 & 0 \\ 0 & c_{44} & 0 & 0 & 0 & 0 & 0 & 0 & 0 \\ c_{55} & 0 & 0 & 0 & 0 & 0 & 0 & 0 & 0 \\ 0 & 0 & 0 & 0 & 0 & 0 & 0 & 0 & 0 \end{bmatrix} \tag{5.76}$$

The eigenvalues of the matrix $\mathbf{A}_1$ (given by Eq. 5.74), $\mathbf{A}_2$ (given by Eq. 5.75), $\mathbf{A}_3$ (given by Eq. 5.76) are given by the following sets represented by Eqs. 5.77, 5.78, 5.79, respectively.

$$\left\{\sqrt{\frac{c_{11}}{\rho}}, -\sqrt{\frac{c_{11}}{\rho}}, \sqrt{\frac{c_{55}}{\rho}}, -\sqrt{\frac{c_{55}}{\rho}}, \sqrt{\frac{c_{66}}{\rho}}, -\sqrt{\frac{c_{66}}{\rho}}, 0, 0, 0\right\} \tag{5.77}$$

$$\left\{\sqrt{\frac{c_{22}}{\rho}}, -\sqrt{\frac{c_{22}}{\rho}}, \sqrt{\frac{c_{44}}{\rho}}, -\sqrt{\frac{c_{44}}{\rho}}, \sqrt{\frac{c_{66}}{\rho}}, -\sqrt{\frac{c_{66}}{\rho}}, 0, 0, 0\right\} \tag{5.78}$$

$$\left\{\sqrt{\frac{c_{33}}{\rho}}, -\sqrt{\frac{c_{33}}{\rho}}, \sqrt{\frac{c_{44}}{\rho}}, -\sqrt{\frac{c_{44}}{\rho}}, \sqrt{\frac{c_{55}}{\rho}}, -\sqrt{\frac{c_{55}}{\rho}}, 0, 0, 0\right\} \tag{5.79}$$

In Fig. 5.1, the interpretation of all wave propagation speeds in a case of orthorhombic anisotropy is presented. Three mutually perpendicular axes of symmetry 1, 2 and 3 are presented. These axes also define a Cartesian coordinate system, in which the elastic constants tensor presented by Eq. 5.15 is given.

Three cases of wave propagating are considered in this Section:

1. Along the 1-direction. The longitudinal waves propagate with the speed $\sqrt{\frac{c_{11}}{\rho}}$. The transverse waves propagate with the speed $\sqrt{\frac{c_{66}}{\rho}}$, the motion of medium in

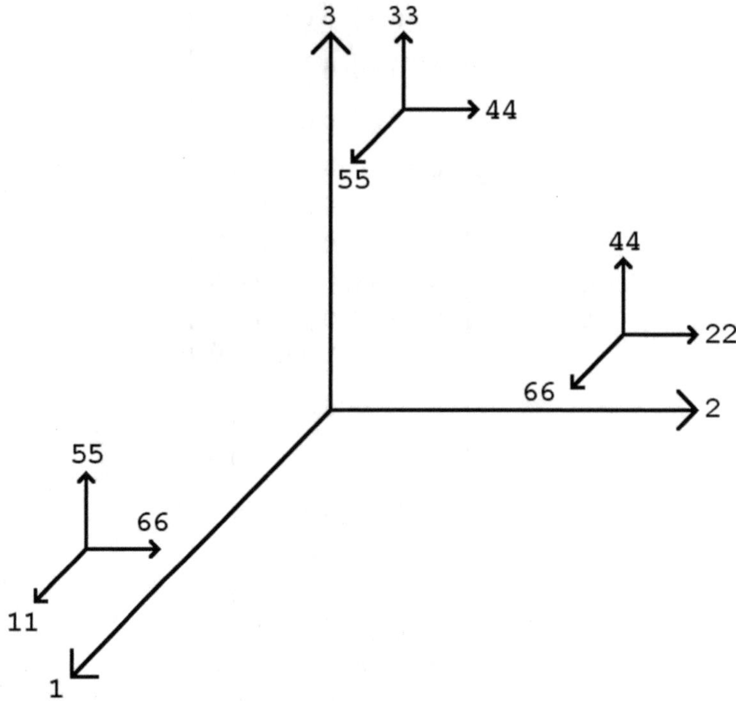

Fig. 5.1 The speeds of propagation of various types of seismic waves in a medium with orthorhombic anisotropy

which is directed along the direction 2. The transverse waves propagate with the speed $\sqrt{\frac{c_{55}}{\rho}}$, the motion of medium in which is directed along the direction 3.

2. Along the 2-direction. The longitudinal waves propagate with the speed $\sqrt{\frac{c_{22}}{\rho}}$. The transverse waves propagate with the speed $\sqrt{\frac{c_{44}}{\rho}}$, the motion of medium in which is directed along the direction 3. The transverse waves propagate with the speed $\sqrt{\frac{c_{66}}{\rho}}$, the motion of medium in which is directed along the direction 1.
3. Along the 3-direction. The longitudinal waves propagate with the speed $\sqrt{\frac{c_{33}}{\rho}}$. The transverse waves propagate with the speed $\sqrt{\frac{c_{55}}{\rho}}$, the motion of medium in which is directed along the direction 1. The transverse waves propagate with the speed $\sqrt{\frac{c_{44}}{\rho}}$, the motion of medium in which is directed along the direction 2.

In the case, when the axes of the coordinate system (ξ_1, ξ_2, ξ_3) do not coincide with the distinguished axes of the orthorhombic anisotropy under consideration, the case of orthorhombic anisotropy will pass to the general case of anisotropy. Vectors $\left(\vec{e}_1, \vec{e}_2, \vec{e}_3\right)$ denote the basis of unit vectors directed along the coordinate axes (ξ_1, ξ_2, ξ_3). Vectors $\left(\vec{a}_1, \vec{a}_2, \vec{a}_3\right)$ denote a basis of unit vectors directed along the distinguished axes of the orthorhombic anisotropy.

Let the following relations given by Eqs. 5.80, 5.81 hold for the vector $\vec{r}$.

$$\vec{r} = r_1^e \vec{e}_1 + r_2^e \vec{e}_2 + r_2^e \vec{e}_3 \tag{5.80}$$

$$\vec{r} = r_1^a \vec{a}_1 + r_2^a \vec{a}_2 + r_2^a \vec{a}_3 \tag{5.81}$$

The following Eq. 5.82 is true in these conditions.

$$\begin{bmatrix} r_1^e \\ r_2^e \\ r_2^e \end{bmatrix} = \mathbf{H} \begin{bmatrix} r_1^a \\ r_2^a \\ r_2^a \end{bmatrix} \tag{5.82}$$

In Eq. 5.82, **H** is the transition matrix having components given by Eq. 5.83.

$$\mathbf{H} = \begin{bmatrix} h_{11} & h_{12} & h_{13} \\ h_{21} & h_{22} & h_{21} \\ h_{31} & h_{32} & h_{33} \end{bmatrix} \tag{5.83}$$

Let the elastic constants tensor be given by Eq. 5.15 in the coordinate system with unit vectors $\left(\vec{a}_1, \vec{a}_2, \vec{a}_3\right)$. This tensor can be written by Eq. 5.84 in the coordinate system (ξ_1, ξ_2, ξ_3) with unit vectors $\left(\vec{e}_1, \vec{e}_2, \vec{e}_3\right)$.

$$c_{\alpha\beta}^{\xi} = \sum_{i=1}^{3}\sum_{j=1}^{3}\sum_{k=1}^{3}\sum_{l=1}^{3} h_{m(\alpha),i} h_{n(\alpha),j} h_{p(\beta),k} h_{q(\beta),l} c_{\gamma(i,j)\psi(k,l)} \tag{5.84}$$

Let the tensor be given by Eq. 5.3 in the general case. Let the structure be given by Eq. 5.4 in the general case. Under these assumptions, the functions of the transition in Eq. 5.84 from the indices of the components of the tensor to the indices into the structure and back are given in accordance with Eqs. 5.5–5.13. Thus, Eq. 5.84 can be expanded as written in Eq. 5.85.

$$
\begin{aligned}
C^e_{ijkl} = & \left(h_{1i}h_{1j}h_{1k}h_{1l}\right)c_{11} + \left(h_{2i}h_{2j}h_{2k}h_{2l}\right)c_{22} + \left(h_{3i}h_{3j}h_{3k}h_{3l}\right)c_{33} \\
& + \left(h_{2i}h_{3j}h_{2k}h_{3l} + h_{2i}h_{3j}h_{3k}h_{2l} + h_{3i}h_{2j}h_{2k}h_{3l} + h_{3i}h_{2j}h_{3k}h_{2l}\right)c_{44} \\
& + \left(h_{1i}h_{3j}h_{1k}h_{3l} + h_{1i}h_{3j}h_{3k}h_{1l} + h_{3i}h_{1j}h_{1k}h_{3l} + h_{3i}h_{1j}h_{3k}h_{1l}\right)c_{55} \\
& + \left(h_{1i}h_{2j}h_{1k}h_{2l} + h_{1i}h_{2j}h_{2k}h_{1l} + h_{2i}h_{1j}h_{1k}h_{2l} + h_{2i}h_{1j}h_{2k}h_{1l}\right)c_{66} \\
& + \left(h_{1i}h_{1j}h_{2k}h_{2l} + h_{2i}h_{2j}h_{1k}h_{1l}\right)c_{12} + \left(h_{1i}h_{1j}h_{3k}h_{3l} + h_{3i}h_{3j}h_{1k}h_{1l}\right)c_{13} \\
& + \left(h_{2i}h_{2j}h_{3k}h_{3l} + h_{3i}h_{3j}h_{2k}h_{2l}\right)c_{23} \\
& + \left(h_{1i}h_{1j}h_{2k}h_{3l} + h_{1i}h_{1j}h_{3k}h_{2l} + h_{2i}h_{3j}h_{1k}h_{1l} + h_{3i}h_{2j}h_{1k}h_{1l}\right)c_{14} \\
& + \left(h_{1i}h_{1j}h_{1k}h_{3l} + h_{1i}h_{1j}h_{3k}h_{1l} + h_{1i}h_{3j}h_{1k}h_{1l} + h_{3i}h_{1j}h_{1k}h_{1l}\right)c_{15} \\
& + \left(h_{1i}h_{1j}h_{1k}h_{2l} + h_{1i}h_{1j}h_{2k}h_{1l} + h_{1i}h_{2j}h_{1k}h_{1l} + h_{2i}h_{1j}h_{1k}h_{1l}\right)c_{16} \\
& + \left(h_{2i}h_{2j}h_{2k}h_{3l} + h_{2i}h_{2j}h_{3k}h_{2l} + h_{2i}h_{3j}h_{2k}h_{2l} + h_{3i}h_{2j}h_{2k}h_{2l}\right)c_{24} \\
& + \left(h_{1i}h_{3j}h_{2k}h_{2l} + h_{2i}h_{2j}h_{1k}h_{3l} + h_{2i}h_{2j}h_{3k}h_{1l} + h_{3i}h_{1j}h_{2k}h_{2l}\right)c_{25} \\
& + \left(h_{1i}h_{2j}h_{2k}h_{2l} + h_{2i}h_{1j}h_{2k}h_{2l} + h_{2i}h_{2j}h_{1k}h_{2l} + h_{2i}h_{2j}h_{2k}h_{1l}\right)c_{26} \\
& + \left(h_{2i}h_{3j}h_{3k}h_{3l} + h_{3i}h_{2j}h_{3k}h_{3l} + h_{3i}h_{3j}h_{2k}h_{3l} + h_{3i}h_{3j}h_{3k}h_{2l}\right)c_{34} \\
& + \left(h_{1i}h_{3j}h_{3k}h_{3l} + h_{3i}h_{1j}h_{3k}h_{3l} + h_{3i}h_{3j}h_{1k}h_{3l} + h_{3i}h_{3j}h_{3k}h_{1l}\right)c_{35} \\
& + \left(h_{1i}h_{2j}h_{3k}h_{3l} + h_{2i}h_{1j}h_{3k}h_{3l} + h_{3i}h_{3j}h_{1k}h_{2l} + h_{3i}h_{3j}h_{2k}h_{1l}\right)c_{36} \\
& + \left(h_{1i}h_{3j}h_{2k}h_{3l} + h_{1i}h_{3j}h_{3k}h_{2l} + h_{2i}h_{3j}h_{1k}h_{3l} + h_{2i}h_{3j}h_{3k}h_{1l} + h_{3i}h_{1j}h_{2k}h_{3l}\right. \\
& \left. + h_{3i}h_{1j}h_{3k}h_{2l} + h_{3i}h_{2j}h_{1k}h_{3l} + h_{3i}h_{2j}h_{3k}h_{1l}\right)c_{45} \\
& + \left(h_{1i}h_{2j}h_{2k}h_{3l} + h_{1i}h_{2j}h_{3k}h_{2l} + h_{2i}h_{1j}h_{2k}h_{3l} + h_{2i}h_{1j}h_{3k}h_{2l} + h_{2i}h_{3j}h_{1k}h_{2l}\right. \\
& \left. + h_{2i}h_{3j}h_{2k}h_{1l} + h_{3i}h_{2j}h_{1k}h_{2l} + h_{3i}h_{2j}h_{2k}h_{1l}\right)c_{46} \\
& + \left(h_{1i}h_{2j}h_{1k}h_{3l} + h_{1i}h_{2j}h_{3k}h_{1l} + h_{1i}h_{3j}h_{1k}h_{2l} + h_{1i}h_{3j}h_{2k}h_{1l} + h_{2i}h_{1j}h_{1k}h_{3l}\right. \\
& \left. + h_{2i}h_{1j}h_{3k}h_{1l} + h_{3i}h_{1j}h_{1k}h_{2l} + h_{3i}h_{1j}h_{2k}h_{1l}\right)c_{56}
\end{aligned} \tag{5.85}
$$

5.3.5 *Grid-Characteristic Methods for Isotropic Elastic Waves*

Consider the direction ξ_1 for definiteness. Let the vector $\vec{n}$ be directed along the chosen direction and the vectors $\vec{n}_1$ and $\vec{n}_2$ (or only the vector $\vec{n}_1$ in a two-dimensional case) form the Cartesian coordinate system with this vector $\vec{n}$. The following symmetric tensors of the second rank are introduced by Eq. 5.86.

$$
\mathbf{N}_{ij} = \frac{1}{2}\left(n_i \otimes n_j + n_j \otimes n_i\right) \tag{5.86}
$$

In Eq. 5.86, vector $\vec{n}_0$ means the vector $\vec{n}$. The vector of unknowns is given by Eq. 5.64 for a three-dimensional case and by Eq. 5.87 for a two-dimensional case.

$$\vec{q} = \begin{bmatrix} \vec{v} \\ \boldsymbol{\sigma} \end{bmatrix} = \begin{bmatrix} v_1 \\ v_2 \\ \sigma_{11} \\ \sigma_{22} \\ \sigma_{12} \end{bmatrix} \tag{5.87}$$

The action of the matrix $\mathbf{A}_1$ on the vector of unknowns can be written by Eq. 5.88 both for 2D and 3D cases.

$$\mathbf{A}_1 \begin{bmatrix} \vec{v} \\ \boldsymbol{\sigma} \end{bmatrix} = - \begin{bmatrix} \frac{1}{\rho} \left(\boldsymbol{\sigma} \cdot \vec{n} \right) \\ \lambda \left(\vec{n} \cdot \vec{v} \right) \mathbf{I} + \mu \left(\vec{n} \otimes \vec{v} + \vec{v} \otimes \vec{n} \right) \end{bmatrix} \tag{5.88}$$

Matrices $\mathbf{A}_1^{3D}$, $\mathbf{A}_2^{3D}$, $\mathbf{A}_3^{3D}$ have the same set of eigenvalues provided by Eq. 5.89.

$$\{c_P, -c_P, c_S, -c_S, c_S, -c_S, 0, 0, 0\} \tag{5.89}$$

Matrices $\mathbf{A}_1^{2D}$, $\mathbf{A}_2^{2D}$ also have the same set of eigenvalues given by Eq. 5.90.

$$\{c_P, -c_P, c_S, -c_S, 0\} \tag{5.90}$$

The action of the matrix $\boldsymbol{\Omega}_1$ on the vector of unknowns can be represented by Eqs. 5.91, 5.92 in 2D and 3D cases, respectively.

$$\begin{bmatrix} \omega_1 \\ \omega_2 \\ \omega_3 \\ \omega_4 \\ \omega_5 \end{bmatrix} = \boldsymbol{\Omega}_1 \begin{bmatrix} \vec{v} \\ \boldsymbol{\sigma} \end{bmatrix} = \begin{bmatrix} \vec{n} \cdot \vec{v} - \frac{1}{c_P \rho} \mathbf{N}_{00} \div \boldsymbol{\sigma} \\ \vec{n} \cdot \vec{v} + \frac{1}{c_P \rho} \mathbf{N}_{00} \div \boldsymbol{\sigma} \\ \vec{n}_1 \cdot \vec{v} - \frac{1}{c_S \rho} \mathbf{N}_{01} \div \boldsymbol{\sigma} \\ \vec{n}_1 \cdot \vec{v} + \frac{1}{c_S \rho} \mathbf{N}_{01} \div \boldsymbol{\sigma} \\ \left(\mathbf{N}_{11} - \frac{\lambda}{\lambda + 2\mu} \mathbf{N}_{00} \right) \div \boldsymbol{\sigma} \end{bmatrix} \tag{5.91}$$

$$\begin{bmatrix} \omega_1 \\ \omega_2 \\ \omega_3 \\ \omega_4 \\ \omega_5 \\ \omega_6 \\ \omega_7 \\ \omega_8 \\ \omega_9 \end{bmatrix} = \boldsymbol{\Omega}_1 \begin{bmatrix} \vec{v} \\ \boldsymbol{\sigma} \end{bmatrix} = \begin{bmatrix} \vec{n} \cdot \vec{v} - \frac{1}{c_P \rho} \mathbf{N}_{00} \div \boldsymbol{\sigma} \\ \vec{n} \cdot \vec{v} + \frac{1}{c_P \rho} \mathbf{N}_{00} \div \boldsymbol{\sigma} \\ \vec{n}_1 \cdot \vec{v} - \frac{1}{c_S \rho} \mathbf{N}_{01} \div \boldsymbol{\sigma} \\ \vec{n}_1 \cdot \vec{v} + \frac{1}{c_S \rho} \mathbf{N}_{01} \div \boldsymbol{\sigma} \\ \vec{n}_2 \cdot \vec{v} - \frac{1}{c_S \rho} \mathbf{N}_{02} \div \boldsymbol{\sigma} \\ \vec{n}_2 \cdot \vec{v} + \frac{1}{c_S \rho} \mathbf{N}_{02} \div \boldsymbol{\sigma} \\ \mathbf{N}_{12} \div \boldsymbol{\sigma} \\ (\mathbf{N}_{11} - \mathbf{N}_{22}) \div \boldsymbol{\sigma} \\ \left(\mathbf{N}_{11} + \mathbf{N}_{22} - \frac{2\lambda}{\lambda + 2\mu} \mathbf{N}_{00} \right) \div \boldsymbol{\sigma} \end{bmatrix} \tag{5.92}$$

In a two-dimensional case, the action of the matrix $(\boldsymbol{\Omega}_1)^{-1}$ on the vector $\vec{\omega}$ can be represented by Eq. 5.93.

$$\begin{aligned} \begin{bmatrix} \vec{v} \\ \boldsymbol{\sigma} \end{bmatrix} &= (\boldsymbol{\Omega}_1)^{-1} \vec{\omega} \\ &= \frac{1}{2} \begin{bmatrix} (\omega_1 + \omega_2)\, \vec{n} + (\omega_3 + \omega_4)\, \vec{n}_1 \\ (\omega_2 - \omega_1)\, (\rho\, (c_P - c_3)\, \mathbf{N}_{00} + \rho c_3 \mathbf{I}) + 2\rho c_S\, (\omega_4 - \omega_3)\, \mathbf{N}_{01} + 2\omega_5\, (\mathbf{I} - \mathbf{N}_{00}) \end{bmatrix} \end{aligned} \tag{5.93}$$

In a three-dimensional case, the action of the matrix $(\boldsymbol{\Omega}_1)^{-1}$ on the vector $\vec{\omega}$ can be represented by Eq. 5.94.

$$\begin{aligned} \begin{bmatrix} \vec{v} \\ \boldsymbol{\sigma} \end{bmatrix} &= (\boldsymbol{\Omega}_1)^{-1} \vec{\omega} \\ &= \frac{1}{2} \begin{bmatrix} (\omega_1 + \omega_2)\, \vec{n} + (\omega_3 + \omega_4)\, \vec{n}_1 + (\omega_5 + \omega_6)\, \vec{n}_2 \\ \rho\, (\omega_2 - \omega_1)\, ((c_P - c_3)\, \mathbf{N}_{00} + c_3 \mathbf{I}) + 2 c_S \rho\, (\omega_4 - \omega_3)\, \mathbf{N}_{01} \\ +2 c_S \rho\, (\omega_6 - \omega_5)\, \mathbf{N}_{02} + 4\omega_7 \mathbf{N}_{12} + \omega_8\, (\mathbf{N}_{11} - \mathbf{N}_{22}) + \omega_9\, (\mathbf{I} - \mathbf{N}_{00}) \end{bmatrix} \end{aligned} \tag{5.94}$$

In Eqs. 5.93, 5.94, the quantity c_3 is denoted in Eq. 5.95 and in Sect. 5.4.3.

$$c_3 = \frac{\lambda}{\lambda + 2\mu} c_P. \tag{5.95}$$

5.3.6 Grid-Characteristic Methods for Acoustic Waves

Also for definiteness, the direction ξ_1 is considered. The vector of unknowns for a three-dimensional case can be written by Eq. 5.96.

$$\vec{q} = \begin{bmatrix} \vec{v} \\ p \end{bmatrix} = \begin{bmatrix} v_1 \\ v_2 \\ v_3 \\ p \end{bmatrix} \tag{5.96}$$

The vector of unknowns for a 3D case can be written by Eq. 5.97.

$$\vec{q} = \begin{bmatrix} \vec{v} \\ p \end{bmatrix} = \begin{bmatrix} v_1 \\ v_2 \\ p \end{bmatrix} \tag{5.97}$$

The action of the matrix $\mathbf{A}_1$ on the vector of unknowns can be written by Eq. 5.98 both for 2D and 3D cases.

$$\mathbf{A}_1 \begin{bmatrix} \vec{v} \\ p \end{bmatrix} = \begin{bmatrix} \frac{p}{\rho}\vec{n} \\ c^2\rho\left(\vec{n}\cdot\vec{v}\right) \end{bmatrix} \tag{5.98}$$

Matrices $\mathbf{A}_1^{3D}$, $\mathbf{A}_2^{3D}$, $\mathbf{A}_3^{3D}$ have the same set of eigenvalues:

$$\{c, -c, 0, 0\}\,. \tag{5.99}$$

Matrices $\mathbf{A}_1^{2D}$, $\mathbf{A}_2^{2D}$ also have the same set of eigenvalues:

$$\{c, -c, 0\}\,. \tag{5.100}$$

In a two-dimensional case, the action of the matrix $\mathbf{\Omega}_1$ on the vector of unknowns can be represented by Eq. 5.101.

$$\begin{bmatrix} \omega_1 \\ \omega_2 \\ \omega_3 \end{bmatrix} = \mathbf{\Omega} \begin{bmatrix} \vec{v} \\ p \end{bmatrix} = \begin{bmatrix} \vec{n}\cdot\vec{v} + \frac{p}{c\rho} \\ \vec{n}\cdot\vec{v} - \frac{p}{c\rho} \\ \vec{n}_1\cdot\vec{v} \end{bmatrix} \tag{5.101}$$

In a three-dimensional case, the action of the matrix $\mathbf{\Omega}_1$ on the vector of unknowns can be represented by Eq. 5.102.

$$\begin{bmatrix} \omega_1 \\ \omega_2 \\ \omega_3 \\ \omega_4 \end{bmatrix} = \mathbf{\Omega} \begin{bmatrix} \vec{v} \\ p \end{bmatrix} = \begin{bmatrix} \vec{n} \cdot \vec{v} + \frac{p}{c\rho} \\ \vec{n} \cdot \vec{v} - \frac{p}{c\rho} \\ \vec{n}_1 \cdot \vec{v} \\ \vec{n}_2 \cdot \vec{v} \end{bmatrix} \tag{5.102}$$

In a two-dimensional case, the action of the matrix $(\mathbf{\Omega}_1)^{-1}$ on the vector $\vec{\omega}$ can be represented by Eq. 5.103.

$$\begin{bmatrix} \vec{v} \\ p \end{bmatrix} = \mathbf{\Omega}^{-1}\vec{\omega} = \frac{1}{2} \begin{bmatrix} (\omega_1 + \omega_2)\,\vec{n} + \omega_3\vec{n}_1 \\ c\rho\,(\omega_1 - \omega_2) \end{bmatrix} \tag{5.103}$$

In a three-dimensional case, the action of the matrix $(\mathbf{\Omega}_1)^{-1}$ on the vector $\vec{\omega}$ can be represented by Eq. 5.104.

$$\begin{bmatrix} \vec{v} \\ p \end{bmatrix} = \mathbf{\Omega}^{-1}\vec{\omega} = \frac{1}{2} \begin{bmatrix} (\omega_1 + \omega_2)\,\vec{n} + \omega_3\vec{n}_1 + \omega_4\vec{n}_2 \\ c\rho\,(\omega_1 - \omega_2) \end{bmatrix} \tag{5.104}$$

5.4 Boundary and Interface Conditions

The family of grid-characteristic methods both on structured and unstructured triangular and tetrahedral grids allows us to apply the most correct computational algorithms on the boundaries and interfaces into the integration domain.

In this Section, main cases of boundary and interface conditions are discussed. Other cases one can find in works [3–5, 9, 10]. The general propositions for boundary and interface conditions calculation are discussed in Sect. 5.4.1. Calculation of the boundary and interface conditions for a case of anisotropic elastic waves equations is discussed in Sect. 5.4.2. The case of isotropic elastic waves is proposed in Sect. 5.4.3. The case of acoustic waves is represented in Sect. 5.4.4.

5.4.1 General Provisions of Boundary and Interface Conditions

Suppose that the boundary conditions are written by Eq. 5.105 in a matrix form.

$$\mathbf{D}\vec{q}\,(\xi_1, \xi_2, \xi_3, t + \tau) = \vec{d} \tag{5.105}$$

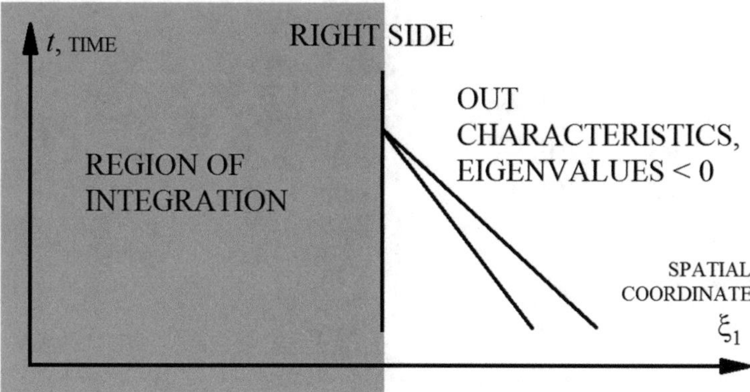

Fig. 5.2 The right side of the region of integration for the direction ξ_1

In Eq. 5.105, $\vec{q}\,(\xi_1, \xi_2, \xi_3, t + \tau)$ are the values of velocity components and stress tensor at the next step of integration at the boundary point.

For each matrix $\mathbf{A}_j$, there are zero, positive, and negative eigenvalues. For each direction there will be two types of correctors: for the left and right borders. This means that in cases of anisotropic and isotropic elastic wave equations, 3 scalar values and 2 scalar values given at the boundaries should be set for a three-dimensional case and a two-dimensional case, respectively. In the cases of acoustic wave equations, 1 scalar value given at the boundaries should be set for both three-dimensional and two-dimensional cases. In the cases of anisotropic and isotropic elastic wave equations, 6 scalar values given at the interfaces should be set for a three-dimensional case and 4 scalar values given at the interfaces should be set for a two-dimensional case. In the cases of acoustic wave equations, 2 scalar values given at the interfaces should be set for both three-dimensional and two-dimensional cases.

Suppose along the direction ξ_1 the characteristics corresponding to *negative* eigenvalues of the matrix go outside the region of integration. This side is called the *right side of the region of integration for the direction* ξ_1 in accordance with Fig. 5.2.

Suppose along the direction ξ_1 the characteristics corresponding to *positive* eigenvalues of the matrix go outside the region of integration. This side is called the *left side of the region of integration for the direction* ξ_1 in accordance with Fig. 5.3.

For definiteness, consider one of the sides of the region of integration and one of directions (ξ_1, ξ_2, ξ_3) for a 3D case and one of directions (ξ_1, ξ_2) in a 2D case. Suppose that along the direction ξ_1 the characteristics corresponding to negative eigenvalues of the matrix go outside the region of integration. This side is called the right side of the region of integration. Then, at the stage of calculating the internal points in accordance with Eqs. 5.52, 5.53, the following vectors given in Eqs. 5.106, 5.107 will be calculated.

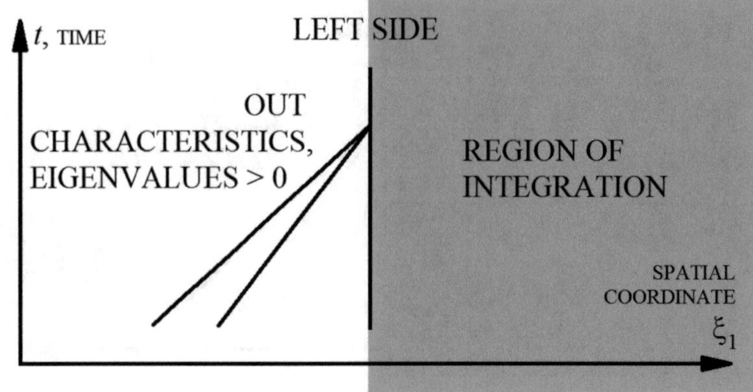

Fig. 5.3 The left side of the region of integration for the direction ξ_1

$$\vec{q}\left(\xi_1, \xi_2, t+\tau\right) = \vec{q}\left(\xi_1, \xi_2, t\right) + \sum_{i \in I^+} \mathbf{X}_i^{2D,1}\left(\vec{q}\left(\xi_1 - c_i^{2D,1}\tau, \xi_2, t\right) - \vec{q}\left(\xi_1, \xi_2, t\right)\right) \tag{5.106}$$

$$\begin{aligned}\vec{q}\left(\xi_1, \xi_2, \xi_3, t+\tau\right) &= \vec{q}\left(\xi_1, \xi_2, \xi_3, t\right) \\ &+ \sum_{i \in I^+} \mathbf{X}_i^{3D,1}\left(\vec{q}\left(\xi_1 - c_i^{2D,1}\tau, \xi_2, \xi_3, t\right) - \vec{q}\left(\xi_1, \xi_2, \xi_3, t\right)\right)\end{aligned} \tag{5.107}$$

The matrix $\mathbf{\Omega}^{*,out}$ is composed of eigenvectors corresponding to negative eigenvalues. The action of the corrector at the boundary point is performed by the following Eq. 5.108, where $\vec{r}_\mathrm{B}$ is the given point at the boundary.

$$\vec{q}\left(\vec{r}_\mathrm{B}, t+\tau\right) = \mathbf{F}\vec{q}^{\,\mathrm{in}}\left(\vec{r}_\mathrm{B}, t+\tau\right) + \mathbf{\Phi}\vec{d} \tag{5.108}$$

The condition given in Eq. 5.105 is satisfied with the same order of convergence as the method used for solving systems of equations under consideration. This method is given by Eqs. 5.48, 5.49 into the inner points of the modelling domain (region of integration) and, accordingly, is using for finding values in Eqs. 5.52, 5.53. In Eq. 5.108, the matrices $\mathbf{\Phi}$ and $\mathbf{F}$ are calculated from the following Eqs. 5.109–5.110.

$$\mathbf{\Phi} = \mathbf{\Omega}^{*,out}\left(\mathbf{D}\mathbf{\Omega}^{*,out}\right)^{-1} \tag{5.109}$$

$$\mathbf{F} = \mathbf{I} - \mathbf{\Phi}\mathbf{D} \tag{5.110}$$

In Eq. 5.109, the matrix $\left(\mathbf{D}\mathbf{\Omega}^{*,out}\right)^{-1}$ is calculated in order the Eq. 5.111 to be true.

$$\left(\mathbf{D}\boldsymbol{\Omega}^{*,out}\right)^{-1}\mathbf{D}\boldsymbol{\Omega}^{*,out} = \mathbf{I}. \tag{5.111}$$

5.4.2 Boundary and Interface Conditions for Anisotropic Elastic Waves

The boundary corrector with a given traction is considered. In this case, the condition given by Eq. 5.105 takes the form represented in Eq. 5.112.

$$\boldsymbol{\sigma} \cdot \vec{p} = \vec{f} \tag{5.112}$$

In Eq. 5.112, $\vec{f}$ is the traction also called as the density of external forces and being a given vector, $\vec{p}$ is the outer normal to the boundary here and below. The matrix **D** from Eq. 5.105 can be written by Eq. 5.113.

$$\mathbf{D} = \begin{bmatrix} 0 & 0 & 0 & p_1 & 0 & 0 & 0 & p_3 & p_2 \\ 0 & 0 & 0 & 0 & p_2 & 0 & p_3 & 0 & p_1 \\ 0 & 0 & 0 & 0 & 0 & p_3 & p_2 & p_1 & 0 \end{bmatrix} = \mathbf{D}\left(\vec{p}\right) \tag{5.113}$$

The vector $\vec{d}$ from Eq. 5.105 can be written by Eq. 5.114.

$$\vec{d} = \begin{bmatrix} f_1 \\ f_2 \\ f_3 \end{bmatrix} = \vec{f} = \vec{d}\left(\vec{f}\right) \tag{5.114}$$

The boundary corrector with the given boundary velocity is considered. In this case, the condition in Eq. 5.105 takes the form represented by Eq. 5.115.

$$\vec{v} = \vec{V} \tag{5.115}$$

In Eq. 5.115, $\vec{V}$ is the velocity of the boundary being a given vector. The matrix **D** from Eq. 5.105 can be written by Eq. 5.116.

$$\mathbf{D} = \begin{bmatrix} 1 & 0 & 0 & 0 & 0 & 0 & 0 & 0 & 0 \\ 0 & 1 & 0 & 0 & 0 & 0 & 0 & 0 & 0 \\ 0 & 0 & 1 & 0 & 0 & 0 & 0 & 0 & 0 \end{bmatrix} \tag{5.116}$$

The vector $\vec{d}$ from Eq. 5.105 can be written by Eq. 5.117.

$$\vec{d} = \begin{bmatrix} V_1 \\ V_2 \\ V_3 \end{bmatrix} = \vec{d}(\vec{V}) = \vec{V} \tag{5.117}$$

The boundary corrector of mixed boundary conditions with the given normal projection of the boundary velocity and the non-zero given tangential component of the traction is considered. In this case, the Eq. 5.105 takes the form given by Eqs. 5.118, 5.119.

$$\vec{v} \cdot \vec{p} = V_p \tag{5.118}$$

$$\boldsymbol{\sigma} \cdot \vec{p} = \vec{f} \tag{5.119}$$

In Eq. 5.119, $\vec{f}$ is given by Eq. 5.120.

$$\vec{f} = \vec{f}_\tau + \left(\left(\vec{f} - \vec{f}_\tau\right) \cdot \vec{p}\right) \vec{p} \tag{5.120}$$

In Eq. 5.118, V_p is the projection of the boundary velocity on the normal being a given scalar. In Eq. 5.120, $\vec{f}_\tau$ is the non-zero tangential component of the traction being a given vector. The matrix $\mathbf{D}$ from Eq. 5.105 can be written by Eq. 5.121.

$$\mathbf{D} = \mathbf{D}\left(\vec{f}_\tau, \vec{p}\right) = \begin{bmatrix} s_{11} & s_{12} & s_{13} & 0 & 0 & 0 & 0 & 0 & 0 \\ 0 & 0 & 0 & s_{11}s_{31} & s_{12}s_{32} & s_{13}s_{33} & s_{12}s_{33} + s_{13}s_{32} & s_{11}s_{33} + s_{13}s_{31} & s_{11}s_{32} + s_{12}s_{31} \\ 0 & 0 & 0 & s_{11}s_{21} & s_{12}s_{22} & s_{13}s_{23} & s_{12}s_{23} + s_{13}s_{22} & s_{11}s_{23} + s_{13}s_{21} & s_{11}s_{22} + s_{12}s_{21} \end{bmatrix} \tag{5.121}$$

The vector $\vec{d}$ from Eq. 5.105 can be written by Eq. 5.122.

$$\vec{d} = \begin{bmatrix} V_p \\ 0 \\ f_\tau \end{bmatrix} = \vec{d}\left(V_p, \vec{f}_\tau, \vec{p}\right) \tag{5.122}$$

In Eq. 5.121, the designation introduced in Eqs. 5.123–5.131 is used.

$$s_{11} = p_1 \tag{5.123}$$

$$s_{12} = p_2 \tag{5.124}$$

$$s_{13} = p_3 \tag{5.125}$$

$$s_{21} = \frac{f_{\tau 1}}{f_\tau} = \frac{f_{\tau 1}}{\sqrt{(f_{\tau 1})^2 + (f_{\tau 2})^2 + (f_{\tau 3})^2}} \tag{5.126}$$

$$s_{22} = \frac{f_{\tau 2}}{f_\tau} = \frac{f_{\tau 2}}{\sqrt{(f_{\tau 1})^2 + (f_{\tau 2})^2 + (f_{\tau 3})^2}} \tag{5.127}$$

$$s_{23} = \frac{f_{\tau 3}}{f_\tau} = \frac{f_{\tau 3}}{\sqrt{(f_{\tau 1})^2 + (f_{\tau 2})^2 + (f_{\tau 3})^2}} \tag{5.128}$$

$$s_{31} = \frac{p_2 f_{\tau 3} - p_3 f_{\tau 2}}{\sqrt{(f_{\tau 1})^2 + (f_{\tau 2})^2 + (f_{\tau 3})^2}} = s_{12}s_{23} - s_{13}s_{22} \tag{5.129}$$

$$s_{32} = \frac{p_3 f_{\tau 1} - p_1 f_{\tau 3}}{\sqrt{(f_{\tau 1})^2 + (f_{\tau 2})^2 + (f_{\tau 3})^2}} = s_{13}s_{21} - s_{11}s_{23} \tag{5.130}$$

$$s_{33} = \frac{p_1 f_{\tau 2} - p_2 f_{\tau 1}}{\sqrt{(f_{\tau 1})^2 + (f_{\tau 2})^2 + (f_{\tau 3})^2}} = s_{11}s_{22} - s_{12}s_{21} \tag{5.131}$$

The boundary corrector of mixed boundary conditions with the given normal projection of the boundary velocity and zero tangential component of the traction is considered. In this case, the Eq. 5.105 takes the form of Eqs. 5.132, 5.133.

$$\vec{v} \cdot \vec{p} = V_p \tag{5.132}$$

$$\boldsymbol{\sigma} \cdot \vec{p} = \vec{f} \tag{5.133}$$

In Eq. 5.133, $\vec{f}$ is given by Eq. 5.134.

$$\vec{f} = \left(\vec{f} \cdot \vec{p}\right) \vec{p} \tag{5.134}$$

In Eq. 5.132, V_p is the projection of the boundary velocity on the normal being a given scalar. The vector $\vec{d}$ from Eq. 5.105 can be written by Eq. 5.135.

$$\vec{d} = \begin{bmatrix} V_p \\ 0 \\ 0 \end{bmatrix} = \vec{d}\left(V_p\right) \tag{5.135}$$

A view of the matrix **D** depends on the vector of the normal. If $p_1 \neq 0$, then this matrix takes the form represented by Eq. 5.136.

$$\mathbf{D} = \mathbf{D}\left(\vec{p}\right) = \begin{bmatrix} p_1 & p_2 & p_3 & 0 & 0 & 0 & 0 & 0 & 0 \\ 0 & 0 & 0 & -p_1 p_2 & p_1 p_2 & 0 & p_1 p_3 & p_2 p_3 & \left(p_1 p_1 - p_2 p_2\right) \\ 0 & 0 & 0 & -p_1 p_3 & 0 & p_1 p_3 & p_1 p_2 & \left(p_1 p_1 - p_3 p_3\right) & p_2 p_3 \end{bmatrix} \tag{5.136}$$

If $p_2 \neq 0$, then the matrix **D** takes the form represented by Eq. 5.137.

$$\mathbf{D} = \mathbf{D}\left(\vec{p}\right) = \begin{bmatrix} p_1 & p_2 & p_3 & 0 & 0 & 0 & 0 & 0 & 0 \\ 0 & 0 & 0 & p_1 p_2 & -p_1 p_2 & 0 & -p_1 p_3 & p_2 p_3 & \left(p_2 p_2 - p_1 p_1\right) \\ 0 & 0 & 0 & 0 & -p_2 p_3 & p_2 p_3 & \left(p_2 p_2 - p_3 p_3\right) & p_1 p_2 & -p_1 p_3 \end{bmatrix} \tag{5.137}$$

If $p_3 \neq 0$, then the matrix **D** takes the form represented by Eq. 5.138.

$$\mathbf{D} = \mathbf{D}\left(\vec{p}\right) = \begin{bmatrix} p_1 & p_2 & p_3 & 0 & 0 & 0 & 0 & 0 & 0 \\ 0 & 0 & 0 & p_1 p_3 & 0 & -p_1 p_3 & -p_1 p_2 & \left(p_3 p_3 - p_1 p_1\right) & p_2 p_3 \\ 0 & 0 & 0 & 0 & p_2 p_3 & -p_2 p_3 & \left(p_3 p_3 - p_2 p_2\right) & -p_1 p_2 & p_1 p_3 \end{bmatrix} \tag{5.138}$$

The non-reflective boundary corrector is considered. Then the differences of the values along the characteristics that go beyond the boundaries of the integration region must be zero. Thus, the matrix **D** from Eq. 5.105 takes the form represented by Eq. 5.139 and vector $\vec{d}$ from Eq. 5.105 can be written by Eq. 5.140.

$$\mathbf{D} = \mathbf{\Omega}^{out} \tag{5.139}$$

$$\vec{d} = 0 \tag{5.140}$$

In Eq. 5.139, $\mathbf{\Omega}^{out}$ is the matrix composed of columns of the matrix $\mathbf{\Omega}$ corresponding to the outgoing characteristics. Equations 5.109, 5.110 take the following form given by Eqs. 5.141, 5.142.

$$\mathbf{\Phi} = \mathbf{\Omega}^{*,out}\left(\mathbf{\Omega}^{out}\mathbf{\Omega}^{*,out}\right)^{-1} \tag{5.141}$$

$$\mathbf{F} = \mathbf{I} - \mathbf{\Omega}^{*,out}\left(\mathbf{\Omega}^{out}\mathbf{\Omega}^{*,out}\right)^{-1}\mathbf{\Omega}^{out} \tag{5.142}$$

Let here and further for the cases of interface correctors there is a medium a and a medium b, $\vec{p}$ is the external normal to the boundary of the medium a.

The interface corrector of continuity of velocity and traction is considered. These interface conditions are written by Eqs. 5.143, 5.144.

$$\vec{v}^{a} = \vec{v}^{b} = \vec{V} \tag{5.143}$$

$$\vec{f}^{a} = -\vec{f}^{b} \tag{5.144}$$

The matrix $\mathbf{P}$ is introduced by Eq. 5.145.

$$\mathbf{P} = \begin{bmatrix} p_1 & 0 & 0 & 0 & p_3 & p_2 \\ 0 & p_2 & 0 & p_3 & 0 & p_1 \\ 0 & 0 & p_3 & p_2 & p_1 & 0 \end{bmatrix} \tag{5.145}$$

The matrices $\mathbf{\Phi}^a$, $\mathbf{F}^a$ for the medium a are obtained in accordance with Eqs. 5.109, 5.110. For the medium b, the matrices $\mathbf{\Phi}^b$, $\mathbf{F}^b$ are obtained in accordance with the analogues of Eqs. 5.109, 5.110 for the positive outgoing characteristics. In Eqs. 5.109, 5.110 and their analogs, the matrix $\mathbf{D}$ is used for the boundary corrector of the given velocity in accordance with Eqs. 5.115–5.117. On the basis of these matrices, 4 new matrices $\mathbf{\Phi}^{a,\sigma}$, $\mathbf{F}^{a,\sigma}$, $\mathbf{\Phi}^{b,\sigma}$, $\mathbf{F}^{b,\sigma}$ can be introduces. New 6×9 matrices $\mathbf{F}^{a,\sigma}$, $\mathbf{F}^{b,\sigma}$ can be found using Eqs. 5.146.

$$\mathbf{F}^{\sigma} = \begin{bmatrix} \mathbf{F}_{41} & \mathbf{F}_{42} & \cdots & \mathbf{F}_{49} \\ \mathbf{F}_{51} & \mathbf{F}_{52} & \cdots & \mathbf{F}_{59} \\ \vdots & \vdots & \ddots & \vdots \\ \mathbf{F}_{91} & \mathbf{F}_{92} & \cdots & \mathbf{F}_{99} \end{bmatrix} \tag{5.146}$$

New 6×3 matrices $\mathbf{\Phi}^{a,\sigma}$, $\mathbf{\Phi}^{b,\sigma}$ can be found using Eq. 5.147.

$$\mathbf{\Phi}^{\sigma} = \begin{bmatrix} \mathbf{\Phi}_{41} & \mathbf{\Phi}_{42} & \mathbf{\Phi}_{43} \\ \mathbf{\Phi}_{51} & \mathbf{\Phi}_{52} & \mathbf{\Phi}_{53} \\ \vdots & \vdots & \vdots \\ \mathbf{\Phi}_{91} & \mathbf{\Phi}_{92} & \mathbf{\Phi}_{93} \end{bmatrix} \tag{5.147}$$

The 3×3 matrix given by Eq. 5.148 is considered as well.

$$\mathbf{P}\left(\mathbf{\Phi}^{b,\sigma} - \mathbf{\Phi}^{a,\sigma}\right) \tag{5.148}$$

The matrix $\left(\mathbf{P}\left(\mathbf{\Phi}^{b,\sigma} - \mathbf{\Phi}^{a,\sigma}\right)\right)^{-1}$ should be calculated using Eq. 5.149.

$$\left(\mathbf{P}\left(\mathbf{\Phi}^{b,\sigma}-\mathbf{\Phi}^{a,\sigma}\right)\right)\left(\mathbf{P}\left(\mathbf{\Phi}^{b,\sigma}-\mathbf{\Phi}^{a,\sigma}\right)\right)^{-1}=\mathbf{I} \tag{5.149}$$

The following Eqs. 5.150, 5.151 set new designations $\mathbf{\Gamma}^a$ and $\mathbf{\Gamma}^b$.

$$\mathbf{\Gamma}^a=\left(\mathbf{P}\left(\mathbf{\Phi}^{b,\sigma}-\mathbf{\Phi}^{a,\sigma}\right)\right)^{-1}\mathbf{P}\mathbf{F}^{a,\sigma} \tag{5.150}$$

$$\mathbf{\Gamma}^b=-\left(\mathbf{P}\left(\mathbf{\Phi}^{b,\sigma}-\mathbf{\Phi}^{a,\sigma}\right)\right)^{-1}\mathbf{P}\mathbf{F}^{b,\sigma} \tag{5.151}$$

Using these new designations one can find $\vec{V}$ by Eq. 5.152.

$$\vec{V}=\mathbf{\Gamma}^a\vec{q}^{\,a,in}\left(t,\vec{r}_{\mathrm{B}}\right)+\mathbf{\Gamma}^b\vec{q}^{\,b,in}\left(t,\vec{r}_{\mathrm{B}}\right) \tag{5.152}$$

Then one can use the corrector of the given boundary velocity with $\vec{d}=\vec{V}$ for two different media a and b in accordance with Eqs. 5.153, 5.154.

$$\vec{q}^{\,a}\left(t+\tau,\vec{r}_{\mathrm{B}}\right)=\mathbf{F}^a\vec{q}^{\,a,in}\left(t,\vec{r}_{\mathrm{B}}\right)+\mathbf{\Phi}^a\vec{V} \tag{5.153}$$

$$\vec{q}^{\,b}\left(t+\tau,\vec{r}_{\mathrm{B}}\right)=\mathbf{F}^b\vec{q}^{\,b,in}\left(t,\vec{r}_{\mathrm{B}}\right)+\mathbf{\Phi}^b\vec{V} \tag{5.154}$$

Let us consider the free sliding interface corrector. The conditions are given by Eqs. 5.155–5.157.

$$\vec{v}_a\cdot\vec{p}=\vec{v}_b\cdot\vec{p}=V_p \tag{5.155}$$

$$f_p^a=-f_p^b \tag{5.156}$$

$$f_\tau^a=f_\tau^b=0 \tag{5.157}$$

The matrix-row $\mathbf{P}$ given by Eq. 5.158 is used.

$$\mathbf{P}=\left[(p_1)^2\;(p_2)^2\;(p_3)^2\;2p_2p_3\;2p_1p_3\;2p_1p_2\right] \tag{5.158}$$

The matrices $\mathbf{\Phi}^a$, $\mathbf{F}^a$ for the medium a are obtained in accordance with the Eqs. 5.109, 5.110. For the medium b, the matrices $\mathbf{\Phi}^b$, $\mathbf{F}^b$ are obtained in accordance with the analogues of Eqs. 5.109, 5.110 for the positive outgoing characteristics. In Eqs. 5.109, 5.110 and their analogs, the matrix $\mathbf{D}$ is used for the boundary corrector of mixed boundary conditions with the given normal projection of the boundary velocity and with zero tangential component of the traction in accordance with Eqs. 5.132–5.138. This boundary corrector is used with the normal $\vec{p}$ for the medium a and the normal $-\vec{p}$ for the medium b. The direction of the normal should be taken into account into Eqs. 5.132–5.138.

On the basis of these matrices, 4 new matrices $\mathbf{\Phi}^{a,\sigma}$, $\mathbf{F}^{a,\sigma}$, $\mathbf{\Phi}^{b,\sigma}$, $\mathbf{F}^{b,\sigma}$ can be introduced. New 6×9 matrices $\mathbf{F}^{a,\sigma}$, $\mathbf{F}^{b,\sigma}$ can be found using Eq. 5.146 but 6×1 matrices $\mathbf{\Phi}^{a,\sigma}$, $\mathbf{\Phi}^{b,\sigma}$ should be calculated using Eq. 5.159.

$$\mathbf{\Phi}^{\sigma} = \begin{bmatrix} \mathbf{\Phi}_{41} \\ \mathbf{\Phi}_{51} \\ \vdots \\ \mathbf{\Phi}_{91} \end{bmatrix} \tag{5.159}$$

For a given interface corrector the expression $\mathbf{P}\left(\mathbf{\Phi}^{b} - \mathbf{\Phi}^{a}\right)$ is scalar. The following notation is introduced by Eqs. 5.160, 5.161.

$$\mathbf{\Gamma}^{a} = -\left(\mathbf{P}\left(\mathbf{\Phi}^{a} + \mathbf{\Phi}^{b}\right)\right)^{-1} \mathbf{P}\mathbf{F}^{a} \tag{5.160}$$

$$\mathbf{\Gamma}^{b} = \left(\mathbf{P}\left(\mathbf{\Phi}^{a} + \mathbf{\Phi}^{b}\right)\right)^{-1} \mathbf{P}\mathbf{F}^{b} \tag{5.161}$$

Then the projection of the velocity on the normal V_p is found from Eq. 5.162.

$$V_p = \mathbf{\Gamma}^{a}\vec{q}^{\,a,in}\left(t, \vec{r}_{\mathrm{B}}\right) + \mathbf{\Gamma}^{b}\vec{q}^{\,b,in}\left(t, \vec{r}_{\mathrm{B}}\right) \tag{5.162}$$

Next, the boundary corrector of mixed boundary conditions with the zero tangential projection of the traction is applied using Eqs. 5.163, 5.164.

$$\vec{d}^{\,a} = \begin{bmatrix} V_p \\ 0 \\ 0 \end{bmatrix} \tag{5.163}$$

$$\vec{d}^{\,b} = \begin{bmatrix} -V_p \\ 0 \\ 0 \end{bmatrix} \tag{5.164}$$

Then one can apply Eqs. 5.165, 5.166.

$$\vec{q}^{\,a}\left(t + \tau, \vec{r}_{\mathrm{B}}\right) = \mathbf{F}^{a}\vec{q}^{\,a,in}\left(t, \vec{r}_{\mathrm{B}}\right) + \mathbf{\Phi}^{a}\vec{d}^{\,a} \tag{5.165}$$

$$\vec{q}^{\,b}\left(t + \tau, \vec{r}_{\mathrm{B}}\right) = \mathbf{F}^{b}\vec{q}^{\,b,in}\left(t, \vec{r}_{\mathrm{B}}\right) + \mathbf{\Phi}^{b}\vec{q}^{\,b}. \tag{5.166}$$

5.4.3 Boundary and Interface Conditions for Isotropic Elastic Waves

The boundary corrector with the given traction is considered. In this case, Eq. 5.105 takes the form of Eq. 5.112. In the case, when the normal $\vec{p}$ is collinear to the direction of splitting $\vec{n}$, Eq. 5.167 is true for the right boundary.

$$\vec{p} = \vec{n} \tag{5.167}$$

Equation 5.168 is true for the left boundary.

$$\vec{p} = -\vec{n} \tag{5.168}$$

The case, when the normal $\vec{p}$ is collinear to the direction of splitting $\vec{n}$, is realized during using the regular meshes. In the case of unstructured triangular or tetrahedral grids, the normal might not being collinear to the split direction.

If the normal is collinear to the direction of splitting, then the boundary corrector will act according to Eqs. 5.169, 5.170.

$$\vec{v}^{n+1} = \vec{v}^{in} - \frac{1}{\rho c_S}\vec{z} + \frac{1}{\rho}\left(\frac{1}{c_S} - \frac{1}{c_P}\right)\left(\vec{z}\cdot\vec{n}\right)\vec{n} \tag{5.169}$$

$$\boldsymbol{\sigma}^{n+1} = \boldsymbol{\sigma}^{in} \pm \left(\vec{z}\otimes\vec{n} + \vec{n}\otimes\vec{z}\right) \pm \frac{\vec{z}\cdot\vec{n}}{\lambda + 2\mu}\left(\lambda\mathbf{I} - 2\left(\lambda + \mu\right)\mathbf{N}_{00}\right) \tag{5.170}$$

In Eqs. 5.169, 5.170 and further, the upper sign corresponds to the left boundary of the integration region and the lower sign corresponds to the right boundary. The following value is denoted by Eq. 5.171.

$$\vec{z} = \mp\left(\boldsymbol{\sigma}^{in}\cdot\vec{n}\right) - \vec{f} \tag{5.171}$$

If the normal is not collinear to the direction of splitting, then the boundary corrector will act according to Eqs. 5.172, 5.173.

$$\vec{v}^{n+1} = \vec{v}^{in} \pm \frac{1}{\rho}\left(\omega_1\vec{n} - \left(\vec{n}\cdot\vec{b}\right)\vec{n} + \vec{b}\right) \tag{5.172}$$

$$\boldsymbol{\sigma}^{n+1} = \boldsymbol{\sigma}^{in} - \left(\left(c_P - c_3\right)\omega_1 - 2c_S\left(\vec{n}\cdot\vec{b}\right)\right)\mathbf{N}_{00} - c_3\omega_1\mathbf{I} - c_S\left(\vec{n}\otimes\vec{b} + \vec{b}\otimes\vec{n}\right) \tag{5.173}$$

In Eqs. 5.172, 5.173, values $\vec{b}$ and ω_1 depend on $\vec{z}$ given by Eq. 5.174 according to Eqs. 5.175, 5.176.

$$\vec{z} = \boldsymbol{\sigma}^{in}\cdot\vec{p} - \vec{f} \tag{5.174}$$

$$\omega_1 = \frac{2\left(\vec{n}\cdot\vec{p}\right)\left(\vec{n}\cdot\vec{z}\right) - \left(\vec{p}\cdot\vec{z}\right)}{\left(c_P + c_3\right)\left(\vec{n}\cdot\vec{p}\right)^2 - c_3\left|\vec{p}\right|^2} \tag{5.175}$$

$$\vec{b} = \frac{1}{c_S\left(\vec{n}\cdot\vec{p}\right)}\left(\vec{z} - \omega_1 c_3\vec{p}\right) \tag{5.176}$$

The boundary corrector with a given boundary velocity is considered. In this case, Eq. 5.105 takes the form of Eq. 5.115. The boundary corrector acts in accordance with Eqs. 5.177, 5.178.

$$\vec{v}^{n+1} = \vec{v}^{in} - \vec{z} = \vec{V} \tag{5.177}$$

$$\boldsymbol{\sigma}^{n+1} = \boldsymbol{\sigma}^{in} \pm \rho \left(\left(\vec{z} \cdot \vec{n} \right) \left(\left(c_P - 2c_S - c_3 \right) \mathbf{N}_{00} + c_3 \mathbf{I} \right) + c_S \left(\vec{n} \otimes \vec{z} + \vec{z} \otimes \vec{n} \right) \right) \tag{5.178}$$

In Eqs. 5.177, 5.178, $\vec{z}$ is given by Eq. 5.179.

$$\vec{z} = \vec{v}^{in} - \vec{V} \tag{5.179}$$

The boundary corrector of the first mixed conditions with given normal projection of the boundary velocity and the tangential component of the traction is considered. In this case, Eq. 5.105 takes the form of Eqs. 5.118–5.120. Note that $\vec{f}_\tau$ is scalar in a 2D case. In the case, when the normal is collinear to the direction of the splitting, the traction can be calculated from Eq. 5.180.

$$\vec{f} = \vec{f}_\tau + \vec{n} \left(\mp \rho c_P V_p + \left(\mp \left(\boldsymbol{\sigma}^{in} \cdot \vec{n} \right) - \rho c_P \vec{v}^{in} \right) \cdot \vec{n} \right) \tag{5.180}$$

Then this traction is substituted into Eqs. 5.169–5.171 for calculation of the boundary corrector with the given traction.

If the normal is collinear to the direction of the splitting, then the force density is computed using Eq. 5.181.

$$\vec{f} = \vec{f}_\tau + f_p \vec{p} \tag{5.181}$$

Further, this traction is substituted into Eqs. 5.172–5.176 for calculation of the boundary corrector with the given traction. In Eq. 5.181, the following quantity f_p is denoted by Eq. 5.182.

$$f_p = -\frac{A_M + B_M \left(\vec{n} \cdot \vec{f}_\tau \right)}{\left(B_M + C_M \right) \left(\vec{n} \cdot \vec{p} \right)} \tag{5.182}$$

In Eq. 5.182, the following notations given by Eqs. 5.183–5.185 are introduced.

$$C_M = c_P - c_S \tag{5.183}$$

$$B_M = \left(2c_S + c_3 - c_P \right) \left(\vec{n} \cdot \vec{p} \right)^2 - c_3 \tag{5.184}$$

$$A_M = \pm\rho c_S \left((c_P + c_3) \left(\vec{n} \cdot \vec{p}\right)^2 - c_3 \right) \cdot \left(V_p - \left(\vec{v}^{in} \cdot \vec{p} \right) \right)$$
$$\pm B_M \left(\boldsymbol{\sigma}^{in} \div \mathbf{N}_{00}\right) \pm C_M \left(\vec{n} \cdot \vec{p}\right) \left(\vec{p} \cdot \left(\boldsymbol{\sigma}^{in} \cdot \vec{n}\right)\right) \quad (5.185)$$

Also, the boundary corrector of the second mixed conditions with the given tangential component of the velocity of the boundary and the normal projection of the traction is considered. In this case, Eq. 5.105 takes the form of Eqs. 5.186–5.188.

$$\vec{V} = \vec{V}_\tau \quad (5.186)$$

$$\boldsymbol{\sigma} \cdot \vec{p} = \vec{f} \quad (5.187)$$

$$\vec{f} \cdot \vec{p} = f_p \quad (5.188)$$

In Eq. 5.186, $\vec{V}_\tau$ is the tangential component of the velocity of the boundary being a given vector. In Eq. 5.188, f_p is the projection of the traction on the normal being a given scalar. Note that $\vec{V}_\tau$ is scalar in a 2D case.

In the case, when the normal is collinear to the direction of the splitting, the velocity $\vec{V}$ is calculated using Eq. 5.189.

$$\vec{V} = \vec{V}_\tau + \vec{n} \left(\vec{v}^{in} \cdot \vec{n} \pm \frac{1}{\rho c_P} \left(\vec{n} \cdot \boldsymbol{\sigma}^{in} \cdot \vec{n} - f_p \right) \right) \quad (5.189)$$

Then this velocity $\vec{V}$ is substituted into Eqs. 5.177–5.179 for calculation of the boundary condition with the given boundary velocity.

If the normal is not collinear to the direction of the splitting, then the velocity is calculated by Eq. 5.190.

$$\vec{v} = \vec{V}_\tau + V_p \vec{p} \quad (5.190)$$

In Eq. 5.190, the quantity that can be found from Eq. 5.191 is introduced.

$$V_p = \frac{D_M}{E_M} \quad (5.191)$$

In Eq. 5.191, the notations D_M, E_M are introduced by Eqs. 5.192, 5.193.

$$D_M = \left((c_P - 2c_S - c_3) \left(\vec{n} \cdot \vec{p}\right)^2 + c_3 \right) \cdot \left(\left(\vec{v}^{in} - \vec{V}_\tau \right) \cdot \vec{n} \right)$$
$$+ 2c_S \left(\vec{n} \cdot \vec{p}\right) \left(\vec{v}^{in} \cdot \vec{p}\right) \mp \frac{1}{\rho} \left(f_p - \vec{p} \cdot \boldsymbol{\sigma}^{in} \cdot \vec{p} \right) \quad (5.192)$$

$$E_M = \left((c_P - 2c_S - c_3) \left(\vec{n} \cdot \vec{p}\right)^2 + 2c_S + c_3 \right) \left(\vec{n} \cdot \vec{p}\right) \quad (5.193)$$

For the nonreflecting boundary corrector, Eqs. 5.109, 5.110 take the form of Eqs. 5.141, 5.142, respectively.

The interface corrector of continuity of velocity and traction is considered. The conditions on the interface are given by Eqs. 5.143, 5.144. In the case, when the normal $\vec{p}$ is collinear to the direction of the splitting $\vec{n}$, the velocity $\vec{V}$ is calculated using Eq. 5.194.

$$\vec{V} = \frac{1}{\rho^a c_S^a + \rho^b c_S^b} \left\{ \rho^a \left(\left(\vec{p} \cdot \vec{v}^{a,in} \right) \left(c_P^a - c_S^a \right) \vec{p} + c_S^a \vec{v}^{a,in} \right) \right.$$
$$+ \rho^b \left(\left(\vec{p} \cdot \vec{v}^{b,in} \right) \left(c_P^b - c_S^b \right) \vec{p} + c_S^b \vec{v}^{b,in} \right) - \left(\boldsymbol{\sigma}^{a,in} - \boldsymbol{\sigma}^{b,in} \right) \cdot \vec{p}$$
$$- \frac{\rho^a \left(c_P^a - c_S^a \right) + \rho^b \left(c_P^b - c_S^b \right)}{\rho^a c_P^a + \rho^b c_P^b}$$
$$\left. \cdot \left(\vec{p} \cdot \left(\left(\rho^a c_P^a \vec{v}^{a,in} + \rho^b c_P^b \vec{v}^{b,in} \right) - \left(\boldsymbol{\sigma}^{a,in} - \boldsymbol{\sigma}^{b,in} \right) \cdot \vec{p} \right) \right) \right\} \vec{p} \tag{5.194}$$

Further, this velocity $\vec{V}$ is substituted into Eqs. 5.177–5.179 for calculation of the boundary condition with the given boundary velocity.

If the normal $\vec{p}$ is not collinear to the direction of the splitting $\vec{n}$, then the velocity $\vec{V}$ should be found using Eq. 5.195.

$$\vec{V} = \frac{\left(2 \left(\vec{n} \cdot \vec{p} \right)^2 \left(\rho^b c_S^b + \rho^a c_S^a + \rho^b c_3^b + \rho^a c_3^a \right) - B_C \right) \left(\vec{A}_C \cdot \vec{n} \right) \vec{n}}{B_C \left(\rho^b c_S^b + \rho^a c_S^a \right) \left(\vec{n} \cdot \vec{p} \right)}$$
$$- \frac{\left(\rho^b c_S^b + \rho^a c_S^a + \rho^b c_3^b + \rho^a c_3^a \right) \left(\vec{A}_C \cdot \vec{p} \right) \vec{n}}{B_C \left(\rho^b c_S^b + \rho^a c_S^a \right)} + \frac{\vec{A}_C}{\left(\rho^b c_S^b + \rho^a c_S^a \right) \left(\vec{n} \cdot \vec{p} \right)}$$
$$+ \frac{\left(\rho^b c_3^b + \rho^a c_3^a \right) \left(\vec{A}_C \cdot \vec{p} \right) \vec{p}}{B_C \left(\rho^b c_S^b + \rho^a c_S^a \right) \left(\vec{n} \cdot \vec{p} \right)} - \frac{2 \left(\rho^b c_3^b + \rho^a c_3^a \right) \left(\vec{A}_C \cdot \vec{n} \right) \vec{p}}{B_C \left(\rho^b c_S^b + \rho^a c_S^a \right)} \tag{5.195}$$

In Eq. 5.195, the following notations given by Eqs. 5.196, 5.197 are introduced.

$$\vec{A}_C = \left(\boldsymbol{\sigma}^{b,in} - \boldsymbol{\sigma}^{a,in}\right) \cdot \vec{p} + \left(\left(\rho^b c_S^b \vec{v}^{b,in} + \rho^a c_S^a \vec{v}^{a,in}\right) \cdot \vec{p}\right) \vec{n}$$
$$+ \left(\vec{n} \cdot \vec{p}\right) \left(\rho^b c_S^b \vec{v}^{b,in} + \rho^a c_S^a \vec{v}^{a,in}\right)$$
$$+ \rho^b \left(\vec{v}^{b,in} \cdot \vec{n}\right) \left(\left(c_P^b - 2c_S^b - c_3^b\right) \left(\vec{n} \cdot \vec{p}\right) \vec{n} + c_{b3} \vec{p}\right)$$
$$+ \rho^a \left(\vec{v}^{a,in} \cdot \vec{n}\right) \left(\left(c_P^a - 2c_S^a - c_3^a\right) \left(\vec{n} \cdot \vec{p}\right) \vec{n} + c_{a3} \vec{p}\right) \tag{5.196}$$

$$B_C = \left(\rho^b c_P^b + \rho^a c_P^a + \rho^b c_3^b + \rho^a c_3^a\right) \left(\vec{n} \cdot \vec{p}\right)^2 - \left(\rho^b c_3^b + \rho^a c_3^a\right) \tag{5.197}$$

The free sliding interface corrector is considered. The conditions on the interface are given by Eqs. 5.155–5.157. In the case, when the normal $\vec{p}$ is collinear to the direction of the splitting $\vec{n}$, one calculates V_p using Eq. 5.198.

$$V_p = \frac{\left(\rho^a c_P^a \vec{v}^{a,in} + \rho^b c_P^b \vec{v}^{b,in} - \left(\boldsymbol{\sigma}^{a,in} - \boldsymbol{\sigma}^{b,in}\right) \cdot \vec{p}\right) \cdot \vec{p}}{\rho^a c_P^a + \rho^b c_P^b} \tag{5.198}$$

Further, V_p and $\vec{f}_\tau = 0$ are substituted into the first mixed conditions given by Eq. 5.180 for medium a with an external normal $\vec{p}$. Values $-V_p$ and $\vec{f}_\tau = 0$ are substituted into the first mixed conditions given by Eq. 5.180 for medium b with an external normal $-\vec{p}$.

If the normal $\vec{p}$ is not collinear to the direction of the splitting $\vec{n}$, then V_p is calculated by Eq. 5.199.

$$V_p = \frac{B_F}{A_F} \tag{5.199}$$

In Eq. 5.199, the following notations are introduced by Eqs. 5.200, 5.201.

$$A_F = \left(B_M^a + C_M^a\right) \rho^b c_S^b \left(\left(c_P^b + c_3^b\right) \left(\vec{n} \cdot \vec{p}\right)^2 - c_3^b\right)$$
$$+ \left(B_M^b + C_M^b\right) \rho^a c_S^a \left(\left(c_P^a + c_3^a\right) \left(\vec{n} \cdot \vec{p}\right)^2 - c_3^a\right) \tag{5.200}$$

$$
\begin{aligned}
B_F = &-\left(B_M^a + C_M^a\right)\rho^b c_S^b\left(\left(c_P^b + c_3^b\right)\left(\vec{n}\cdot\vec{p}\right)^2 - c_3^b\right)\left(\vec{v}^{\,b,in}\cdot\vec{p}\right)\\
&-\left(B_M^b + C_M^b\right)\rho^a c_S^a\left(\left(c_P^a + c_3^a\right)\left(\vec{n}\cdot\vec{p}\right)^2 - c_3^a\right)\left(\vec{v}^{\,a,in}\cdot\vec{p}\right)\\
&+\left(B_M^b + C_M^b\right)B_M^a\left(\boldsymbol{\sigma}^{a,in} \div \mathbf{N}_{00}\right) - \left(B_M^a + C_M^a\right)B_M^b\left(\boldsymbol{\sigma}^{b,in} \div \mathbf{N}_{00}\right)\\
&+\left(B_M^b + C_M^b\right)C_M^a\left(\vec{n}\cdot\vec{p}\right)\left(\vec{p}\cdot\left(\boldsymbol{\sigma}^{a,in}\cdot\vec{n}\right)\right)\\
&-\left(B_M^a + C_M^a\right)C_M^b\left(\vec{n}\cdot\vec{p}\right)\left(\vec{p}\cdot\left(\boldsymbol{\sigma}^{b,in}\cdot\vec{n}\right)\right)
\end{aligned}
\tag{5.201}
$$

Further, values V_p and $\vec{f}_\tau = 0$ are substituted into the first mixed conditions given by Eqs. 5.181–5.185 for medium a with an external normal $\vec{p}$ and values $-V_p$ and $\vec{f}_\tau = 0$ are substituted into the first mixed conditions given by Eqs. 5.181–5.185 for medium b with an external normal $-\vec{p}$. In Eq. 5.201, parameters B_M^a, C_M^a, and B_M^b, C_M^b were introduced into Eqs. 5.183, 5.184 for media a and b, respectively.

5.4.4 Boundary and Interface Conditions for Acoustic Waves

The boundary condition with a given pressure is considered. The Eq. 5.105 takes the form of Eq. 5.202.

$$p^{n+1} = p \tag{5.202}$$

The boundary corrector will act according to Eqs. 5.203, 5.204.

$$\vec{v}^{\,n+1} = \vec{v}^{\,in} + \frac{p^{in} - p}{c\rho}\vec{p} \tag{5.203}$$

$$p^{n+1} = p. \tag{5.204}$$

The boundary condition with a given normal projection of the velocity is considered. Equation 5.105 takes the form of Eq. 5.205.

$$\vec{v}^{\,n+1}\cdot\vec{p} = V_p \tag{5.205}$$

The boundary corrector will act according to Eqs. 5.206, 5.207.

$$\vec{v}^{\,n+1} = \vec{v}^{\,in} - \left(\vec{v}^{\,in}\cdot\vec{p} - V_p\right)\vec{p} \tag{5.206}$$

$$p^{n+1} = p^{in} + c_1\rho\left(\vec{v}^{\,in}\cdot\vec{p} - V_p\right) \tag{5.207}$$

Free sliding interface corrector is considered and given by Eqs. 5.208, 5.209.

$$\vec{v}_a \cdot \vec{p} = \vec{v}_b \cdot \vec{p} = V_p \tag{5.208}$$

$$p^a = p^b \tag{5.209}$$

Firstly, one can find p using Eq. 5.210.

$$p = \frac{c^a \rho^a p^{in,b} + c^b \rho^b p^{in,a} - c^a \rho^a c^b \rho^b \left(\left(\vec{v}^{in,b} \cdot \vec{p} \right) - \left(\vec{v}^{in,a} \cdot \vec{p} \right) \right)}{c^a \rho^a + c^b \rho^b} \tag{5.210}$$

Substituting p into Eqs. 5.203, 5.204 for both media, one can obtain the action of this interface corrector.

5.5 Conclusions

The family of grid-characteristic methods for solving hyperbolic systems of equations is considered in this chapter. This family of numerical methods adapts successfully to the basic systems of equations used to solve applied problems of seismic oil and gas exploration, seismic resistance of various structures, ultrasonic non-destructive testing of railway tracks, ultrasonic studies of various materials, human body, and ultrasonic operations. In this chapter, cases of isotropic and anisotropic elastic waves and acoustic waves are considered in detail. The structure of the wave processes for an anisotropic elastic medium is also obtained and considered.

Acknowledgements This work has been performed at Non-state Educational Institution "Educational Scientific and Experimental Center of Moscow Institute of Physics and Technology" and supported by the Russian Science Foundation, grant no. 17-71-20088. This work has been carried out using computing resources of the federal collective usage center Complex for Simulation and Data Processing for Mega-science Facilities at NRC "Kurchatov Institute", http://ckp.nrcki.ru/.

References

1. Favorskaya A, Petrov I, Grinevskiy A (2017) Numerical simulation of fracturing in geological medium. Procedia Comput Sci 112:1216–1224
2. Favorskaya AV, Petrov IB, Vasyukov AV, Ermakov AS, Beklemysheva KA, Kazakov AO, Novikov AV (2014) Numerical simulation of wave propagation in anisotropic media. Dokl Math 90(3):778–780
3. Favorskaya A, Petrov I, Golubev V, Khokhlov N (2017) Numerical simulation of earthquakes impact on facilities by grid-characteristic method. Procedia Comput Sci 112:1206–1215
4. Favorskaya AV, Petrov IB (2016) Wave responses from oil reservoirs in the Arctic shelf zone. Dokl Earth Sci 466(2):214–217

5. Petrov IB, Favorskaya AV, Sannikov AV, Kvasov IE (2013) Grid-characteristic method using high order interpolation on tetrahedral hierarchical meshes with a multiple time step. Math Models Comput Simul 5(5):409–415
6. Biryukov VA, Miryakha VA, Petrov IB, Khokhlov NI (2016) Simulation of elastic wave propagation in geological media: intercomparison of three numerical methods. Comput Math Math Phys 56(6):1086–1095
7. Golubev VI, Petrov IB, Khokhlov NI (2015) Simulation of seismic processes inside the planet using the hybrid grid-characteristic method. Math Models Comput Simul 7(5):439–445
8. Favorskaya AV, Petrov IB (2016) A study of high-order grid-characteristic methods on unstructured grids. Numer Anal Appl 9(2):171–178
9. Beklemysheva KA, Favorskaya AV, Petrov IB (2014) Numerical simulation of processes in solid deformable media in the presence of dynamic contacts using the grid-characteristic method. Math Models Comput Simul 6(3):294–304
10. Vasyukov AV, Ermakov AS, Potapov AP, Petrov IB, Favorskaya AV, Shevtsov AV (2014) Combined grid-characteristic method for the numerical solution of three-dimensional dynamical elastoplastic problems. Comput Math Math Phys 54(7):1176–1189
11. Magomedov KM, Kholodov AS (1988) Grid characteristic methods. Nauka, Moscow
12. LeVeque R (2002) Finite volume methods for hyperbolic problems. Cambridge University Press, Cambridge
13. Carcione JM, Herman GC, Kroode APE (2002) Seismic modeling. Geophysics 67(4):1304–1325
14. Nikitin IS (2008) Dynamic models of layered and block media with slip, friction and separation. Mech Solids 43(4):652–661
15. Nikitin IS (2011) Constitutive relations for a growing masonry with setting mortar. Mech Solids 46(5):669–677
16. Nikitin I, Burago N (2016) A refined theory of the layered medium with the slip at the interface. Chapter 6 in book: continuous Media with Microstructure 2. In: Albers B, Kuczma M (eds) Springer International Publishing Switzerland, pp 77–94
17. Burago NG, Nikitin IS (2016) Improved model of a layered medium with slip on the contact boundaries. J Appl Math Mech 80(2):164–172
18. Nikitin IS, Burago NG, Nikitin AD (2017) Continuum model of the layered medium with slippage and nonlinear conditions at the interlayer boundaries. Solid State Phenom 258:137–140
19. Ainsworth M, Wajid HA (2009) Dispersive and dissipative behavior of the spectral element method. SIAM J Numer Anal 47(5):3910–3937
20. Ainsworth M, Monk P, Muniz W (2006) Dispersive and dissipative properties of discontinuous Galerkin finite element methods for the second-order wave equation. J Sci Comput 27(1):5–40
21. Alford RM, Kelly KR, Boore DM (1974) Accuracy of finite-difference modeling of the acoustic wave equation. Geophysics 39(6):834–842
22. Bohlen T, Saenger EH (2006) Accuracy of heterogeneous staggered-grid finite-difference modeling of Rayleigh waves. Geophysics 71(4):T109–T115
23. Chen JB (2014) A 27-point scheme for a 3D frequency-domain scalar wave equation based on an average-derivative method. Geophys Prospect 62(2):258–277
24. Cohen GC, Gaunaurd GC (2002) Higher-order numerical methods for transient wave equations. Scientific computation. Appl Mech Rev 55:B85
25. Seriani G, Oliveira SP (2008) Dispersion analysis of spectral element methods for elastic wave propagation. Wave Motion 45(6):729–744
26. Saenger EH, Gold N, Shapiro SA (2000) Modeling the propagation of elastic waves using a modified finite-difference grid. Wave Motion 31(1):77–92
27. Moczo P, Kristek J, Galis M, Pazak P (2010) On accuracy of the finite-difference and finite-element schemes with respect to P-wave to S-wave speed ratio. Geophys J Int 182(1):493–510
28. Moczo P, Kristek J, Gális M (2014) The finite-difference modelling of earthquake motions: waves and ruptures. Cambridge University Press, Cambridge
29. Alpert B, Greengard L, Hagstrom T (2002) Nonreflecting boundary conditions for the time-dependent wave equation. J Comput Phys 180(1):270–296

30. Alpert B, Greengard L, Hagstrom T (2000) Rapid evaluation of nonreflecting boundary kernels for time-domain wave propagation. SIAM J Numer Anal 37(4):1138–1164
31. Bécache E, Givoli D, Hagstrom T (2010) High-order absorbing boundary conditions for anisotropic and convective wave equations. J Comput Phys 229(4):1099–1129
32. Appelö D, Kreiss G (2006) A new absorbing layer for elastic waves. J Comput Phys 215(2):642–660
33. Belotserkovskii OM (2000) Modern solution methods for nonlinear multidimensional problems. The Edwin Mellen Press, Mathematics. Mechanics. Turbulence
34. Thomsen L (1995) Elastic anisotropy due to aligned cracks in porous rock. Geophys Prospect 43:805–829
35. Hsu C-J, Schoenberg M (1993) Elastic waves through a simulated fractured medium. Geophysics 58(7):964–977
36. Thomsen L (1986) Weak elastic anisotropy. Geophysics 51(10):1954–1966
37. Winterstein DF (1990) Velocity anisotropy terminology for geophysicists. Geophysics 55:1070–1088
38. Landau LD, Lifshitz EM (1959) Fluid mechanics. Pergamon Press, Oxford
39. Petrov IB, Favorskaya AV, Khokhlov NI, Miryakha VA, Sannikov AV, Golubev VI (2015) Monitoring the state of the moving train by use of high performance systems and modern computation methods. Math Models Comput Simul 7(1):51–61

Chapter 6
Numerical Modelling of Composite Delamination and Non-destructive Testing

Katerina A. Beklemysheva, Alexey V. Vasyukov, Alexander O. Kazakov and Alexey S. Ermakov

Abstract Delamination caused by low-velocity strike is considered as one of the most dangerous failure types. The destruction of contact between the plies or composite components significantly lowers the residual strength of the material but cannot be determined by visual inspection. These failures can mostly be determined by ultrasound testing, however, it requires a long time and cannot be carried out on site, which increases the maintenance cost. Both delamination emergence and ultrasound diagnostic results are determined by wave processes in viscoelastic media. The grid-characteristic method used in this chapter shows good results verified on various experimental data. The results of numerical modelling of delamination and its diagnostics are given in this chapter.

Keywords Numerical modelling · Grid-characteristic method · Delamination Non-destructive testing · Elastic waves · Ultrasound testing

6.1 Introduction

Massive use of composites in the bearing structures causes a number of problems. One of the most crucial problems is a development of fast and reliable method for non-destructive testing [1, 2]. Low-velocity strikes with the strike energy approximately less than 200 J occur very often during the maintenance, repair, takeoff, and landing

K. A. Beklemysheva (✉) · A. V. Vasyukov · A. O. Kazakov · A. S. Ermakov
Moscow Institute of Physics and Technology, 9 Intitutsky per., Dolgoprudniy 141700, Russian Federation
e-mail: amisto@yandex.ru

A. V. Vasyukov
e-mail: vasyukov@gmail.com

A. O. Kazakov
e-mail: alexanderkazak@yandex.ru

A. S. Ermakov
e-mail: fufler@gmail.com

A. V. Favorskaya and I. B. Petrov (eds.), *Innovations in Wave Processes Modelling and Decision Making*, Smart Innovation, Systems and Technologies 90,
https://doi.org/10.1007/978-3-319-76201-2_6

of an aircraft. A high-velocity strike causes visible deformations of material but a low-velocity strike leads to an internal damage, which is barely visible. After being exposed to a low-velocity strike, the composites behave differently than the homogenous materials. Even if a composite shows a high strength in standard tests comparable to the traditional aviation materials, it can be damaged in situations that are not modeled in standard tests. The destruction during a low-velocity strike can take place in the volume of material, between composite plies, or between composite components [3, 4]. It lowers the residual strength of the material without almost any visible damage that impedes its diagnostics.

The non-destructive testing assumes the detection of damaged areas without any irreversible actions towards the material. The visual examination is usually enough for homogenous materials because the considerable decline of material strength occurs only after visible deformations.

One of the composite peculiarities is a fast lowering of material strength with a failure of the internal composite structure (delamination between plies or composite components, matrix cracking). The exposure of a material with such failures to a dynamic or static operational load leads to its final destruction, even if the load strength was not enough to break an intact composite or a homogenous material of the same strengh. While significantly lowering the composite material strength, these internal failures are mostly invisible to the naked eye. This means that the strength norms and standards for the traditional aviation materials are ineffective for composites.

One of the most popular testing methods is the ultrasound testing. In homogenous material, the emitted elastic waves are propagated linearly in the material, then reflected from a crack or a backside, and returned to the sensor. The structure of the composite complicates the picture, because the signal of the sensor has a high level of noise caused by elastic waves that are reflected and refracted on internal contacts between materials [5, 6]. Different composite destruction types can absorb the ultrasound pulse (multiple matrix cracking) or let it through without any interaction (closed delaminations between plies).

The modelling of the ultrasound requires a numerical method that allows to obtain a full elastic wave pattern, considering the borders and contacts behavior with a high temporal and spatial resolution. We use the grid-characteristic numerical method [7–9] based on the characteristic parameters of the original system of equations that allows to model the elastic waves' propagation in the deformable solids and interactions of wavefronts with material borders and obtain the full solution of nonstationary contact problems. The method allows to obtain the components of the stress tensor and strain rate vector with the high temporal and spatial resolution in the modelling area.

Usually the aviation composite consists of several (from 2 to 6) subpackets, each of which consists of 11 orthotropic plies [10]. The subpacket behaves closely to an isotropic material because plies are aligned at different angles. In this chapter, we model the composite as several layers of an isotropic material because the plies number is enough to neglect the subpacket anisotropy during modelling. To obtain the accuracy for a quantitative comparison with an experiment, the additional measurements of the subpacket anisotropic behavior parameters are necessary.

In this chapter, both the formation of internal damage in the composite panel during impact and detection of model defects using ultrasonic testing are considered. The problem statement of delamination caused by multiple low-velocity strike is discussed in Sect. 6.2. Section 6.3 presents the numerical results of delamination caused by multiple low-velocity strike. The problem statement and numerical results of the direct problem of non-destructive testing are given in Sects. 6.4 and 6.5, respectively. Section 6.6 concluded the chapter.

6.2 Delamination Caused by Multiple Low-Velocity Strike: Problem Statement

Numerical modelling results for a single low-velocity strike are presented in [10, 11]. During a multiple strike, the elastic waves' pattern is more complex because the wave interference can alter the destruction pattern or lead to the appearance of new destruction areas. Prediction of these effects without modelling the full process only by single strike data is an extremely difficult task. Modelling of the full elastic waves' pattern in a composite during the multiple low-velocity strike can explain many effects and predict their appearance in experiments.

In this chapter, different number of the strikers is considered (one, two, and four). Each striker is modeled as a sphere. The case of a collision at an angle is also considered. Different modes of colliding, such as the simultaneous strike, strike with a short delay, and strike with a long delay, are examined. The length of a short delay was estimated from a characteristic time for the wave processes in a plate. That time is necessary for a longitudinal wave from the striker to reach the backside of the plate. In this case, the waves from multiple strikers still have time to interact. The length of a long delay was estimated from viscous decay processes: the second strike occurs, when the waves from the first one have already dissipated. As a result, the second strike is applied to a previously destructed material without any interaction with wave processes between different strikes.

Thereinafter, the delamination model is considered in Sect. 6.2.1. The calculation parameters are pointed in Sect. 6.2.2.

6.2.1 Delamination Model

One of the characteristic failure modes of the composite is a delamination that means the destruction of contact between the composite plies. This process can be modeled by modifying a contact condition without grid reorganization. The existing delamination models for Finite Element Modelling (FEM) [12–17] face the certain complexities, which require introducing the inner model parameters and lead to the non-physical oscillations. Grid-characteristic method allows for a much simpler contact and border problem statement. In this chapter, we suggest the following model.

Each contacting node can be either intact or failed. During the calculation, we consider only the state of the real node ignoring the virtual ones [8]. For a failed node, the friction contact algorithm is always used [18]. For an intact node, we follow these several steps:

Step 1. The preliminary step. We assume that a node is in the state of complete adhesion and calculate the components of the stress tensor and strain rate vector $(\overrightarrow{v}^*, \sigma^*)$ obtaining the force acting on the contacting node $(\vec{f}^* = \sigma^* \cdot \vec{n})$.

Step 2. The destruction criterion is $\left|\vec{f}^*\right| \geq f_0$, where f_0 is an adhesion strength. If the criterion is not met, then the preliminary step is considered to be correct and we proceed to the next node. Otherwise, we move to Step 3.

Step 3. The recalculation. If the destruction criterion is met, the node is marked as failed. The preliminary step is considered to be incorrect and we calculate the components of the stress tensor and strain rate vector according to the friction contact algorithm.

6.2.2 Calculation Parameters

Several problem statements are mentioned below:

- One striker:
  a. Perpendicular strike.
  b. Strike at an angle.
- Two strikers:
  a. Simultaneous strike.
  b. Strike with a short delay.
  c. Strike with a long delay.
- Four strikers:
  a. Simultaneous strike.
  b. Strike with a short delay.
  c. Strike with a long delay.

The composite is modeled as follows. Three layers are considered to be the elastic deformable bodies made of Carbon Fiber Reinforced Polymer (CFRP). The striker material is steel (Table 6.1). It is assumed that the contacts between layers are destructible with the strength of 42 MPa. Contacts between strikers and plate are friction contacts, $k = 0.1$. Three values of strike energy 50, 100, and 150 J are considered.

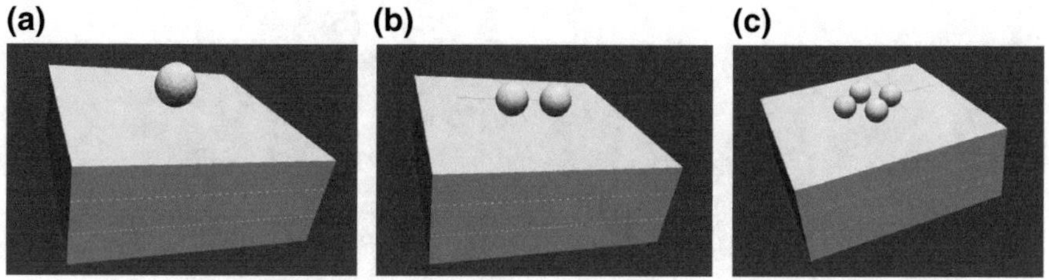

Fig. 6.1 General view: **a** one striker, **b** two strikers, **c** four strikers

The energy is distributed evenly between the strikers. Radius of the single striker is $r_{single} = 1.5$ mm, each of the two strikers is $r_{double} = 1.19$ mm, each of the four strikers is $r_{quad} = 0.94$ mm.

The general view of the calculation area is presented in Fig. 6.1. Time step was determined from Courant condition, $\tau = 1.91 \times 10^{-8}$ s. Length of the short delay is $t_m = 1.91 \times 10^{-6}$ s (100 time steps) and of the long delay is $t_b = 5.73 \times 10^{-6}$ s (300 time steps). For the strike at an angle, three angle values 10°, 30°, and 60° were considered.

6.3 Delamination Caused by Multiple Low-Velocity Strike: Numerical Results

Figures 6.2 and 6.3 present the calculation results for the single striker.

Figure 6.2 presents the strain rate vector modulus and delamination areas at different times in XZ section passing through the collision point. Strike energy is 50 J. Figure 6.3 presents the delamination areas for different values of strike energy.

The characteristic annular delamination form that can be seen in every image is caused by shear stresses on the contact. The front of the longitudinal wave emitted from the collision point is spherical. The compression wave directly beneath the collision point passes through the contact without damaging it. Moving away from the collision point, a force on the contact obtains tangential component increasing until the shear stress becomes high enough to destroy the contact. At the same time, the wave dissipates with the distance and distributes to a larger area lowering its amplitude until it is not enough to destroy the contact. The amplitude of tension waves from the backside and striker bouncing is not enough to destroy the contact.

Table 6.1 Elastic parameters of the plate and strikers

Material	E, GPa	ν	ρ, kg/m^3	λ, GPa	μ, GPa	c_p, m/s	c_s, m/s
CFRP (subpacket)	8.5	0.32	1580	5.72	3.22	2775	1425
Steel	200	0.28	7800	99.43	78.13	5725	3165

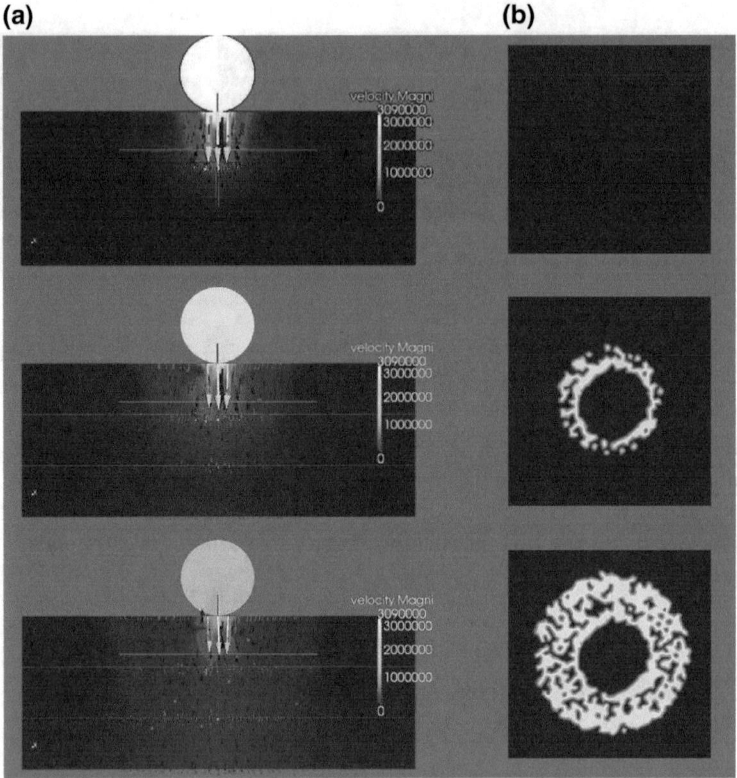

Fig. 6.2 Numerical results for the single striker, strike energy 50 J, section XZ passing through the collision point, and time steps 100, 200, and 300: **a** strain rate vector modulus, **b** delamination areas between upper and middle layers

Edge effects observed in experiments [2] can be seen on the upper layer near sides of the sample (Fig. 6.3).

Numerical results for two strikers with the delamination areas for different collision modes are presented in Fig. 6.4. In the case of a simultaneous collision with two strikers, the compression waves interfere. Shear stresses on the contact in the area between strikers compensate each other and the contact remains intact.

Other modes demonstrate an asymmetry caused by the time delay between strikes. Compression wave from the second striker does not reach the contact and the wave from the first striker destroys a larger area than after the simultaneous strike.

In the case of a long delay, the compression wave from the second striker reaches the contact after the formation of the destruction area from the first collision. Waves do not interfere and the second striker interacts with an already destroyed plate. The destroyed area is significantly higher than both in the case of simultaneous strike and in the case of a single strike of the same total energy.

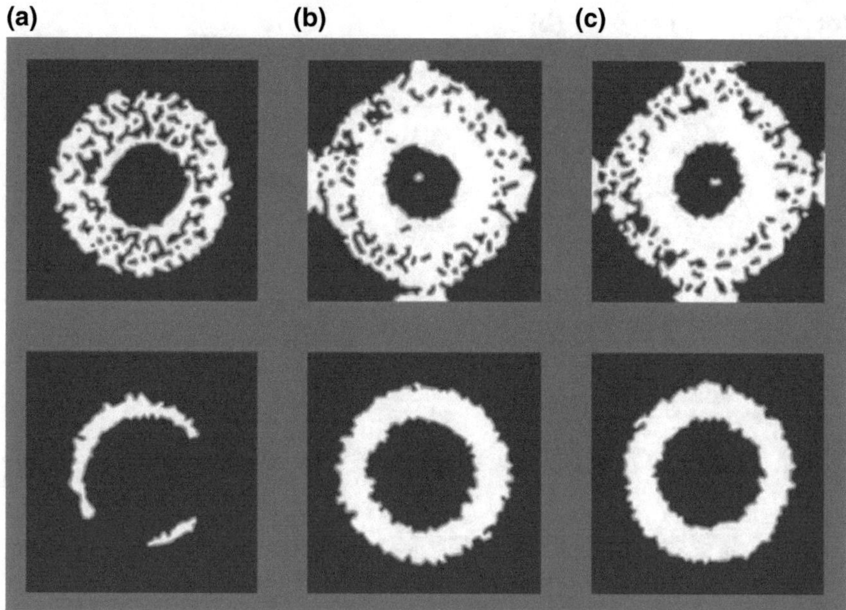

Fig. 6.3 Delamination areas for the single striker with strike energy (at the top—delamination between upper and middle layers, at the bottom—between middle and lower layers): **a** 50 J, **b** 100 J, **c** 150 J

Numerical results for four strikers with the delamination areas for different collision modes are presented in Fig. 6.5. These areas are significantly lower than in the case of two strikers. It is caused by the larger distance between strikers and wave interference between different strikes.

Delamination areas for a single striker and different strike angles are presented in Fig. 6.6. Increasing the strike angle leads to the reduction and shift of delamination area. At the angle of 30°, a considerable asymmetry can be observed. At the angle of 60°, the pattern changes completely. The amplitude of the compression wave is too low to cause delamination and the damage is mostly caused by the shear wave.

The foregoing results show that the delamination pattern after a multiple low-velocity strike is defined by the interference of elastic waves from the collision points. Both form and size of the delamination area is defined not only by the mass and velocity of strikers but also from the time delay between strikes. The collision of a composite plate with a part of complex shape can lead to similar effects.

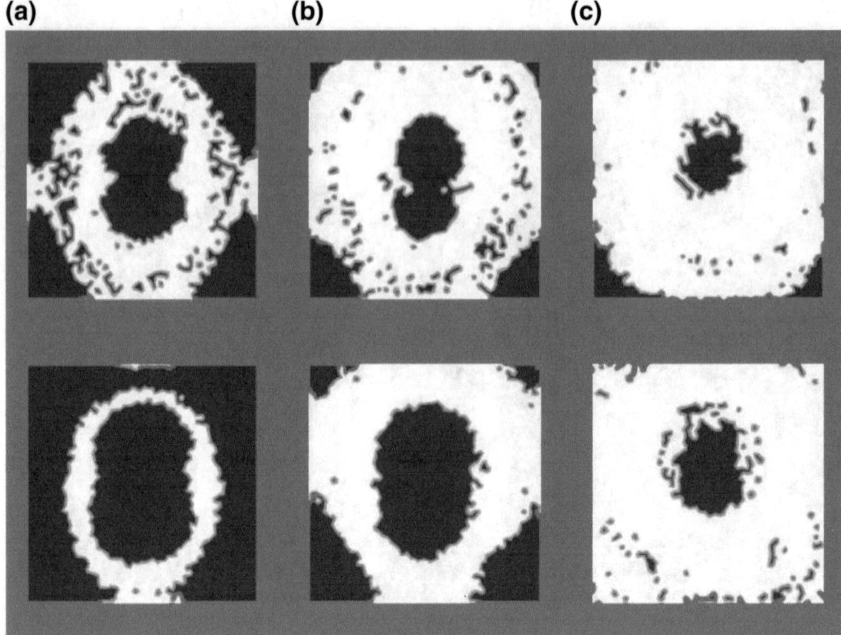

Fig. 6.4 Delamination areas for two strikers, strike energy 100 J (at the top—delamination between upper and middle layers, at the bottom—between middle and lower layers): **a** simultaneous strike, **b** small delay, **c** long delay

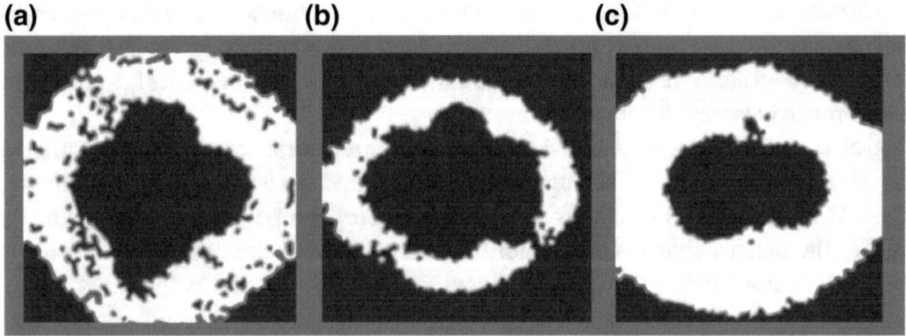

Fig. 6.5 Delamination areas between upper and middle layers for four strikers, strike energy 50 J: **a** simultaneous strike, **b** short delay, **c** long delay

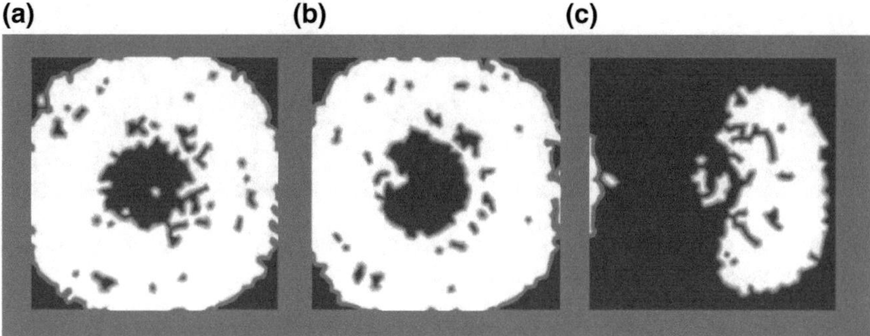

Fig. 6.6 Delamination areas between upper and middle layers for the angled strike, strike energy 50 J: **a** angle 10°, **b** angle 30°, **c** angle 60°

6.4 Direct Problem of Non-destructive Testing: Problem Statement

For a model problem, we took a two-layered composite with the layer width 3 mm and material parameters presented in Table 6.1. Three cases are considered: an intact composite plate, a plate with the radius delamination 2 mm, and a plate with the radius delamination 4 mm. General view of the lower layer is presented in Fig. 6.7, the cross-section of the calculation area is depicted in Fig. 6.8. Note that the axis Z is aligned perpendicular to the plate surface.

Fig. 6.7 Delamination areas, lower layer: **a** no delamination, **b** delamination radius 2 mm, **c** delamination radius 4 mm

Fig. 6.8 Delamination areas, XZ cross-section: **a** no delamination, **b** delamination radius 2 mm, **c** delamination radius 4 mm

Fig. 6.9 Initial state on the upper surface: **a** perpendicular strike (along the axis Z), **b** tangential strike along the axis X, **c** tangential strike along the axis Y

Fig. 6.10 Line of sensors

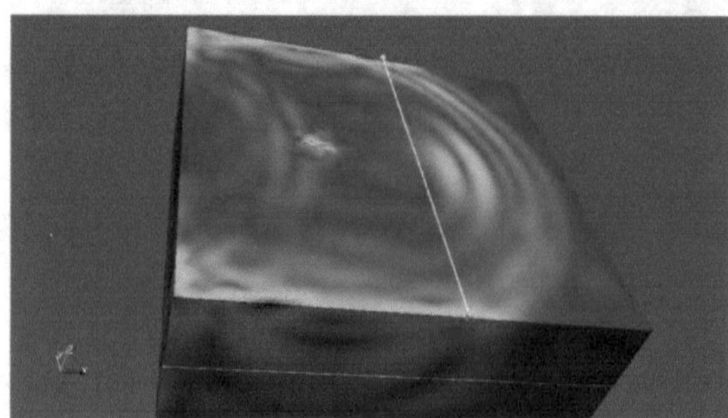

The diagnostic pulse is generated by the applying of an initial state of velocity in a small area of the upper surface of the plate. In real devices, the diagnostic pulse has a complex shape but the given statement is enough to obtain all types of waves generated by the surface strike and analyze their behavior with respect to the delamination area. Three strike directions are considered: a perpendicular strike (along the axis Z) and two tangential strikes (along the axes X and Y). Initial states on the upper surface are presented in Fig. 6.9. The line of sensors (aligned with the axis Y) on the upper surface and general view of the calculation area are presented in Fig. 6.10.

6.5 Direct Problem of Non-destructive Testing: Numerical Results

Consider the numerical results of the perpendicular strike, tangential strike along the axis X, and tangential strike along the axis Y in Sects. 6.5.1–6.5.3, respectively. Sensor line data is given in Sect. 6.5.4. The analysis of numerical results is presented in Sect. 6.5.5.

6.5.1 Perpendicular Strike

Distribution of velocity vector components on the upper surface for 300 time step is presented in Figs. 6.11, 6.12 and 6.13. The distribution of Z velocity vector component in the XZ cross-section for the consecutive time steps and all three considered types of delaminated area are presented in Fig. 6.14.

6.5.2 Tangential Strike Along the Axis X

The distribution of velocity vector components on the upper surface for 300 time step is presented on Figs. 6.15, 6.16 and 6.17. The distribution of X velocity vector component in the XZ cross-section for the consecutive time steps is presented on Fig. 6.18. The distribution of X velocity vector component on the upper surface for the consecutive time steps is presented on Fig. 6.19.

6.5.3 Tangential Strike Along the Axis Y

The distribution of velocity vector components on the upper surface for 300 time step is presented on Figs. 6.20, 6.21 and 6.22. The distribution of Y velocity vector component in the XZ cross-section for the consecutive time steps is depicted in Fig. 6.23. The distribution of Y velocity vector component on the upper surface for the consecutive time steps is presented in Fig. 6.24.

6.5.4 Sensor Line Data

Plots of velocity vector components along the sensor line for the consecutive time steps are presented in Fig. 6.25.

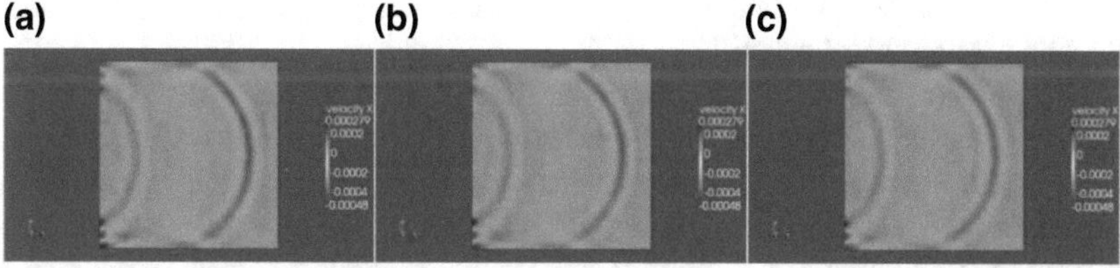

Fig. 6.11 Distribution of X velocity vector component on the upper surface for 300 time step: **a** intact material, **b** delamination area 2 mm, **c** delamination area 4 mm

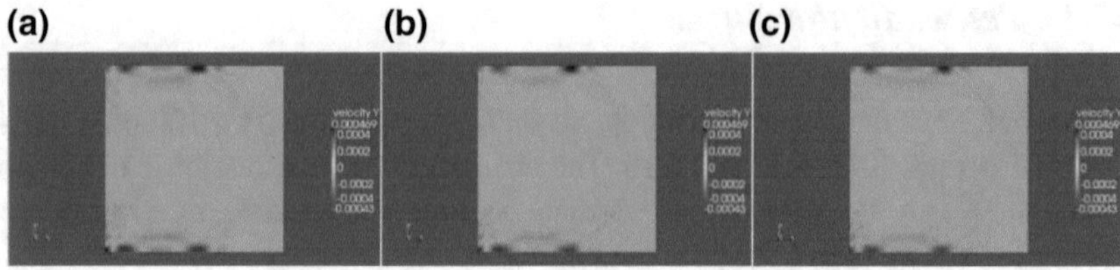

Fig. 6.12 Distribution of Y velocity vector component on the upper surface for 300 time step: **a** intact material, **b** delamination area 2 mm, **c** delamination area 4 mm

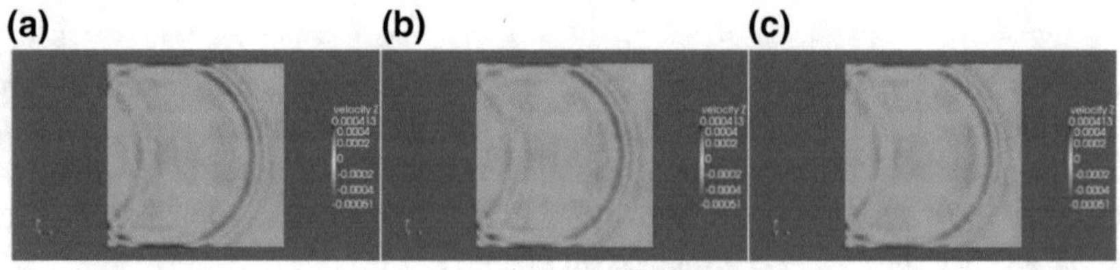

Fig. 6.13 Distribution of Z velocity vector component on the upper surface for 300 time step: **a** intact material, **b** delamination area 2 mm, **c** delamination area 4 mm

6.5.5 Analysis of Numerical Results

Figures 6.11, 6.12 and 6.13 show that the Z velocity vector component is the most suitable for the further analysis of data for the perpendicular strike. Its amplitude is higher and the difference between the intact and delaminated cases is more prominent than for other components. Similarly, Figs. 6.15, 6.16, 6.17, 6.20, 6.21 and 6.22 show that for a tangential strike it is most suitable to take the same velocity vector component as the strike direction.

Figure 6.14 shows the elastic waves propagating from the point of initial state application. The Rayleigh wave has the highest amplitude and is comparatively slow. It propagates along the plate upper surface. Two longitudinal waves (the compression and tension) propagate with the highest speed in a volume of the material and reflect from the plate backside. In the case of delamination, the compression wave passes through the contact and the tension wave reflects from it. Unlike the open fracture, which reflects all longitudinal waves, the delamination as a closed fracture reflects only a half longitudinal waves.

Figure 6.18 shows the wave patterns in the XZ cross-section for the tangential strike along the axis X formed by two wave groups. The longitudinal waves propagate to and from the sensor line. The slower shear wave propagates in the volume of the material and reflects from the backside. Both types of wave give the reflection from the delamination. Firstly, the longitudinal waves' reflection comes to the sensor. Figure 6.19 shows the reflection on the upper surface of the plate.

Figure 6.23 shows the wave patterns in the XZ cross-section for the tangential strike along the axis Y. Since the material is isotropic, the three-dimensional wave

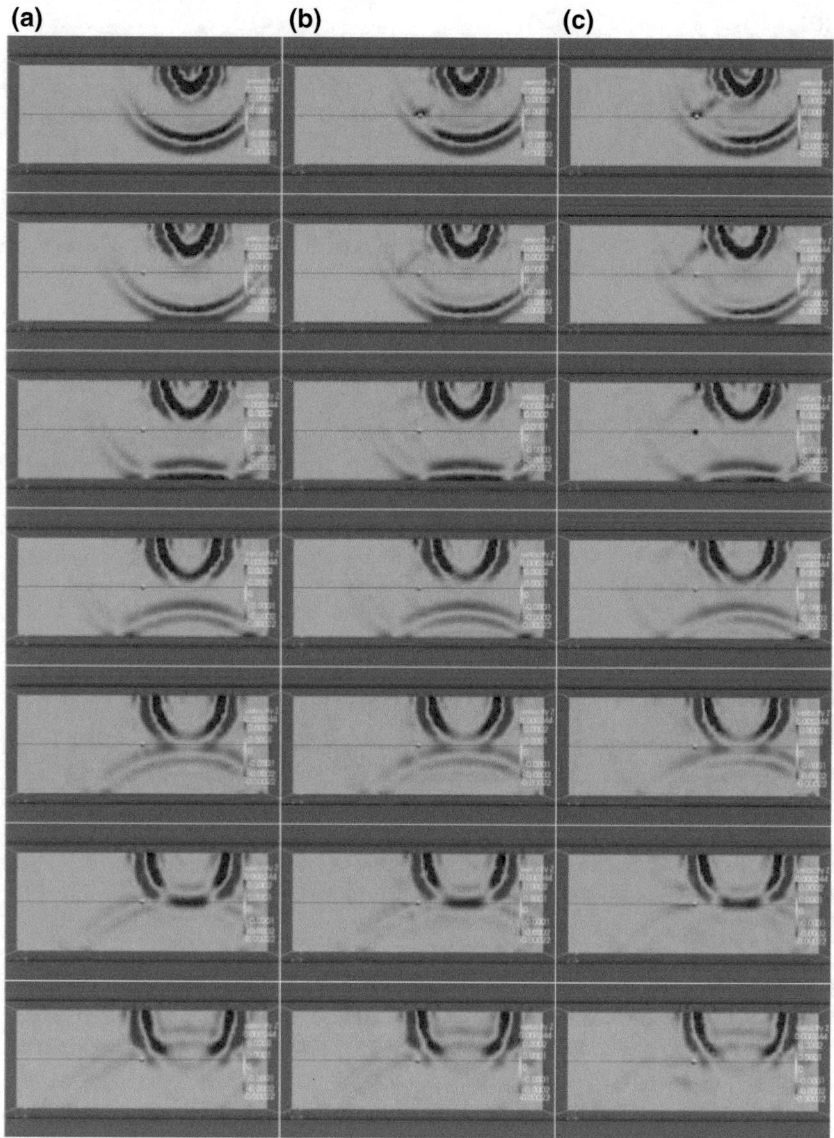

Fig. 6.14 Distribution of Z velocity vector component in the XZ cross-section, from top to bottom—time steps 50, 60, 70, 80, 90, 100, and 110: **a** intact material, **b** delamination area 2 mm, **c** delamination area 4 mm

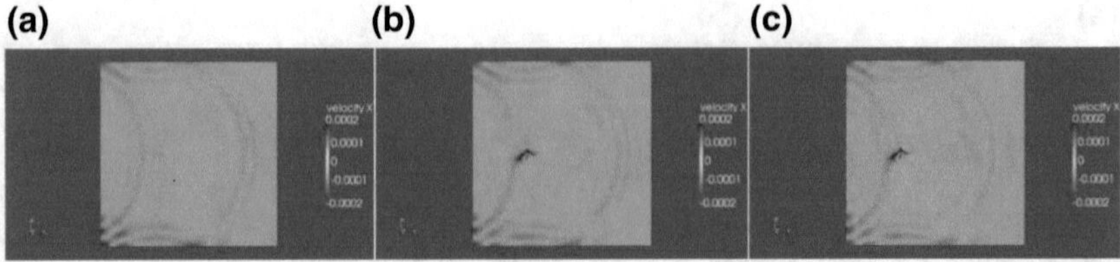

Fig. 6.15 Distribution of X velocity vector components on the upper surface for 300 time step: **a** intact material, **b** delamination area 2 mm, **c** delamination area 4 mm

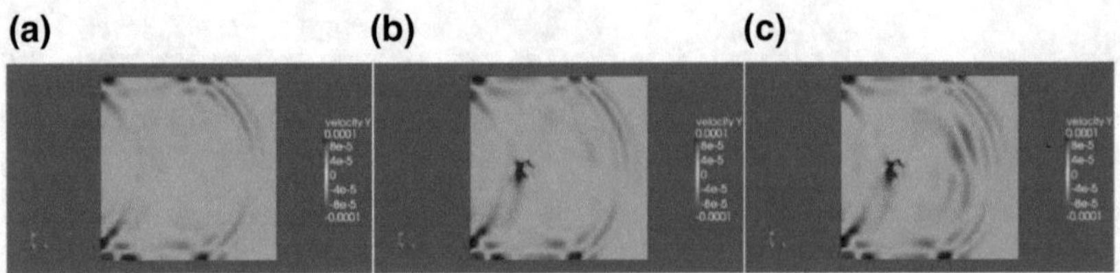

Fig. 6.16 Distribution of Y velocity vector components on the upper surface for 300 time step: **a** intact material, **b** delamination area 2 mm, **c** delamination area 4 mm

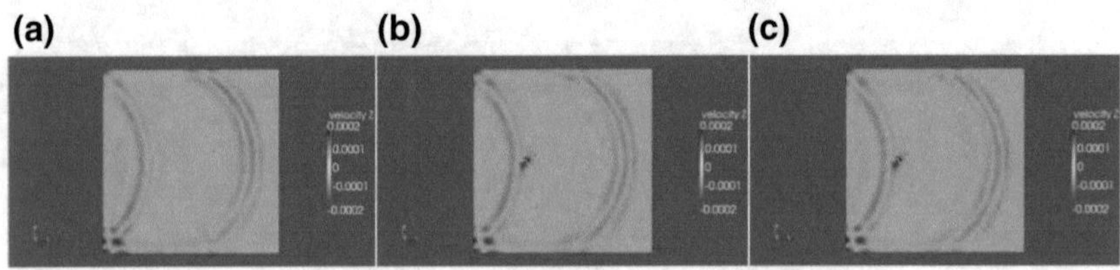

Fig. 6.17 Distribution of Z velocity vector components on the upper surface for 300 time step: **a** intact material, **b** delamination area 2 mm, **c** delamination area 4 mm

pattern in this case coincides with the pattern from the previous case. For better understanding, we look at the cross-section that is perpendicular to the sensor line. Unlike both of the previous cases, we can see only one group of waves in this cross-section.

Consider in detail the three-dimensional wave structure for this case. The shear wave propagates along the axis Z (to the backside) and along the axis X (to the line of sensors). The same wavefront fades into the longitudinal wave in the direction of the axis Y making the wavefront form elliptical. In other directions, the wavefront breaks into the longitudinal and shear groups propagating with different speeds. The amplitude ratio depends on the angle. The closer to the axis X, the higher is the amplitude of shear waves. A surface wave propagating along the axis Y can also be seen in Fig. 6.18, though with a lower amplitude than for the perpendicular strike (Fig. 6.14).

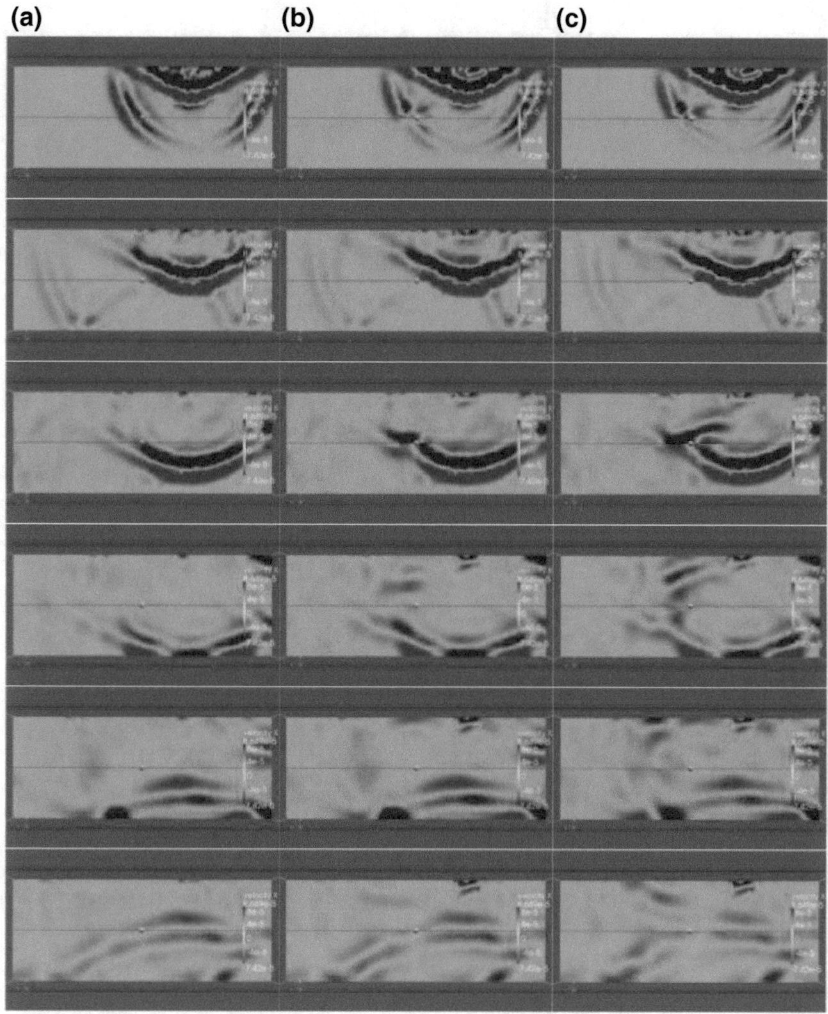

Fig. 6.18 Distribution of X velocity vector components in the XZ cross-section, from top to bottom—time steps 50, 100, 150, 200, 250, and 300: **a** intact material, **b** delamination area 2 mm, **c** delamination area 4 mm

All mentioned waves can be seen in Fig. 6.25. The first column of plots corresponds to the perpendicular strike. At time steps 50, 60, 70, and 80, we can see the direct wave from the diagnostic pulse. At the 90 time step, we can see the fracture response—a positive amplitude pulse that corresponds to the reflected tension wave. Afterward, we can see the passing of the Rayleigh wave. The second column of plots corresponds to the tangential strike along the axis X. At time steps 50, 60, 70, 80, 90, and 100, we can see the longitudinal wave. Further, at time steps 100–220 we

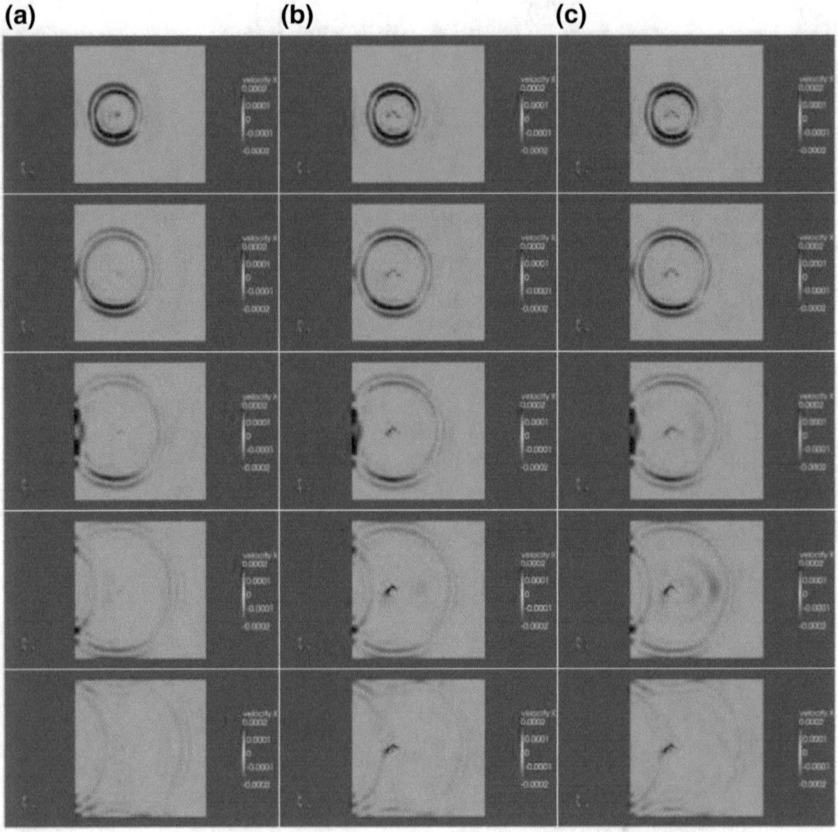

Fig. 6.19 Distribution of X velocity vector components on the upper surface, from top to bottom—time steps 50, 100, 150, 200, 250, and 300: **a** intact material, **b** delamination area 2 mm, **c** delamination area 4 mm

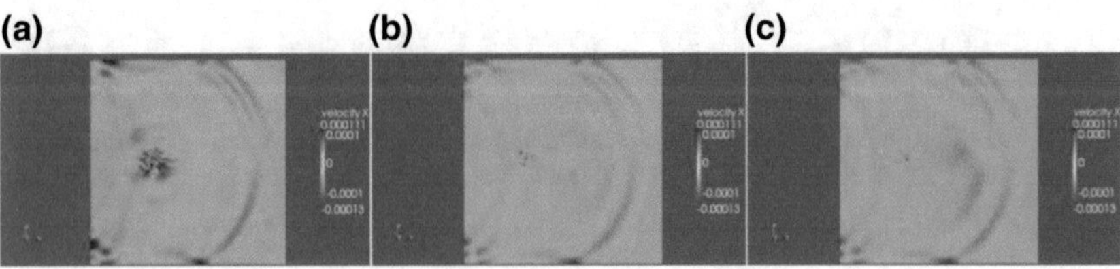

Fig. 6.20 Distribution of X velocity vector components on the upper surface for 300 time step: **a** intact material, **b** delamination area 2 mm, **c** delamination area 4 mm

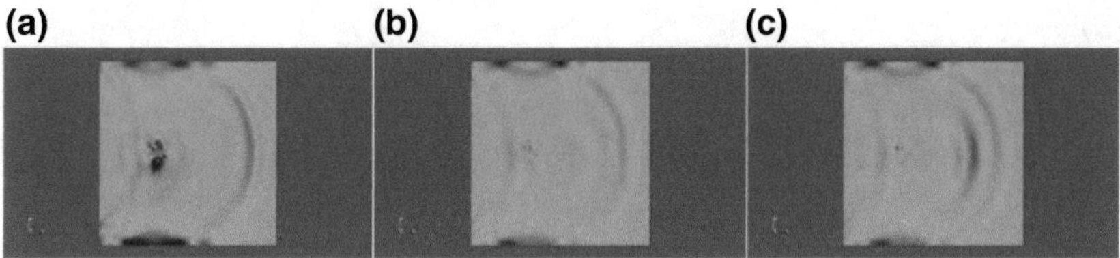

Fig. 6.21 Distribution of Y velocity vector components on the upper surface for 300 time step: **a** intact material, **b** delamination area 2 mm, **c** delamination area 4 mm

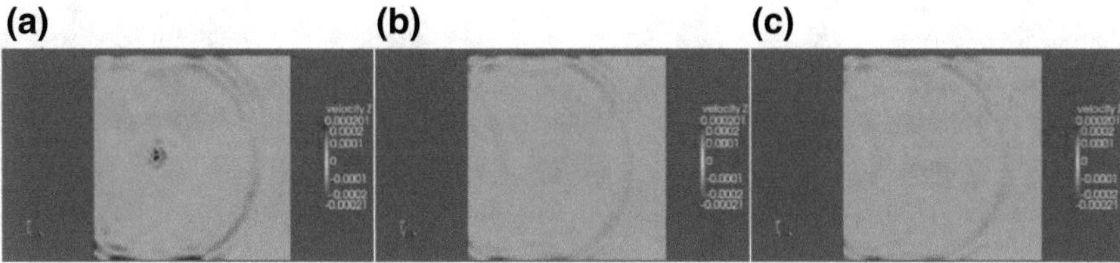

Fig. 6.22 Distribution of Z velocity vector components on the upper surface for 300 time step: **a** intact material, **b** delamination area 2 mm, **c** delamination area 4 mm

can see the passing of the surface wave. The reflection from the delamination can be seen at time steps 220–290. The third column of plots corresponds to the tangential strike along the axis Y. A considerable signal appears only from 120 time step—the passing of the surface wave. The reflection from the delamination can be seen at time steps 210–250.

All three cases demonstrate the significant influence of surface waves on the sensors signal. In some cases, this influence can be removed by placing the sensor on the opposite side but this method has a number of drawbacks. In addition to the technological complexity of adjusting the position and synchronization of the emitter and sensors, the difficulties arise in analyzing the obtained signals. A peculiarity of elastic waves noticed in seismological modelling [19–22] is that they restore the wavefront after passing the damaged area. In our task, this effect can be noticed in Figs. 6.14, 6.18 and 6.23. Consequently, the fracture diagnosis requires a thorough quantitative analysis, which can be performed in numerical experiments but meets a lot of complications in the case of a real material. The lack of data on the inner structure of the composite, especially on the random mistakes in a particular cover part, makes this analysis very inaccurate and unreliable.

The case of the perpendicular strike corresponds to real ultrasound devices. The difference between the intact and delaminated materials is much lower than the surface wave amplitude. Also, the reflected pulse has a relatively short duration. The tangential strikes and shear waves give higher reflected signal amplitude and longer duration, which reduces requirements for the ultrasound sensor. Also, the reflection is necessarily arrived the sensor after the surface wave, thus they cannot overlap.

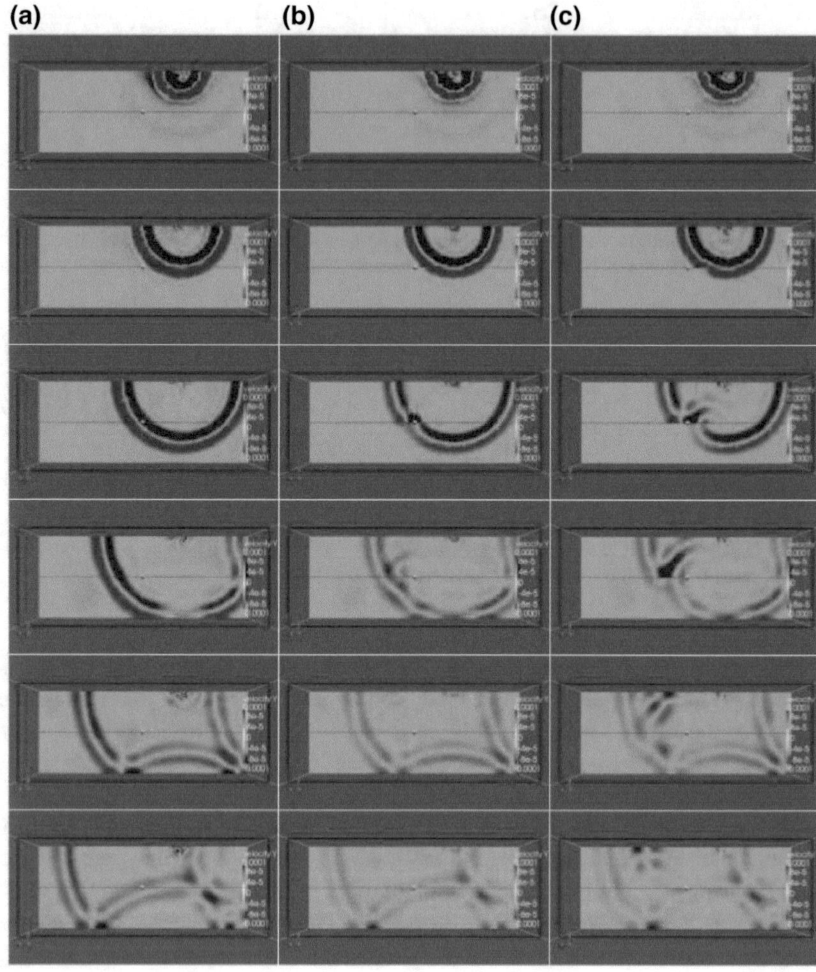

Fig. 6.23 Distribution of Y velocity vector components in the XZ cross-section, from top to bottom—time steps 50, 100, 150, 200, 250, and 300: **a** intact material, **b** delamination area 2 mm, **c** delamination area 4 mm

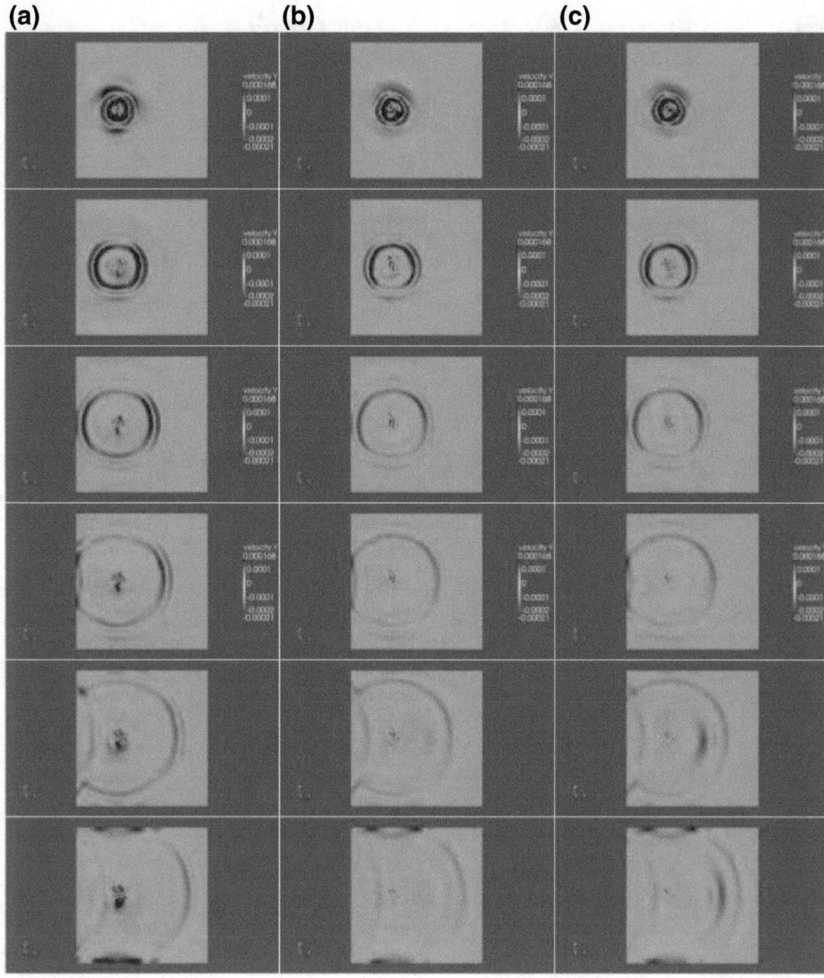

Fig. 6.24 Distribution of Y velocity vector components on the upper surface, from top to bottom—time steps 50, 100, 150, 200, 250, and 300: **a** intact material, **b** delamination area 2 mm, **c** delamination area 4 mm

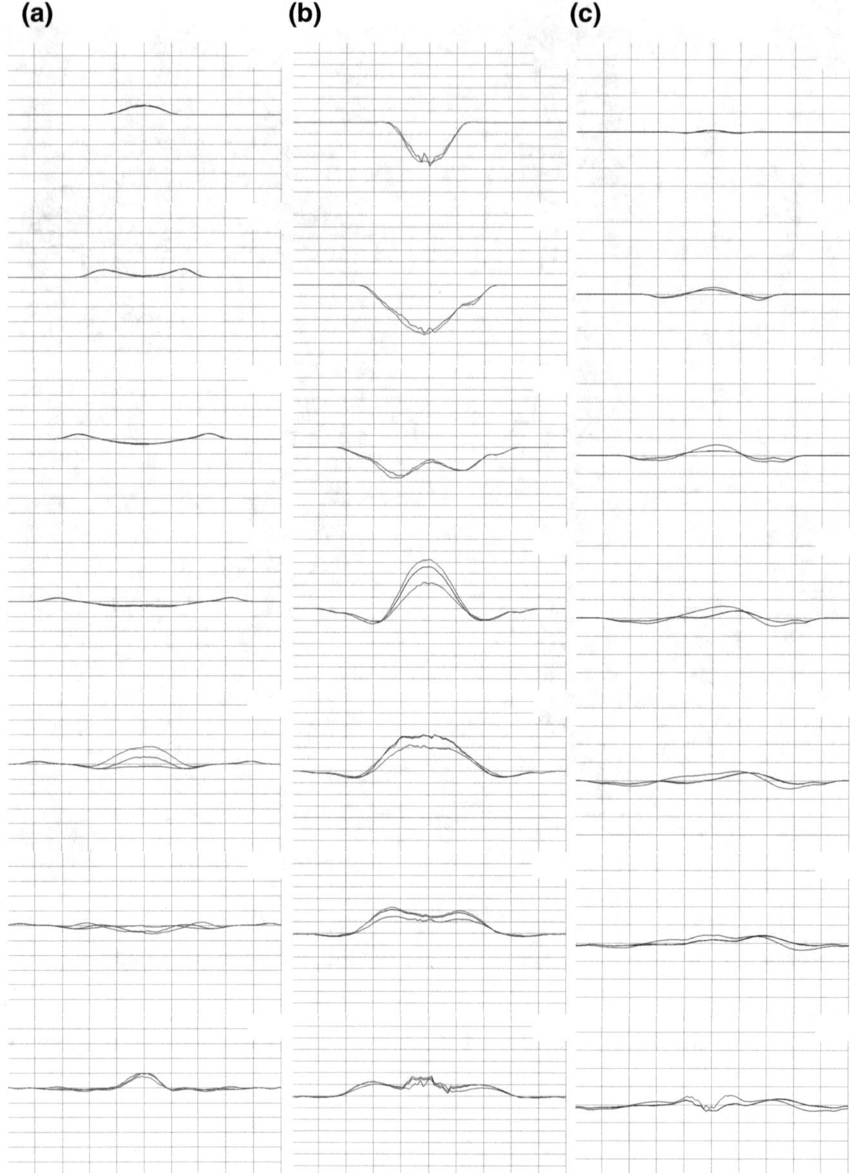

Fig. 6.25 Distribution of the velocity vector components on the sensor line for 50–300 time steps (blue line—the intact material, red line—the delamination 2 mm, orange line—the delamination 4 mm): **a** perpendicular strike (Z velocity vector component), **b** tangential strike along the axis X (X velocity vector component), **c** tangential strike along the axis Y (Y velocity vector component)

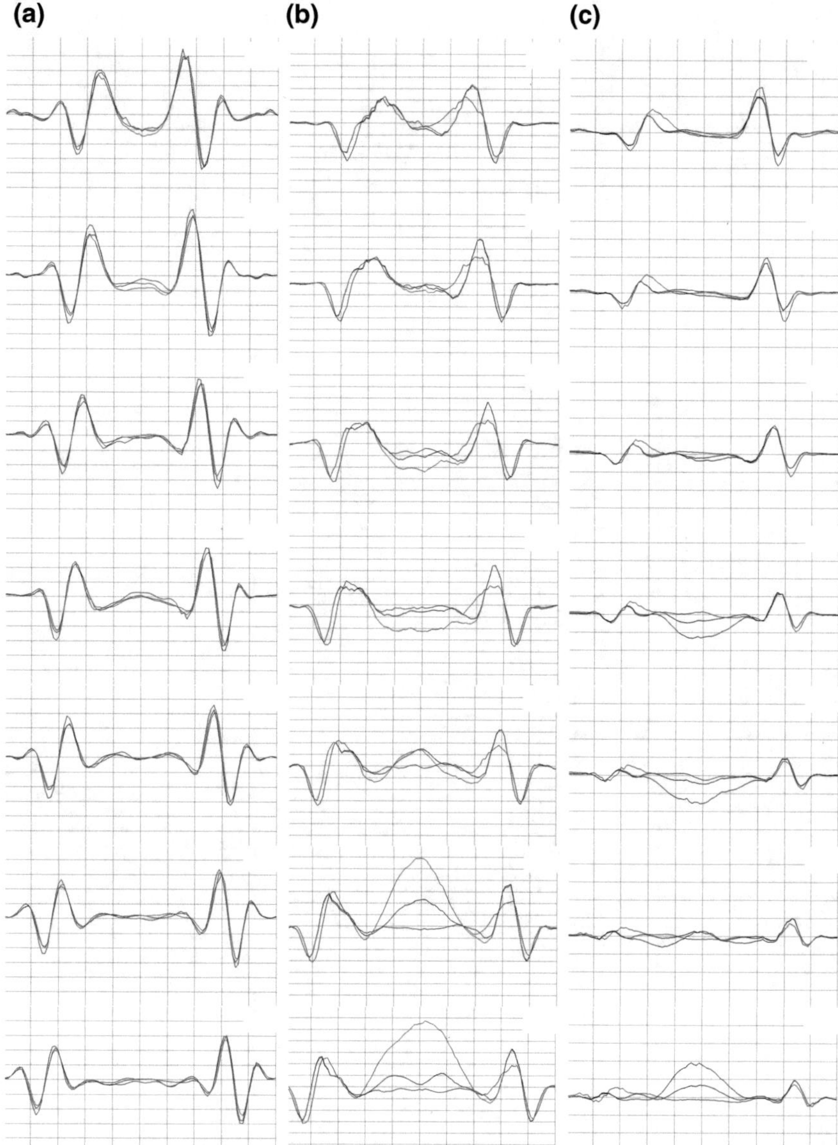

Fig. 6.25 (continued)

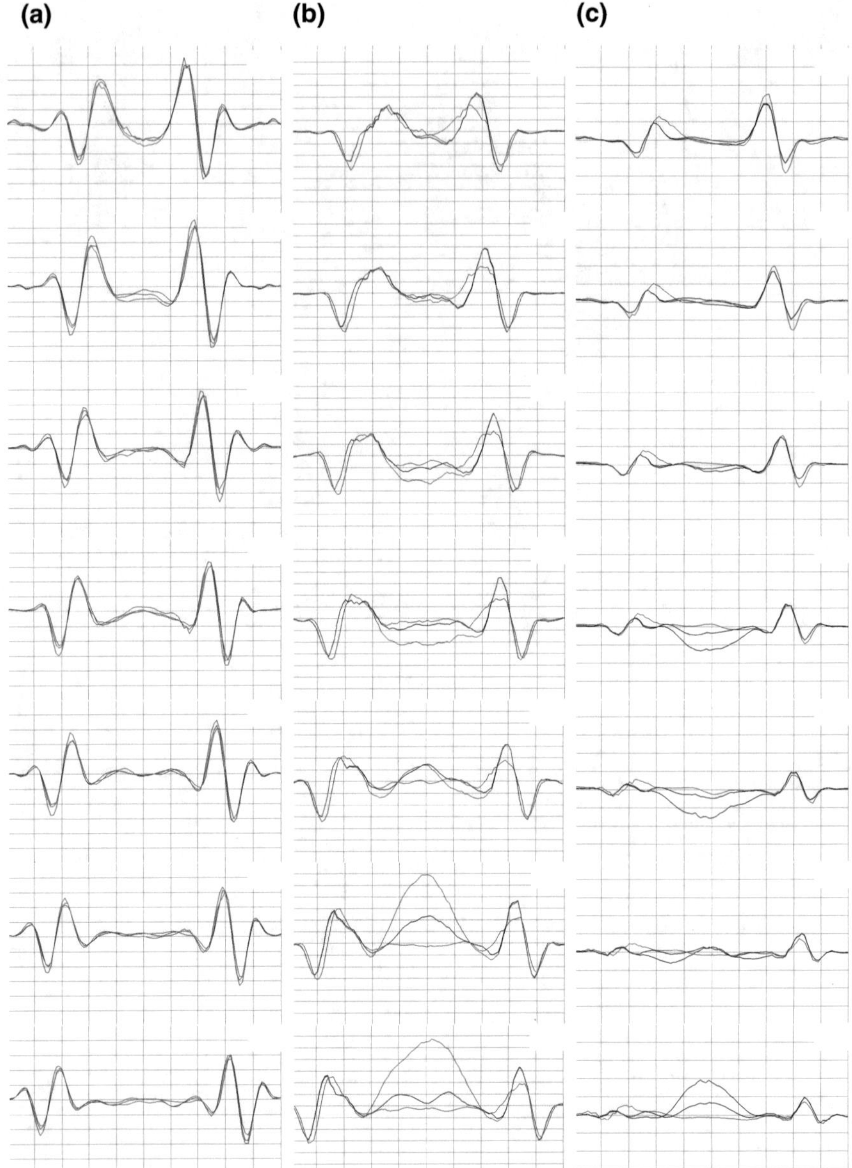

Fig. 6.25 (continued)

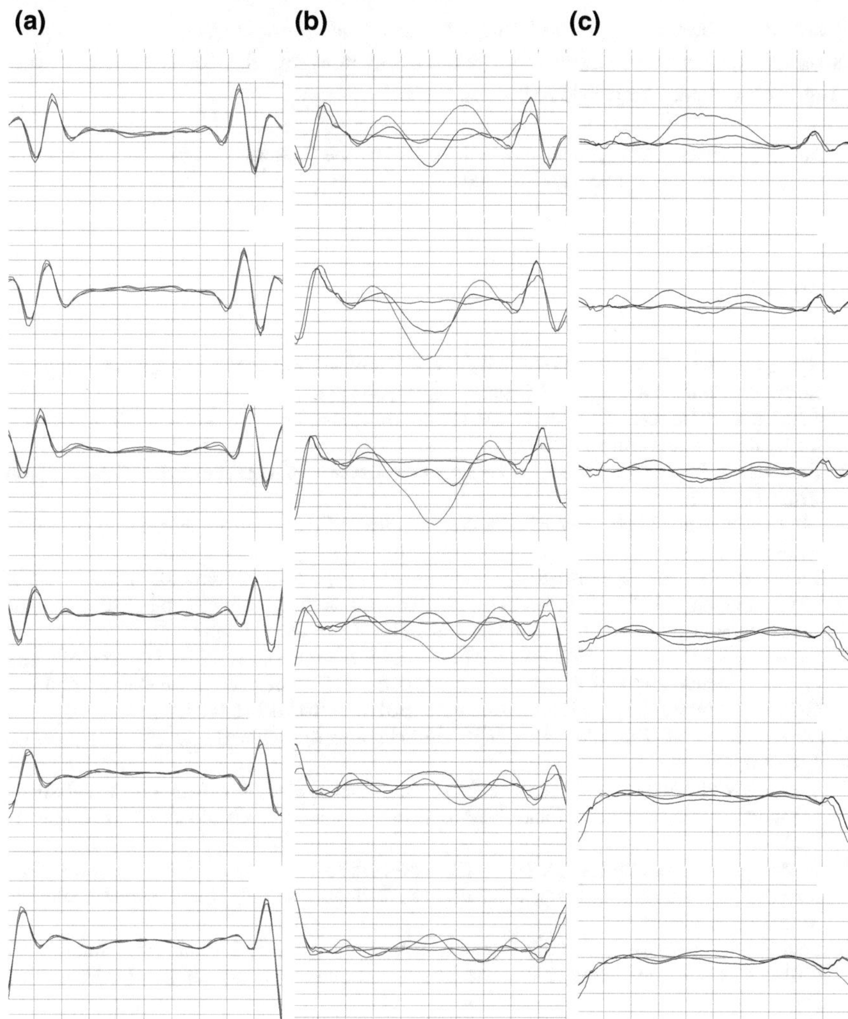

Fig. 6.25 (continued)

6.6 Conclusions

Results of the numerical modelling of delamination due to multiple low-velocity strike show that the delamination pattern depends on the number of strikers. This also means that strikers of the same mass and velocity but with a different shape (for example, different repair tools) can cause delamination of different size.

Results of the numerical modelling of ultrasound diagnostics show that the application of shear waves is advisable for the ultrasound non-destructive testing, especially for the detection of delamination areas.

Acknowledgements The reported study was funded by the Russian Fund for Basic Researches according to the research project № 16-07-00884 A.

References

1. Richardson MOW, Wisheart MJ (1996) Review of low-velocity impact properties of composite materials. Compos A-Appl Sci Manuf 27(12):1123–1131
2. Sjoblom PO, Hartness JT, Cordell TM (1988) On low-velocity impact testing of composite materials. J Compos Muter 22:30–52
3. Cantwell WJ, Morton J (1991) The impact resistance of composite materials. Compos 22(5):347–362
4. Liu D, Malvem LE (1987) Matrix cracking in impacted glass/epoxy plates. J Compos Mater 21:594–609
5. Raillon R, Toullelan G, Darmon M, Lonné S (2013) Experimental study for the validation of CIVA predictions in TOFD inspections. In: 10th international conference on NDE in relation to structural integrity for nuclear and pressurized components, 660–667
6. De Almeida PD, Rebello JMA, Pereira GR, Soares SD, Fernandez R (2013) Ultrasonic inspection of adhesive joints of composite pipelines. In: AIP conference proceedings, vol 1581, QNDE, Baltimore, USA, p 1069, 2014. http://dx.doi.org/10.1063/1.4864939
7. Magomedov KM, Kholodov AS (1988) Grid-characteristic numerical methods. Nauka (in Russian), Moscow
8. Chelnokov FB (2006) Explicit representation of grid-characteristic schemes for the elasticity equations in two-dimensional and three-dimensional spaces. Math Model 18(6):96–108 (in Russian)
9. Beklemysheva KA, Danilov AA, Petrov IB, Vassilevsky YV, Vasyukov AV (2015) Virtual blunt injury of human thorax: age-dependent response of vascular system. Russ J Numer Anal Math Model 30(5):259–268
10. Petrov IB, Vasyukov AV, Beklemysheva KA, Ermakov AS, Dziuba AS, Golovan VI (2014) Numerical modeling of low energy strike at composite stringer panel. Matem Mod 26(9):96–110
11. Beklemysheva KA, Ermakov AS, Petrov IB, Vasyukov AV (2016) Numerical simulation of the failure of composite materials by using the grid-characteristic method. Math Models Comput Simul 8(5):557–567
12. Hu N, Zemba Y, Okabe T, Yan C, Fukunaga H, Elmarakbi A (2008) A new cohesive model for simulating delamination propagation in composite laminates under transverse loads. Mech Mater 40(11):920–935
13. Hu N, Zemba Y, Fukunaga H, Elmarakbi A, Wangh HH (2007) Stable numerical simulations of propagations of complex damages in composite structures under transverse loads. Compos Sci Technol 67:752–765
14. Hou JP, Petrinic N, Ruiz C (2001) A delamination criterion for laminated composites under low-velocity impact. Compos Sci Technol 61(14):2069–2074
15. Geubelle PH, Baylor JS (1998) Impact-induced delamination of composites: a 2D simulation. Compos B: Eng 29(5):589–602
16. Reddy ED Jr, Mello FJ, Guess TR (1996) Modeling the initiation and growth of delaminations in composite structures. J Compos Mater 31(8):812–831

17. Mi Y, Crisfield MA, Davis GAO (1998) Progressive delamination using interface element. J Compos Mater 32(14):1246–1272
18. Petrov IB, Favorskaya AV, Beklemysheva KA (2014) Numerical simulation of processes in solid deformable media in the presence of dynamic contacts using the grid-characteristic method. Math Models Comput Simul 6(3):294–304
19. Petrov IB, Kvasov IE (2012) High-performance computer simulation of wave processes in geological media in seismic exploration. Comput Math Math Phys 52(2):302–313
20. Favorskaya AV, Petrov IB (2016) Wave responses from oil reservoirs in the Arctic shelf zone. Dokl Earth Sci 466(2):214–217
21. Favorskaya A, Petrov I, Grinevskiy A (2017) Numerical simulation of fracturing in geological medium. Procedia Comput Sci 112:1216–1224
22. Favorskaya A, Petrov I, Khokhlov N (2016) Numerical modeling of wave processes during shelf seismic exploration. Procedia Comput Sci 96:920–929

Chapter 7
Wave Processes Modelling in Geophysics

Alena V. Favorskaya, Nikolay I. Khokhlov, Vasiliy I. Golubev, Anton V. Ekimenko, Yurij V. Pavlovskiy, Inga Yu. Khromova and Igor B. Petrov

Abstract In this chapter, the application of the grid-characteristic method to solving seismic prospecting problems is considered. The characteristic seismogeological models, including Marmousi and SEG/EAGE Salt Model, are considered, wave patterns and seismograms are presented. The cases of 2D and 3D modelling, curvilinear boundaries between geological layers, fractured layers taking into account the topology of the Earth's surface, construction of seismograms for both 2D and

A. V. Favorskaya (✉)
Non-State Educational Institution "Educational Scientific and Experimental Center of Moscow Institute of Physics and Technology", 9, Institutsky Pereulok st., Dolgoprudny, Moscow Region 141700, Russian Federation
e-mail: aleanera@yandex.ru

A. V. Favorskaya · N. I. Khokhlov · V. I. Golubev · I. B. Petrov
Department of Computer Science and Numerical Mathematics, Moscow Institute of Physics and Technology, 9, Institutsky Pereulok st., Dolgoprudny, Moscow Region 141700, Russian Federation
e-mail: k_h@inbox.ru

V. I. Golubev
e-mail: w.golubev@mail.ru

I. B. Petrov
e-mail: petrov@mipt.ru

A. V. Favorskaya · N. I. Khokhlov · V. I. Golubev · I. B. Petrov
Scientific Research Institute for System Studies of the Russian Academy of Sciences, 36(1), Nahimovskij ave., Moscow 117218, Russian Federation

A. V. Ekimenko · Y. V. Pavlovskiy
Gazpromneft Science & Technology Centre, 75-79 liter D Moika River emb., St Petersburg 19000, Russian Federation
e-mail: ekimenko.AV@gazpromneft-ntc.ru

Y. V. Pavlovskiy
e-mail: pavlovskiy.YuV@gazpromneft-ntc.ru

I. Y. Khromova
Independent Adviser, Expert of RF State Reserves Committee, 50-3-188 Novocheremushkinskaya st., Moscow 117418, Russian Federation
e-mail: ingakhr@inbox.ru

A. V. Favorskaya and I. B. Petrov (eds.), *Innovations in Wave Processes Modelling and Decision Making*, Smart Innovation, Systems and Technologies 90,
https://doi.org/10.1007/978-3-319-76201-2_7

3D seismic survey cases, and vertical seismic profiling are considered. The investigation of the performance of a software complex developed on the basis of a grid-characteristic method for modelling hydrocarbon deposits of various computational complexity was performed. Also, faults zones of a different nature, both about the length of the faults and the type of geological environment inside these faults, was studied. A detailed analysis of spatial dynamic wave patterns is carried out, and predictive conclusions are made about the nature of the seismograms obtained, which were actually confirmed in the respective seismograms. It will be shown in this Chapter that typical analytical tests cannot guarantee that software gives an opportunity for the geologist to develop right conclusions. This problem can be solved only by understanding the physical basis of the phenomena under consideration and the peculiarities of the operation of the difference methods used in the software, simultaneously. This suggests the method called Wave Logica, fragments of which are also given in the Chapter.

Keywords Grid-characteristic method · Numerical method
Linear-elastic media · Elastic waves · Geological media · Seismic waves
Waves modelling · Seismic prospecting · Seismic exploration · Oil · Gas · Faults

7.1 Introduction

Numerical solution of seismic exploration problems has a large-scale application for the exploration of the Earth's and interior of other planets of Solar system [1–3]. These problems are the tasks of increased computational complexity in complete three-dimensional definitions, taking into account the contact boundaries of complex shapes, such as the seabed, the surface of the well, the surface of the Earth, and various inclusions such as fractured zones and faults. It is important that the numerical method takes into account all types of waves, such as Rayleigh and Love waves [4], Krauklis waves [5–7], SP-waves, PS-waves, etc. Because these types of waves are observed in the field, and if the numerical method does not allow them to be modeled, it is obvious that the results of modelling can differ significantly from physical experiments, which will substantially limit the use of computer simulation for the study of seismic phenomena and their patterns in a case of various geophysical and hydrocarbon objects.

For the calculation of synthetic seismograms, the family of ray-tracing methods is widely used [8–10]. An opportunity of applying finite-difference approach for seismograms modelling is discussed in [11]. Some other methods are used as well [12–15]. A shot review of different methods for full-wave numerical simulation one can read in Chap. 5. Also one can find a comparison of different methods in [16, 17]. Geological faults modelling is discussed in [18, 19]. The applications of finite-difference grid-characteristic method, which was described in Chap. 5 for the seismic waves modelling, are considered in [20–24].

The study of geophysical faults of different topology is discussed in Sect. 7.2. The results of seismic waves and seismograms simulation using different 2D and 3D seismogeological model, including Marmousi and SEG/EAGE Salt Model, are situated in Sect. 7.3. Section 7.4 gives the conclusions of the Chapter.

7.2 Study of Geophysical Faults

In this Section, the emphasis is on demonstrating the possibilities of analyzing dynamic spatial seismic wave patterns called Wave Logica [20–26]. Such analysis permits to localize useful information in seismograms. The recommendations obtained on the basis of the analysis of dynamic wave patterns are verified on an array of synthetic seismograms. The synthetic seismograms help to model the multilayered geological media and clusters of geological faults of various topologies.

In order to solve these problems, the faults in the geological environment of various lengths are considered. Wave processes occurring during the seismic exploration of this type of faults were modeled by solving a system of equations, which describes an elastic wave field using a grid-characteristic method by full wave numerical modelling with the explicit allocation of other geological material within these geological faults. One can find a detailed description of this system of equation and grid-characteristic numerical method into Chap. 5. Also, dynamic wave patterns are analyzed. Some conclusions were drawn about the localization of useful information in seismograms, which are confirmed by the seismograms constructed from the results of calculations.

In Sect. 7.2.1, a formulation of the problem is discussed, and Sect. 7.2.2 presents some conclusions illustrated by appropriate figures.

7.2.1 Formulation of the Problem

Seven calculations were considered, each containing a multi-layered geological environment (Fig. 7.1) consisting of 8 layers with different parameters of the medium. The system of the geological faults was placed into the in the sixth layer.

The seismic characteristics of these layers one can find in Table 7.1. The difference between models was in the different values of the coefficient K (0.5, 0.6, 0.75, 0.9, 1.0), which connects the parameters inside the faults with the parameters of the circling geological media (Fig. 7.1). In Eqs. 7.1–7.3, $c_{\mathrm{P}}^{\mathrm{Faults}}$, $c_{\mathrm{P}}^{\mathrm{Geo}}$ are the speeds of P-waves, $c_{\mathrm{S}}^{\mathrm{Faults}}$, $c_{\mathrm{S}}^{\mathrm{Geo}}$ are the speeds of S-waves, ρ^{Faults}, ρ^{Geo} are densities into the geological faults and into the geological environment around these faults, respectively.

$$c_{\mathrm{P}}^{\mathrm{Faults}} = K \cdot c_{\mathrm{P}}^{\mathrm{Geo}} \tag{7.1}$$

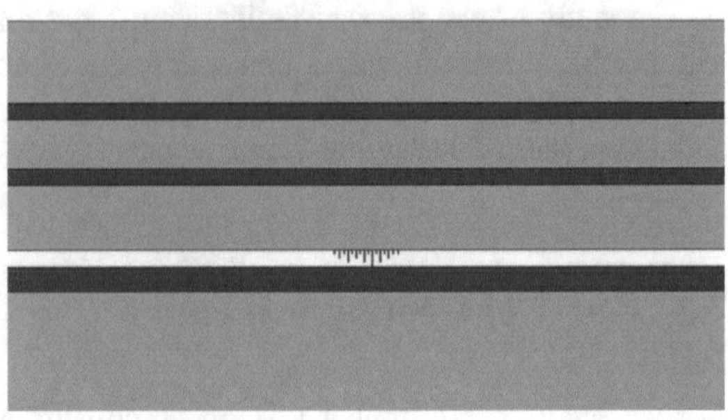

Fig. 7.1 The layers into geological environment and geological faults

Table 7.1 Seismic characteristics of layers

Layer number	Speed of P-waves (m/s)	Speed of S-waves (m/s)	Density (kg/m^3)	Layer thickness (m)
1	2170	674	2000	500
2	2130	795	2300	100
3	2500	1090	2200	300
4	2680	1220	2300	100
5	3000	1385	2400	400
6 (layer with faults)	5550	3144	2700	100
7	6000	1250	2800	150
8	6000	1550	2850	4000

$$c_{\mathrm{S}}^{\mathrm{Faults}} = K \cdot c_{\mathrm{S}}^{\mathrm{Geo}} \tag{7.2}$$

$$\rho^{\mathrm{Faults}} = K \cdot \rho^{\mathrm{Geo}} \tag{7.3}$$

The seismograms V_x (horizontal) and V_y (vertical) and module V were constructed. As wave patterns, the dependence of the modulus V on the coordinates x, y is shown. Values of the module V are depicted in colors. The red color corresponds to some preselected maximum value on the scale, the same everywhere. The blue color corresponds to the zero value of the module V. In each of the calculations, 8 receiver systems were considered, shown in Table 7.2. Note that the source is always fixed, located in the center, and situated above the array of the faults.

Also three types of geometry of the faults region were considered. They are denoted with the following names: "ORIGINAL", "MEDIUM", and "MAXIMUM" fault zones with origin, medium, and maximum lengths of these faults, respectively. The faults of the original length are geometrically represented in Fig. 7.2, and their exact parameters can be found in Table 7.3.

The faults of medium length are geometrically represented in Fig. 7.3, and their exact parameters can be found in Table 7.4. Note that the general topological character of the fault zone remains the same in comparison with the geological faults "ORIGINAL".

Table 7.2 Characteristics of the receivers systems

Number of receiver system	Distance between receivers (m)	Number of receivers
0	10	320
1	25	128
2	50	64
3	75	42
4	100	32
5	150	21
6	200	16
7	400	8

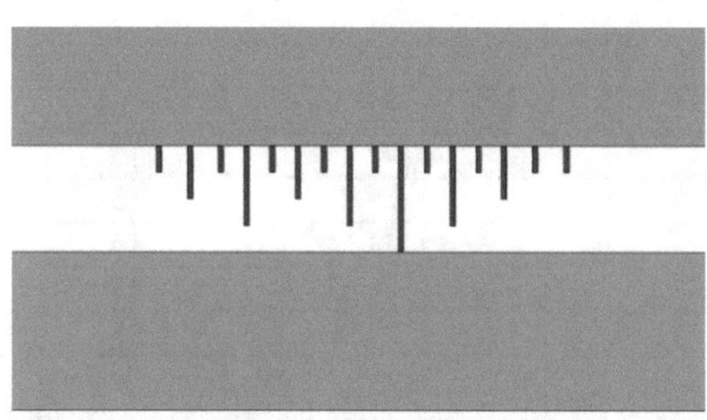

Fig. 7.2 Geological faults of original length, ORIGINAL

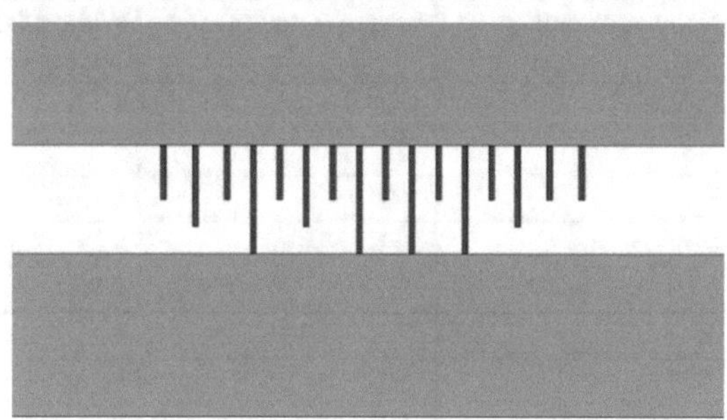

Fig. 7.3 Geological faults of medium length, MEDIUM

The faults of the maximum length considered are represented geometrically in Fig. 7.4, and their exact parameters can be found in Table 7.5. Note that the general topological character of the fault zone has changed slightly in comparison with the faults "ORIGINAL" and "MEDIUM", although a number of topological features are conserved.

Table 7.3 Characteristics of the geological faults, ORIGINAL

Number of fault	Length of fault (m)	Width of fault (m)	Distance between this fault and the next one
1	25	5	25
2	50	5	25
3	25	5	20
4	75	5	20
5	25	5	20
6	50	5	20
7	25	5	20
8	75	5	20
9	25	5	20
10	100	5	20
11	25	5	20
12	75	5	20
13	25	5	20
14	50	5	25
15	25	5	25
16	25	5	Non defined

Table 7.4 Characteristics of the geological faults, MEDIUM

Number of fault	Length of fault (m)	Width of fault (m)	Distance between this fault and the next one
1	50	5	25
2	75	5	25
3	50	5	20
4	100	5	20
5	50	5	20
6	75	5	20
7	50	5	20
8	100	5	20
9	50	5	20
10	100	5	20
11	50	5	20
12	100	5	20
13	50	5	20
14	75	5	25
15	50	5	25
16	50	5	Non defined

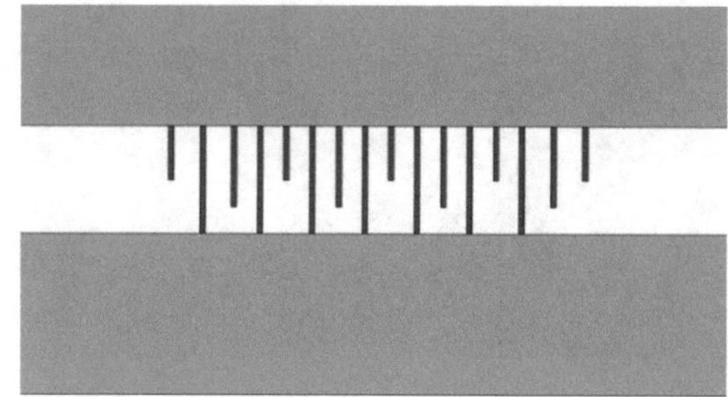

Fig. 7.4 Geological faults of maximum length, MAXIMUM

Table 7.5 Characteristics of the geological faults, MAXIMUM

Number of fault	Length of fault (m)	Width of fault (m)	Distance between this fault and the next one
1	50	5	25
2	100	5	25
3	75	5	20
4	100	5	20
5	50	5	20
6	100	5	20
7	75	5	20
8	100	5	20
9	50	5	20
10	100	5	20
11	75	5	20
12	100	5	20
13	50	5	20
14	100	5	25
15	75	5	25
16	50	5	Non defined

7.2.2 *Main Conclusions Obtained by Wave Logica*

In this section, a study of the influence of the faults length in a geological environment in the seismograms obtained, as well as the causes of these changes in seismograms due to the difference in spatial dynamical wave fields was made.

It can be seen from the dynamics of wave patterns that in addition to standard PP, PS, SP, and PP waves another class of seismic waves is formed in the cluster of the faults by multiple reflections inside the cluster of the incident P-wave. This class of waves propagates at an angle to the faults cluster, and does not strictly upward. In Fig. 7.5, one can see this class of waves occurring after original P-wave passing. This gives the most significant contribution to the difference between the seismograms for the cases of faults of different lengths (Figs. 7.6 and 7.7).

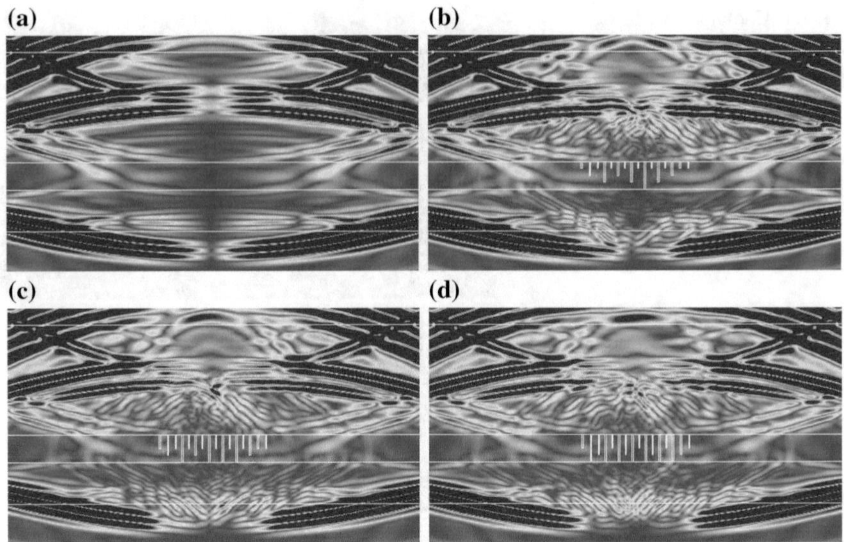

Fig. 7.5 The reflection of original P-wave from the geological faults zone: **a** without geological faults, **b** geological faults "ORIGINAL", **c** geological faults "MIDDLE", **d** geological faults "MAXIMUM"

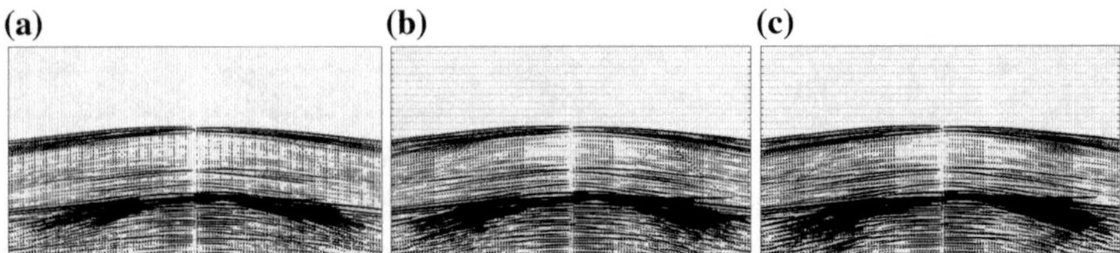

Fig. 7.6 System of receivers № 1, 25 m between receivers, seismograms of horizontal V_x, difference between the case without geological faults and: **a** geological faults "ORIGINAL", **b** geological faults "MIDDLE", **c** geological faults "MAXIMUM"

Fig. 7.7 System of receivers № 1, 25 m between receivers, seismograms of vertical V_y, difference between the case without geological faults and: **a** geological faults "ORIGINAL", **b** geological faults "MIDDLE", **c** geological faults "MAXIMUM"

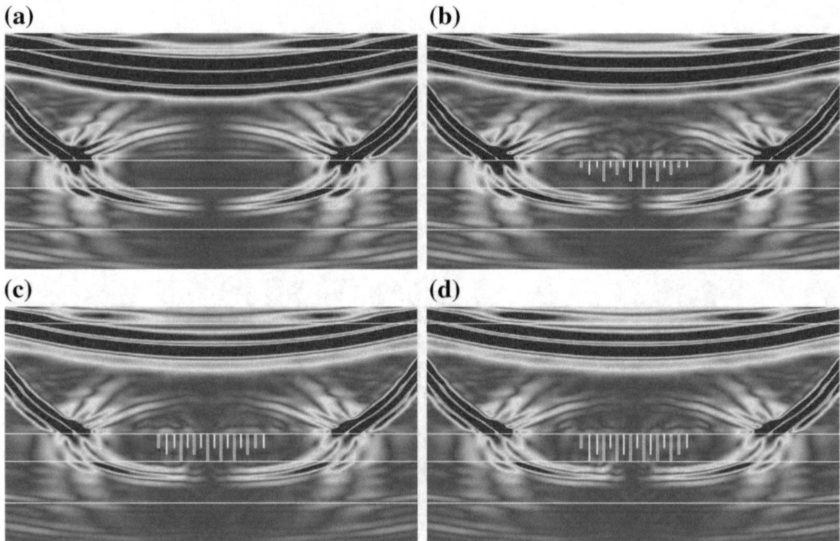

Fig. 7.8 The reflection of S-wave, the approaching of secondary P-wave: **a** without geological faults, **b** geological faults "ORIGINAL", **c** geological faults "MIDDLE", **d** geological faults "MAXIMUM"

Because of the source of seismic waves is located strictly above the cluster of faults, the largest contribution to the waves reflected from the cluster is made by the incident P-wave. This regularity is due to the presence of a central decrease in the amplitude in the incident S-wave (see Chap. 9), in which the cluster of faults occurs. Next waves in importance after the primary P-wave are called S-wave (see Fig. 7.8). And the next one is the secondary P-wave, formed as a result of reflection from the Earth's surface of the PP-wave reflected from the geological medium around the faults zone. Note that the approaching of this secondary P-wave is shown in Fig. 7.8. Reflected from the Earth's surface of the PS-, SP-, and SS-waves does not make such a significant contribution also due to the presence in them of a central decrease in the amplitude. Thus, the focus in this Section is on the P-wave reflection analysis. Notice that these conclusions are true only for the case then the fault zone is strictly under the source of seismic waves in the case of typical seismic source is used.

One can see the difference in the wave patterns of reflections of seismic waves between faults as a function of their length even starting from the instant of time 0.62, 0.64, 0.66, 0.68, 0.7, and 0.74 s in Figs. 7.9, 7.10, 7.11, 7.12, 7.13 and 7.14, respectively.

Beginning from 0.66 s, a "lattice" of re-reflected waves is formed under the faults cluster maximally extended horizontally for the faults "MAXIMUM". This "lattice" has an average length for the faults "MIDDLE" and minimal for the faults "ORIGINAL" (Fig. 7.11).

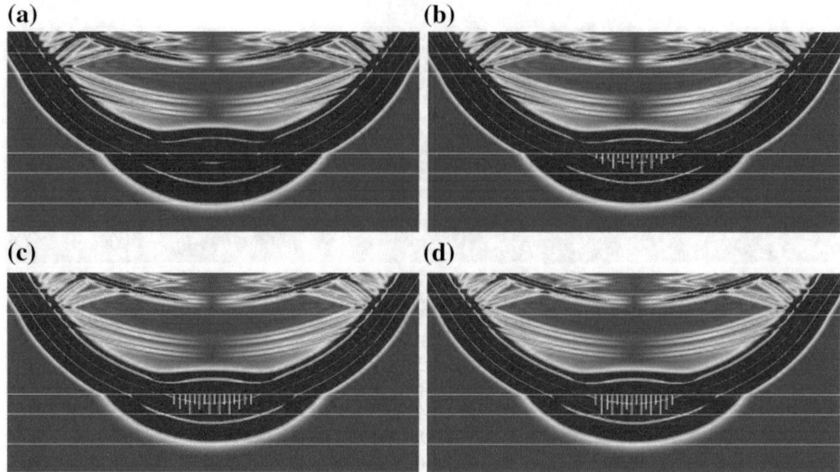

Fig. 7.9 Spatial dynamical wave pattern, time 0.62 s: **a** without geological fault; **b** geological faults "ORIGINAL", **c** geological faults "MIDDLE", **d** geological faults "MAXIMUM"

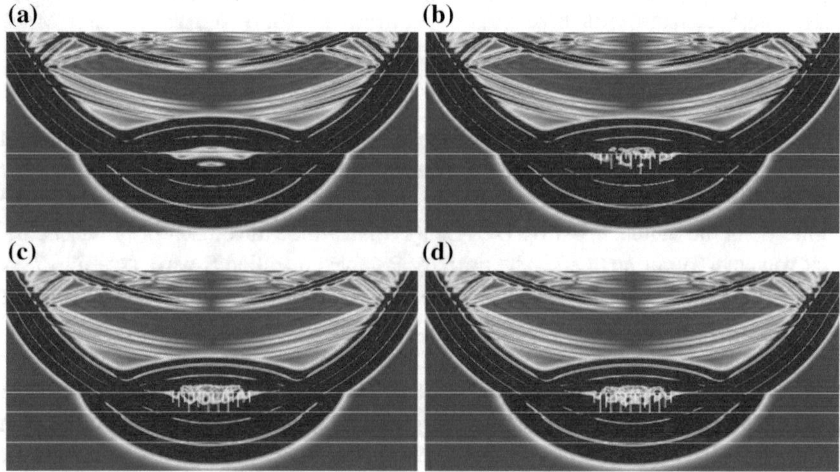

Fig. 7.10 Spatial dynamical wave pattern, time 0.64 s: **a** without geological fault, **b** geological faults "ORIGINAL", **c** geological faults "MIDDLE", **d** geological faults "MAXIMUM"

While wave patterns in the volume above the cluster of faults at 0.66 and at 0.68 s (Fig. 7.12) are topologically similar for the faults "MAXIMUM" and for the faults "ORIGINAL" (they do not have a central amplitude drop) and differ from the wave pattern for the faults "MIDDLE" (the central amplitude drop is).

In the future, at a time of 0.7 s (Fig. 7.13), two decays of the amplitude over the faults are formed for the case of faults "ORIGINAL" and "MIDDLE". In the case

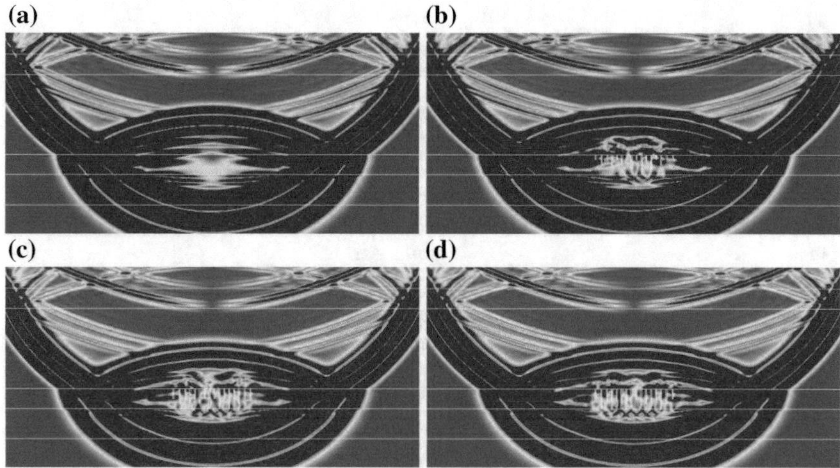

Fig. 7.11 Spatial dynamical wave pattern, time 0.66 s: **a** without geological fault, **b** geological faults "ORIGINAL", **c** geological faults "MIDDLE", **d** geological faults "MAXIMUM"

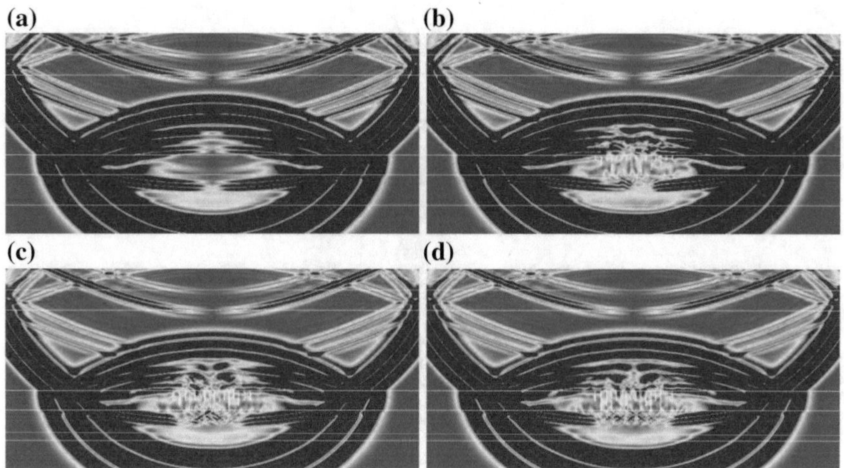

Fig. 7.12 Spatial dynamical wave pattern, time 0.68 s: **a** without geological fault, **b** geological faults "ORIGINAL", **c** geological faults "MIDDLE", **d** geological faults "MAXIMUM"

of faults "MAXIMUM", there is no amplitude decay. However, the decrease in the faults "MIDDLE" is more significant than for the faults "ORIGINAL".

At 0.74 s, formation of lateral responses in the geological layer with a layer with faults can be observed. They are rather weak for the faults "ORIGINAL", and are most clearly expressed for the faults "MAXIMUM" (Fig. 7.14).

At 0.78 s (Fig. 7.15), a "lattice" is formed over the faults cluster. This "lattice" is characterized by a central zone of brightness for the case of the faults "ORIGINAL"

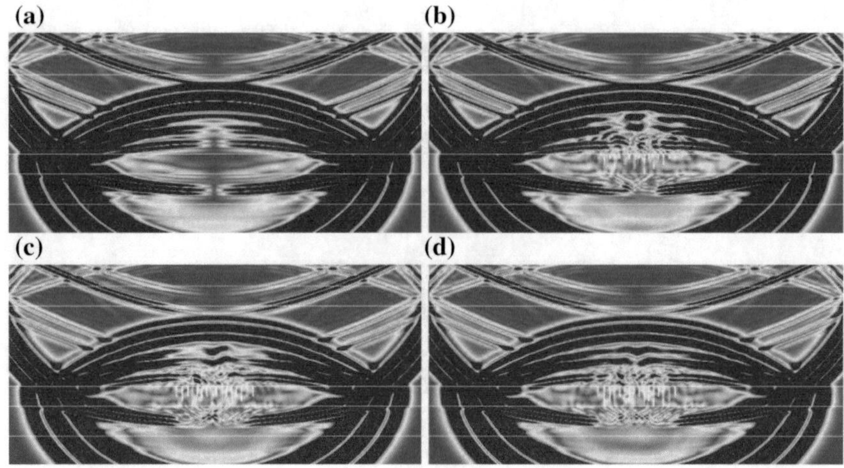

Fig. 7.13 Spatial dynamical wave pattern, time 0.7 s: **a** without geological fault, **b** geological faults "ORIGINAL", **c** geological faults "MIDDLE", **d** geological faults "MAXIMUM"

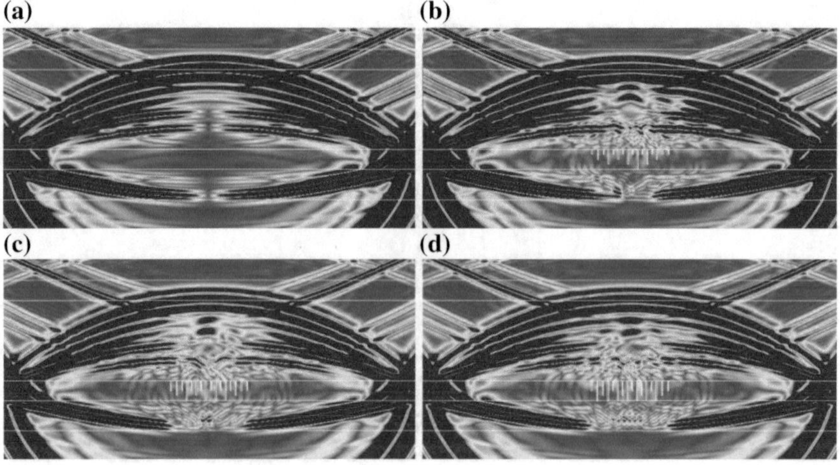

Fig. 7.14 Spatial dynamical wave pattern, time 0.74 s: **a** without geological fault, **b** geological faults "ORIGINAL", **c** geological faults "MIDDLE", **d** geological faults "MAXIMUM"

and "MAXIMUM" of a symmetrical character. Note that it is clearly asymmetric for the case of the faults "MIDDLE". Also, the brightest zones along the edges of the "lattice" are observed in the case of the faults "ORIGINAL". These zones smoothly transform into an analogue of the head P-wave in the case of the absence of faults. These zones become less pronounced for cases of the faults "MIDDLE" and "MAXIMUM". The central zone of the "lattice" is asymmetrical in the faults "ORIGINAL" and approximately symmetrical in the case of the faults "MAXIMUM". It should be noted that this type of waves follows after the exchange PS-wave reflected from the bottom of faults cluster. At the time 0.92 s, it loses a clearly expressed "lattice" structure for all cases of faults length. In general, the resulting "lattice" layer is divided into two fronts, propagating at an angle of approximately 45° along the axes intersecting in the upper part of the geological layer located above the layer with faults. That is, if on these fronts on seismograms to try to restore the location of the faults cluster, in the earth surface OXY plane it will be approximately in the right zone, but along the OZ axis (vertical to the earth surface) the position of the faults cluster will turn out to be overestimated by approximately 300–400 m. The propagation speeds of these types of waves formed from the "lattice" coincide for all cases of the geometry of a faults cluster.

It should be noted that a degree of attenuation of this type of waves greatly exceeds a degree of attenuation of waves of PP-, PS-, SP-, and SS-waves. However, at this stage of the research, it is not possible to guarantee, which part of the attenuation of the wave processes corresponds to physical reality and which is due to the peculiarity of the method used and the quality of the discretization. In other words, in connection with the presence of anisotropy of geological layers, a more pronounced level of this type of waves can be observed on seismograms than is calculated and studied in this section. When developing conclusions and recommendations, it should be borne in mind that the software allows to accurately understand the topological form of the wave fronts and the time and place of their registration by geological equipment. However, it is not possible to accurately estimate their amplitude and the ratio of the given amplitude to the amplitude of the other types of waves.

It should be noted that this feature is a characteristic of numerical simulation and, in addition to the method used, depends also on the grid used and many other factors. Therefore, geologists should be wary of the relative amplitudes obtained in industrial software for direct modelling with a user-friendly interface. Because such a user-friendly interface puts the distribution of the grid into dependence on software algorithms, which makes it difficult to analyze the influence of numerical errors on the attenuation of various types of waves. Notice that typical analytical tests cannot guarantee that industrial software gives an opportunity for the geologist to develop right conclusions. This problem can be solved only by understanding the physical basis of the phenomena under consideration and the peculiarities of the operation of the difference methods used in the software, simultaneously. This suggests the method called Wave Logica.

One might propose that the attenuation of this additional type of waves is occurring due to the separation of energy into waves of different types and the accompanying observed increase in the total length of the front of these additional seismic waves. One can observe this increasing of the total length of the wave front by comparison Figs. 7.13 and 7.16.

Also it should be noted that in the zone of reflected waves above the faults cluster, there is no increase in amplitude compared to the case of absence of faults but its weakening is observed. It is possible to analyze this feature of the topology of wave fronts in more detail from the moment of 1.16 s (Fig. 7.16). That is, for the case

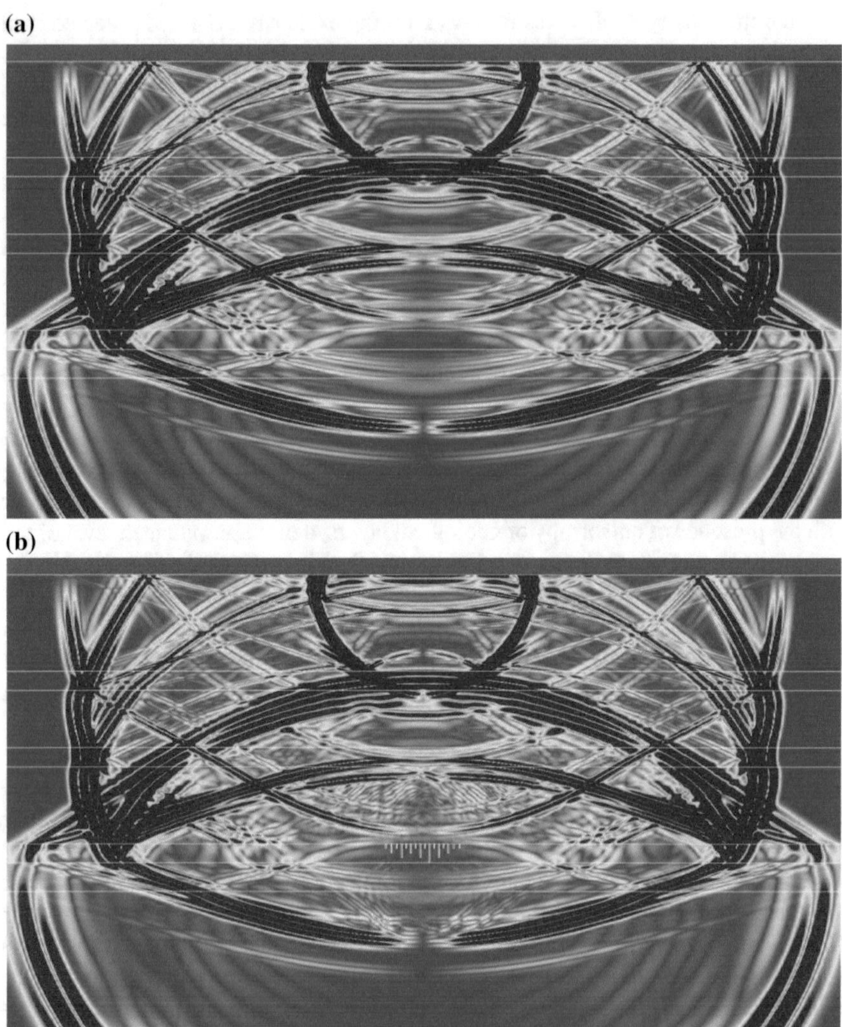

Fig. 7.15 Spatial dynamical wave pattern, time 0.92 s: **a** without geological fault, **b** geological faults "ORIGINAL", **c** geological faults "MIDDLE", **d** geological faults "MAXIMUM"

of no faults, the central amplitude of the PS-wave is quite bright and attenuation is observed, which is maximal for the case of faults "MAXIMUM".

On the basis of Wave Logica, even without studying seismograms, it can be asserted that seismograms-differences should demonstrate greater asymmetry for faults "ORIGINAL" and "MAXIMUM" than for the case of faults "MIDDLE", which is actually observed (Figs. 7.6, 7.7, 7.17 and 7.18).

It should also be noted that the characteristic size of the wavefront of these additional wave reflections reflected from the cluster is approximately 400–500 m, so

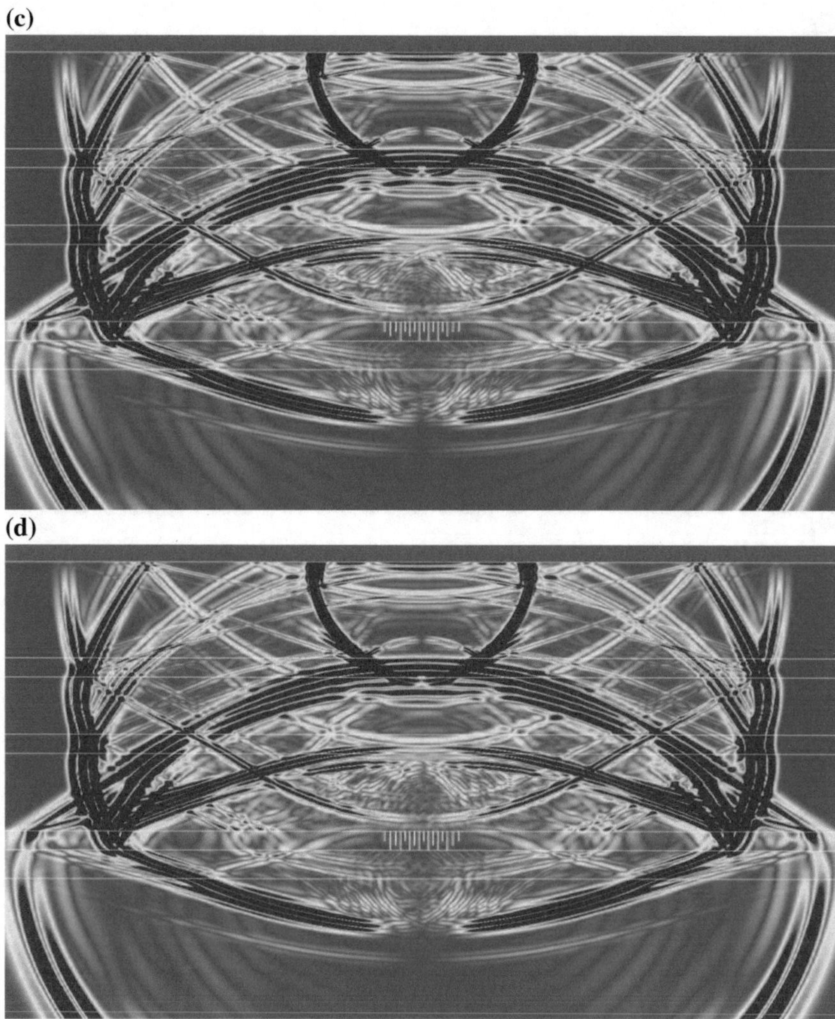

Fig. 7.15 (continued)

if the receivers are too far away from each other, the faults of the fault zone with different lengths will be indistinguishable on seismograms-sacristies, which is also confirmed at the seismograms. See Figs. 7.19, 7.20, 7.21 and 7.22.

It can also be argued that the maximum differences in the response will be on the seismograms of the line with a slope slightly less than the slope of the P-waves located in the lateral parts of the seismograms at a distance approximately equal to the depth of the fracture cluster minus 300–400 m. That is at a distance of 1 km, which is also confirmed by seismograms (Figs. 7.6 and 7.7). Taking into account that the maximum removal of the receiver from the source was taken to be 1600 m, it

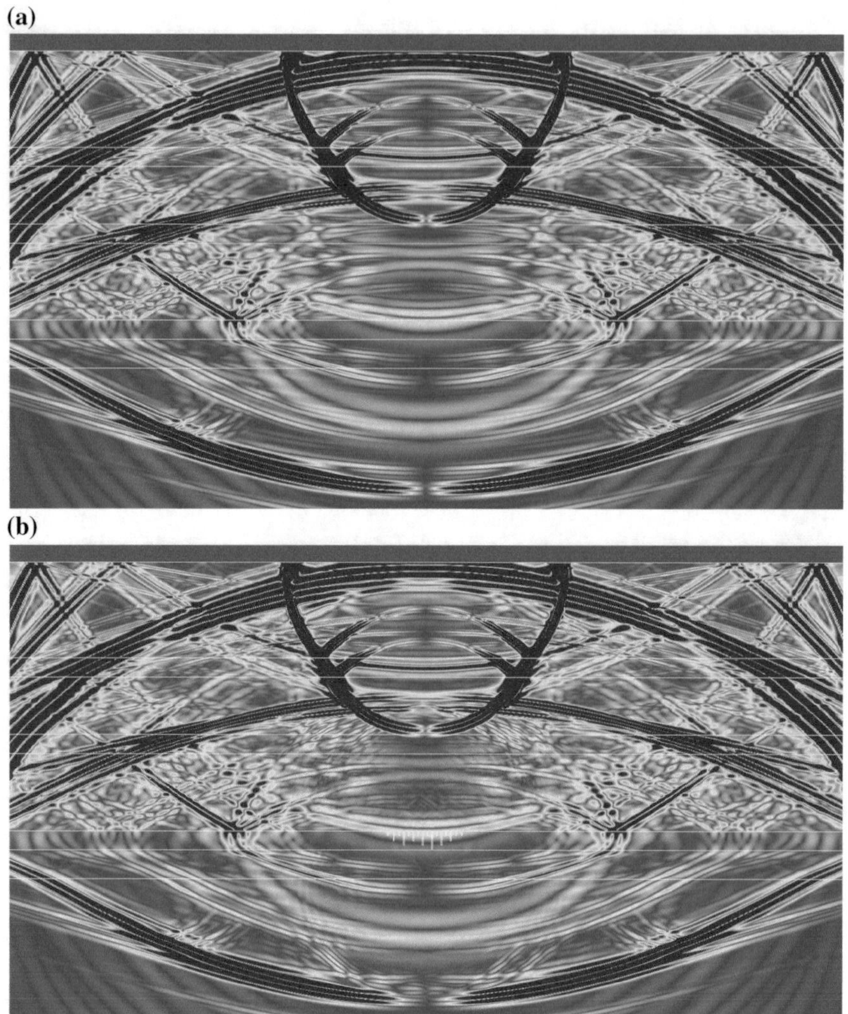

Fig. 7.16 Spatial dynamical wave pattern, time 1.16 s: **a** without geological fault, **b** geological faults "ORIGINAL", **c** geological faults "MIDDLE", **d** geological faults "MAXIMUM"

is necessary to expect the position of these differences at a distance of 1/6 from the edge of the seismogram.

Also note that the speed of this additional type of reflected waves is slightly less than the propagation speed of P-waves in the geological environment. Consequently, it can be expected that they are predominantly P-wave but partly of a mixed nature, propagating at an angle of 45° to the Earth's surface. In this connection, one should expect the maximum visible difference in seismograms representing the horizontal velocity component V_x, while in seismograms V_y obtained from the faults clusters

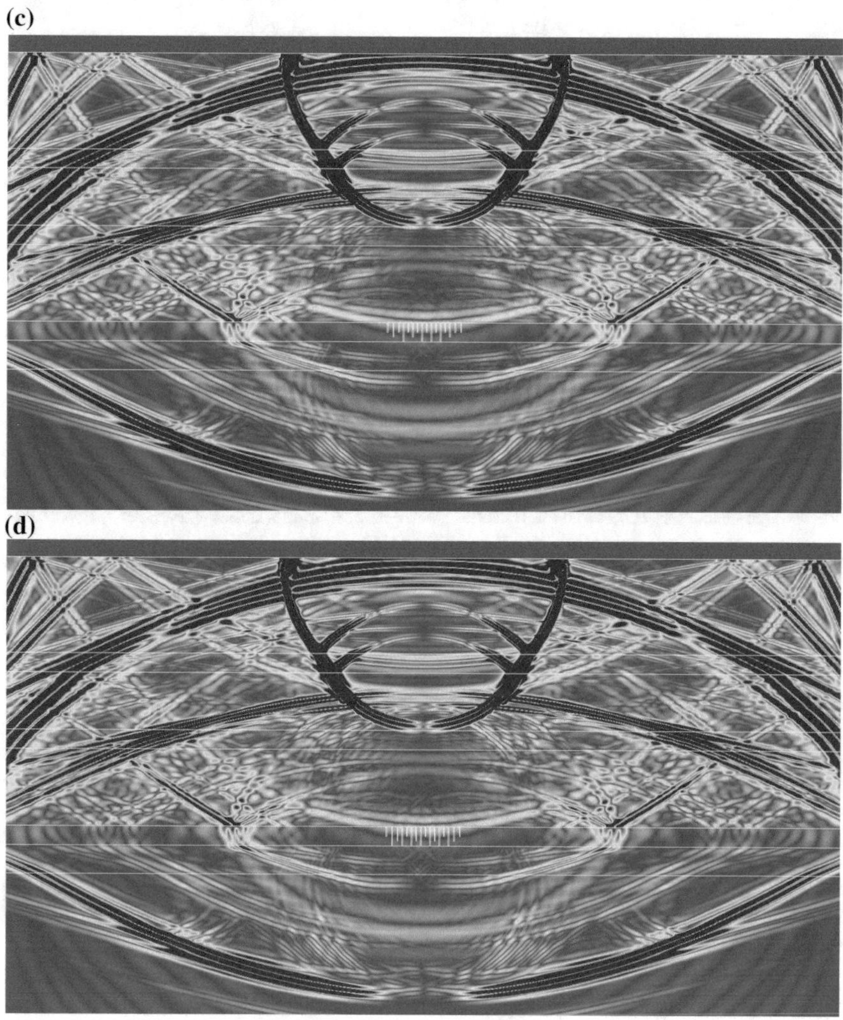

Fig. 7.16 (continued)

of different the length will look identical. This feature is also clearly visible in seismograms (Figs. 7.6 and 7.7).

In general, it can be educed that applying Wave Logica allows to make many conclusions being useful in the subsequent analysis of seismograms. This technology of analysis of the process occurring during seismic prospecting of hydrocarbons and its results allows to analyze the complex effects that are hardly noticeable by usual methods, such as the difference in the length of faults in the faults clusters. The analysis of the spatial dynamical wave patterns permits to determine the seismic

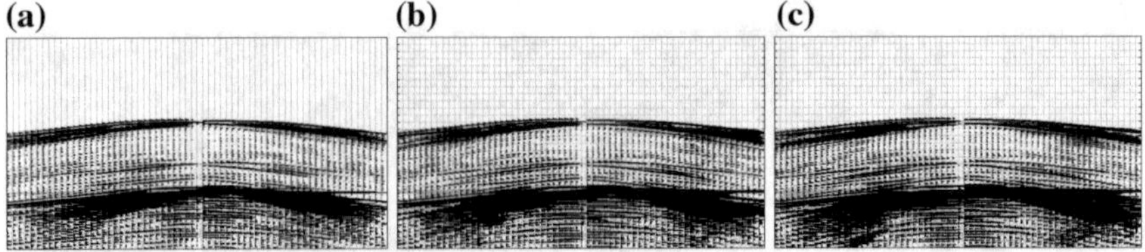

Fig. 7.17 System of receivers №2, 50 m between receivers, seismograms of horizontal V_x, difference between the case without geological faults and: **a** geological faults "ORIGINAL", **b** geological faults "MIDDLE", **c** geological faults "MAXIMUM"

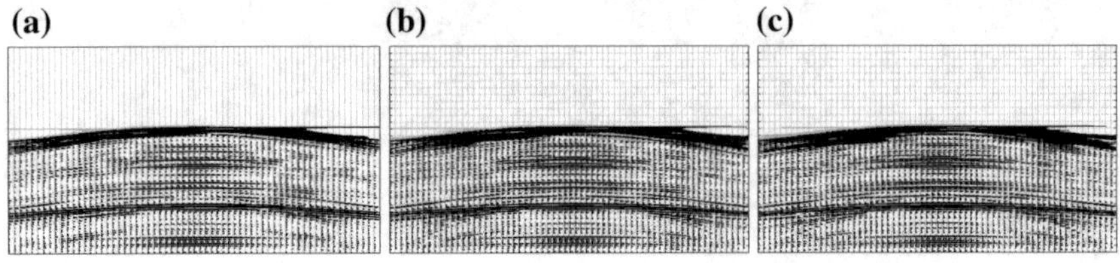

Fig. 7.18 System of receivers №2, 50 m between receivers, seismograms of vertical V_y, difference between the case without geological faults and: **a** geological faults "ORIGINAL", **b** geological faults "MIDDLE", **c** geological faults "MAXIMUM"

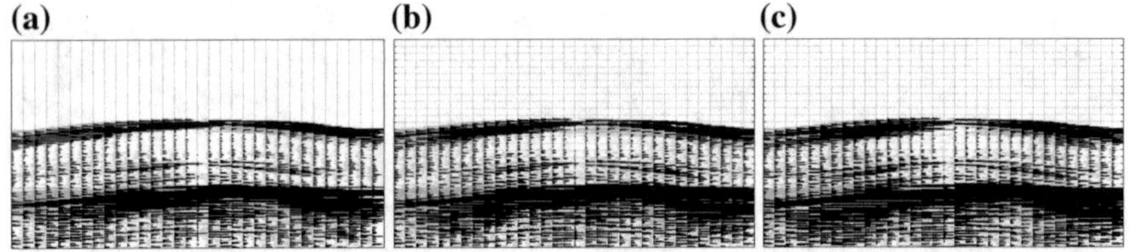

Fig. 7.19 System of receivers №4, 100 m between receivers, seismograms of horizontal V_x, difference between the case without geological faults and: **a** geological faults "ORIGINAL", **b** geological faults "MIDDLE", c geological faults "MAXIMUM"

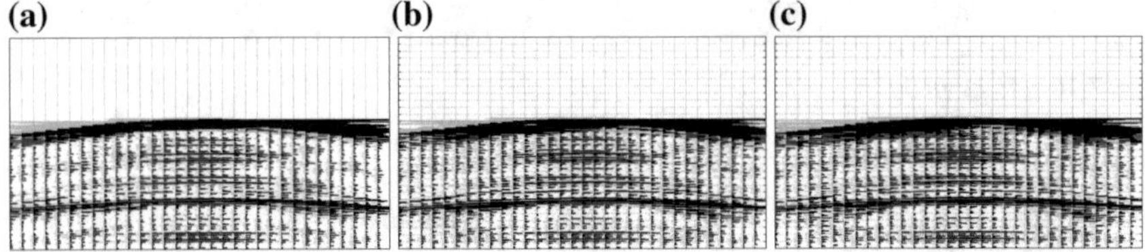

Fig. 7.20 System of receivers №4, 100 m between receivers, seismograms of vertical V_y, difference between the case without geological faults and: **a** geological faults "ORIGINAL", **b** geological faults "MIDDLE", **c** geological faults "MAXIMUM"

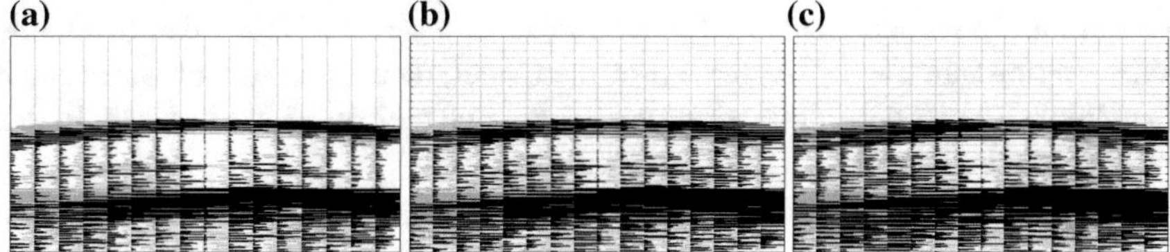

Fig. 7.21 System of receivers №6, 200 m between receivers, seismograms of horizontal V_x, difference between the case without geological faults and: **a** geological faults "ORIGINAL", **b** geological faults "MIDDLE", **c** geological faults "MAXIMUM"

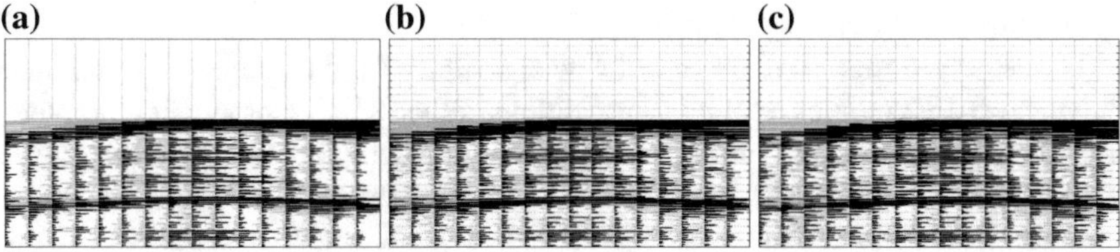

Fig. 7.22 System of receivers №6, 200 m between receivers, seismograms of vertical V_y, difference between the case without geological faults and: **a** geological faults "ORIGINAL", **b** geological faults "MIDDLE", **c** geological faults "MAXIMUM"

survey parameters, which are necessary for obtaining the required information about the geological media and hydrocarbon deposits.

7.3 Wave Simulation in Geological Media

Consider a wave simulation in geological media. Section 7.3.1 presents several seismogeological models for 2D and 3D testing. The calculation of dynamical wave patterns is discussed in Sect. 7.3.2. The calculated synthetic seismograms obtained from these dynamical wave patterns are presented in Sect. 7.3.3. Section 7.3.4 gives some results of the software performance evaluation.

7.3.1 Seismogeological Model for 2D and 3D Testing

During creating algorithms and software for seismic data processing and wave field modelling, reference seismogeological models are widely used [27]. These models are freely available and any researcher can use them in their work. One of these model is the Marmousi model [28]. This model was developed by Institut Francais du Petrole (IFP) in 1988 and imitates the geological structure of the Hansa basin. The model was created to calculate the wave field and test new algorithms for data processing. The model contains 158 horizons and a series of faults that divide the entire section into blocks. At the top of the model, there is a layer of water about 32 m thick. The model's dimensions are 9.2 km in length and 3 km in depth. In this Section, this model is called 2D-Marmousi.

For the three-dimensional calculations, the salt diapir (SEG/EAGE Salt Model) model was used. This model was developed by the SEG Research Committee and was created as a part of the Advanced Computational Technology Initiative in partnership with United States Department of Energy National Laboratories and Technology Centers [29]. The model includes 7 horizons of 12 faults, sand lenses. The main anomalous object in this model is the salt diapir of a complex geometric shape. In this Section, this model is called 3D-SEG.

The described models allow to test new algorithms for data processing and modelling waves for complex geological conditions. When solving production problems of exploration geophysics, the construction of such complex models is not always expedient. Most of the tasks can be solved using simpler models. The most common types of data in the construction of seismogeological models are the distribution maps of the depths of seams and wave velocities.

The following model describes the generalized geological model of the Volga-Ural province. The basis of the model is 12 horizons, the speed of wave propagation varies from 2500 to 6000 m/s. Within the framework of the simulation, the model was gradually complicated and supplemented: the surface of the Earth's relief, the upper part of the section with variable speed, and a series of rupture disorders and salt diapir. Based on this geological model, four models 3D-SIMPLE, 3D-FAULTS, 3D-DEPTH, and 3D-SALT are considered in Sect. 7.3. The ability to include a variety of objects in the model allows to imitate a wide range of geological settings.

The seismogeological model Marmousi was used with the following mathematical characteristics. The size of the model is 13,601 $\times$ 2801 nodes, the grid spacing is 1.25 m. The density data are given in tons per m^3. The model was reinterpolated to a grid of 3401 $\times$ 701 t/m^3 was converted to kg/m^3, all cells with no transverse wave propagation speed were set to 1 m/s. Figure 7.23 shows the resulting distribution of the propagation speed of P-waves in the constructed model.

A seismic geological model 3D-SEG was used with the following mathematical characteristics. The size of the model is 676 $\times$ 676 $\times$ 201 nodes, the grid spacing is 20 m. To create an elastic model, the ratio $c_S = 0.6c_P$ was adopted. The density of the medium was assumed to be constant and equal to 2500 kg/m^3. Figure 7.24 shows the resulting distribution of the propagation speed of P-waves in the constructed model.

The seismic geological model 3D-SIMPLE was used as well. This model includes the altitude maps, P-wave speed in each geological layer, information on the location, and characteristics of the faults. The dimensions of the model are 540 $\times$ 560 $\times$ 275 nodes, the grid pitch is 20 m. The physical size of the model is 10,800 $\times$ 11,200 $\times$ 5500 m. The ratio $c_S = 0.6c_P$ was considered as valid. The density of the medium was assumed to be constant and equal to 2500 kg/m^3. Figure 7.25 shows an example of a converted depth map (after filtering the non-physically large values in the source data).

Fig. 7.23 Distribution of the speed of P-waves in the loaded model Marmousi

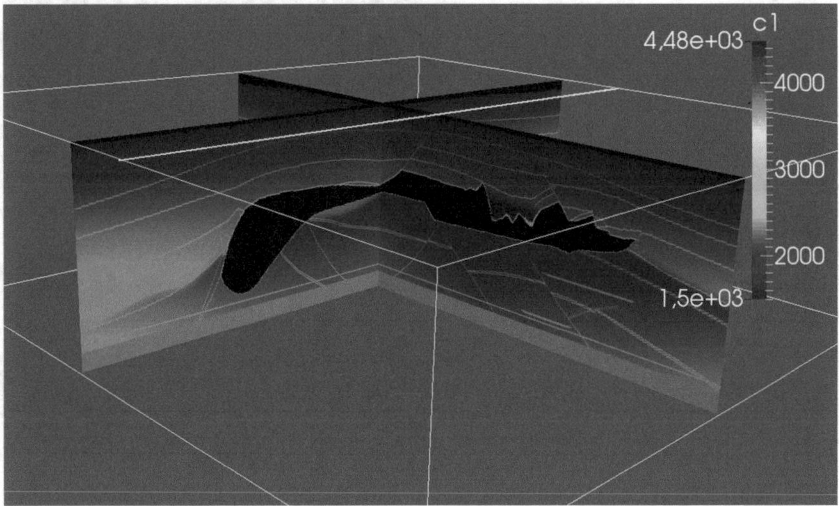

Fig. 7.24 Distribution of the speed of P-waves in the loaded 3D-SEG geological model

Figure 7.26 depicts the constructed seismogeological model. Also, the faults (zones with 10% reduced elastic properties) from the data provided were taken into account (see Fig. 7.27) and this variant of the 3D-SIMPLE model is called 3D-FAULTS.

The 3D-SIMPLE model was complicated by introducing an account of the relief of the earth surface and the variability of the velocities of the longitudinal and transverse waves in the upper layer along the horizontal coordinates. Figure 7.28 shows the speed distribution and in Fig. 7.29 represents the type of model taking into account

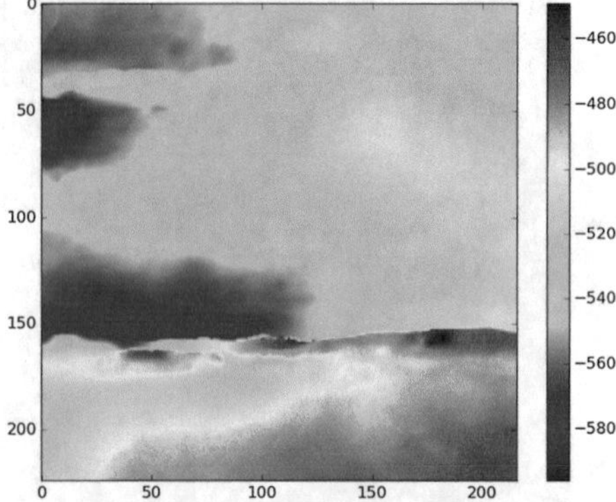

Fig. 7.25 An example of a processed depth map

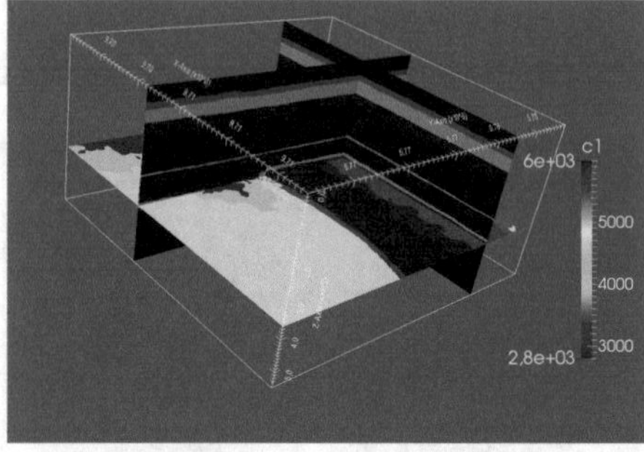

Fig. 7.26 Distribution of the propagation speed of P-waves in the loaded model 3D-SIMPLE

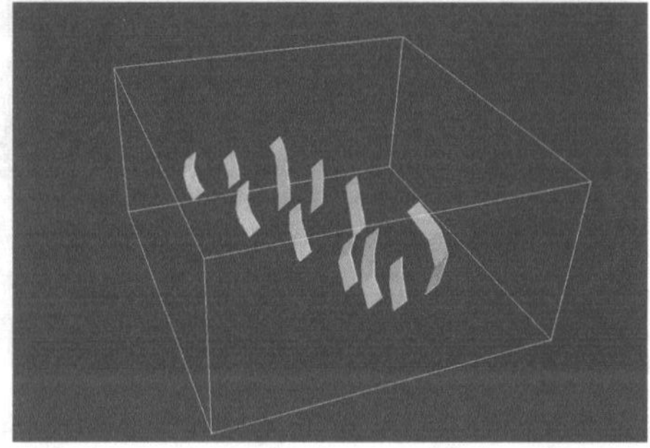

Fig. 7.27 Position of faults in the loaded model 3D-FAULTS

the topography in the complicated version of model 3D-SIMPLE. In this Section, this model is called 3D-DEPTH.

In addition, some calculations were conducted on the model obtained by complicating the previously constructed 3D-SIMPLE model. Figure 7.30 is a sectional view of the final geological model. It includes a salt dome (object of rotation). The speed of P-waves in the object is 4700 m/s. Hereinafter, this model is called 3D-SALT in this Section.

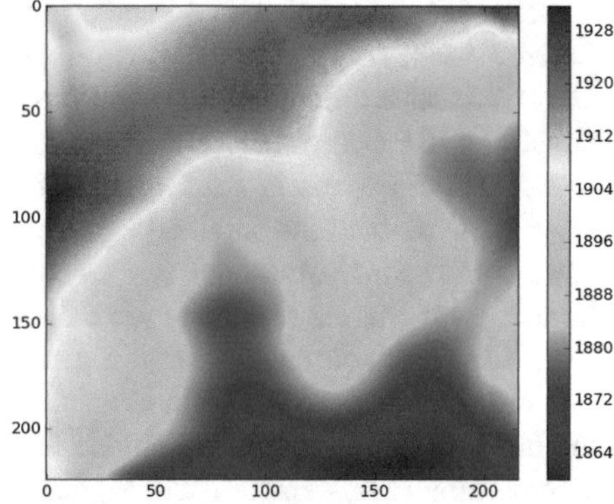

Fig. 7.28 Map of speeds in the near-surface layer of model 3D-DEPTH

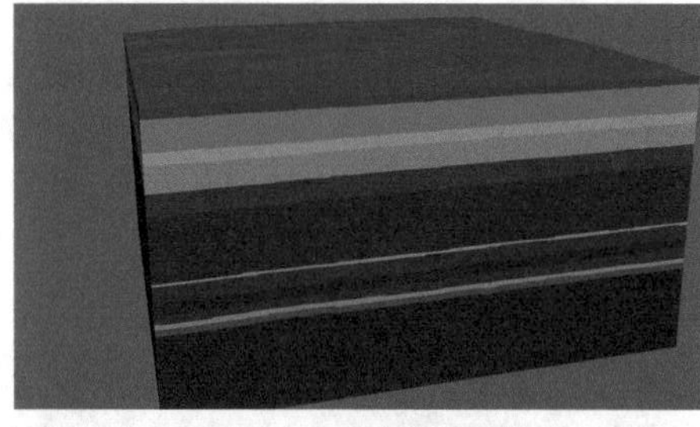

Fig. 7.29 Model 3D-SIMPLE with regard to topography

Fig. 7.30 Geological model 3D-SALT

7.3.2 *Calculation of Seismic Wave Patterns*

For the models constructed in the previous Sect. 7.3.1, the calculations have been made simulating the seismic prospecting process and obtained wave patterns. For the 2D-Marmousi model, two calculations were made: with a point source and with a flat front of a P-wave (analog of zero-offset seismograms). Receiving points were set at a depth of 450 m with a step of 12.32 m across the entire integration area (a total of 1379 geophones). The time function of the source is a 30 Hz Ricker wavelet. The time step was 1 ms, and the total number of steps was 5000. Figure 7.31 shows the wave patterns obtained in the course of numerical calculations.

For the 3D-SEG model, two calculations were made: with a point source and with a flat front of the compressional wave (analog of zero-offset seismograms). Receiving points were set at a depth of 20 m in steps of 20 m across the entire integration area (a total of 676 seismic receivers). The time function of the source is a 30 Hz Ricker wavelet. The time step is 1 ms, and the total number of steps is 4000. Figure 7.32 shows the wave patterns obtained in the set of numerical calculations.

For the 3D-SIMPLE model, two calculations were made with and without faults (the thickness was 20 m). They are called 3D-SIMPLE and 3D-FAULTS, respectively. The point source was set at a depth of 20 m. The receiving points were set in accordance with the transmitted data, simulating the borehole observations. The time function of the source is a 30 Hz Ricker wavelet. The time step is 3 ms, and the total number of steps is 1000. Figure 7.33 shows the wave patterns obtained by numerical calculations for 3D-SIMPLE geological model, while Fig. 7.34 represents wave patterns for 3D-FAULTS geological model.

For the model 3D-SALT, the calculation of the wave field with a point source of 30 Hz (Riker wavelet), buried at 20 m, was performed. The receiving points were also buried 20 m and were located throughout the Earth's surface. Figure 7.35 shows

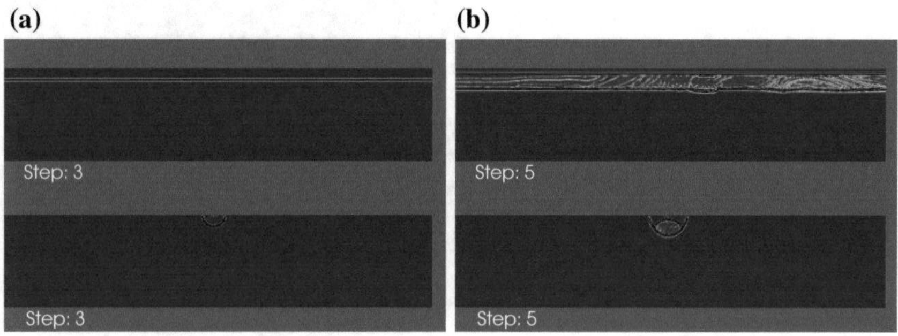

Fig. 7.31 The distribution of the velocity modulus in the medium for the case of a plane wave (the top half of the figures) and for the case of the point source (the bottom half of the figures) at successive instants of time, model 2D-Marmousi: **a** step 3, **b** step 5

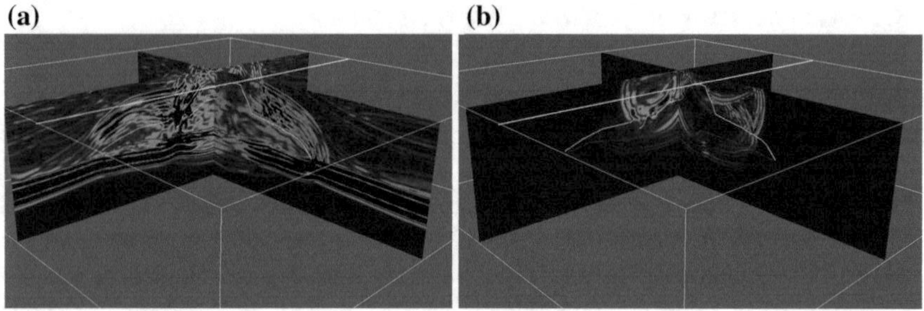

Fig. 7.32 Distribution of the velocity modulus in the medium for 3D-SEG model in the case of: **a** plane wave, **b** point source

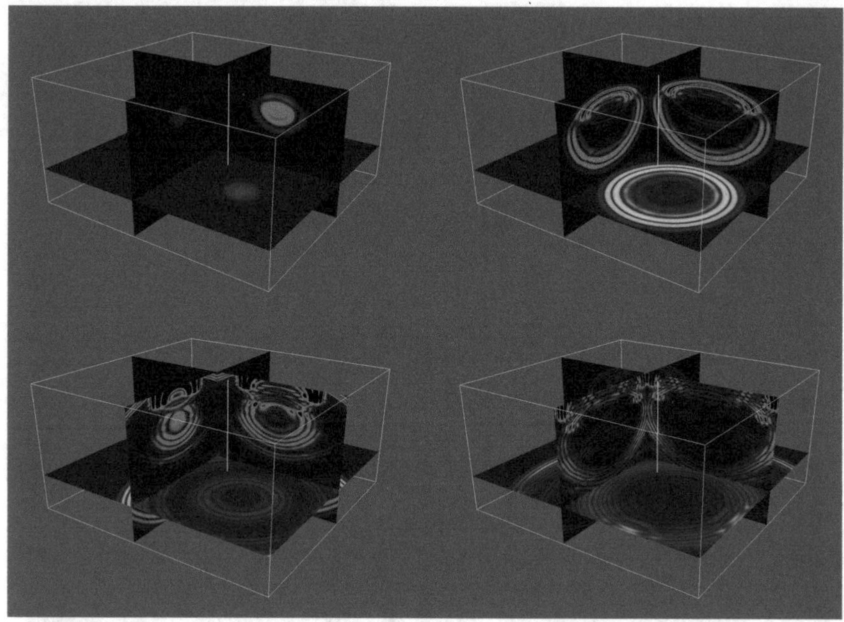

Fig. 7.33 Examples of distribution of the velocity modulus in the medium for 3D-SIMPLE model

one section OXZ, illustrating the propagation of waves in a geological array. 1000 steps are taken with an interval of 3 ms. Based on the calculated wave patterns, an area seismogram was generated.

7.3.3 Calculation of Seismograms

For the models constructed using seismic wave fields calculated in the previous Sect. 7.3.2, synthetic seismograms were obtained. Figures 7.36 and 7.37 show the seismograms plotted along the vertical components of the velocity vector for the seismogeological model 2D-Marmousi.

In Figs. 7.38 and 7.39, the seismograms plotted along the vertical components of the velocity vector for the model 3D-SEG are depicted.

Figure 7.40a, b show the seismograms plotted along the vertical components of the velocity vector for the models 3D-SIMPLE and 3D-FAULTS. The difference of these two seismograms was calculated and represented in Fig. 7.40c in scale 1:100.

Also, a calculation was carried out in a more complicated model taking into account the velocity gradient in the upper layer and topography (3D-DEPTH model). The source is a plane wave of a Ricker wavelet with a frequency of 30 Hz. The receivers are buried 20 m. The time step is 3 ms, the calculation time is 3 s. Figure 7.41

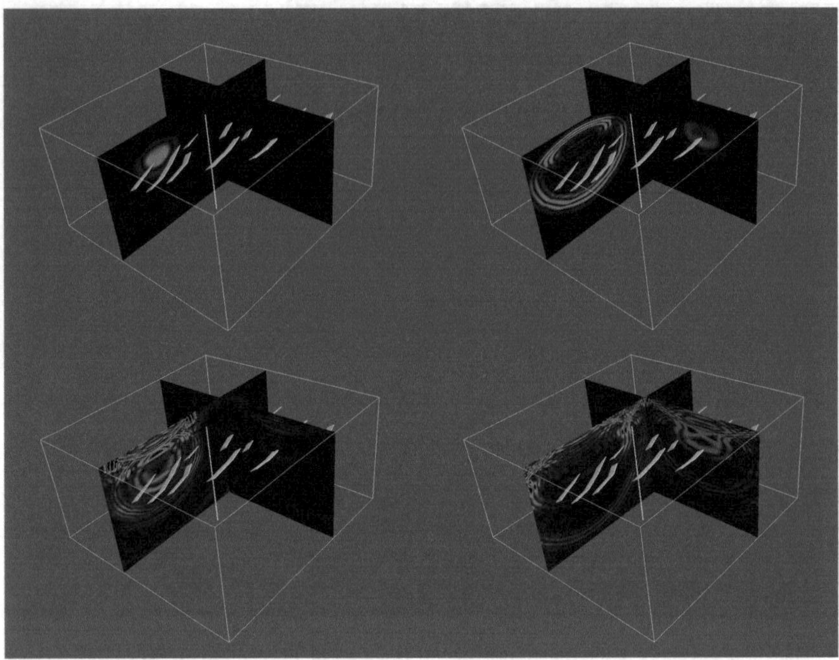

Fig. 7.34 Examples of distribution of the velocity modulus in the medium for 3D-FAULTS model

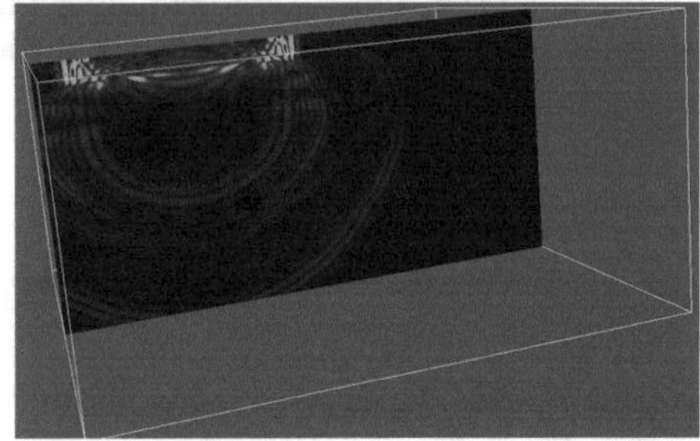

Fig. 7.35 Elastic waves in the model 3D-SALT

shows a seismogram constructed based on the results of calculations for the 3D-DEPTH model.

For the 3D-SALT geological model, the calculation of the 2D seismic survey was performed, in which the sources were set in 50 m increments. The monitoring system is central and symmetrical. Maximum removal of the PP-PV equals 3500 m, while the minimum removal of the PP-PV is 40 m. The record length is 5 s, the sampling step is 2 ms. Coordinates of the first source in the OXY plane were 4000, 4000 m. 70 sources were further simulated in 40 m steps in each of the axes. A total of 70 calculations were made. Figure 7.42 shows the resulting seismogram for a fixed source.

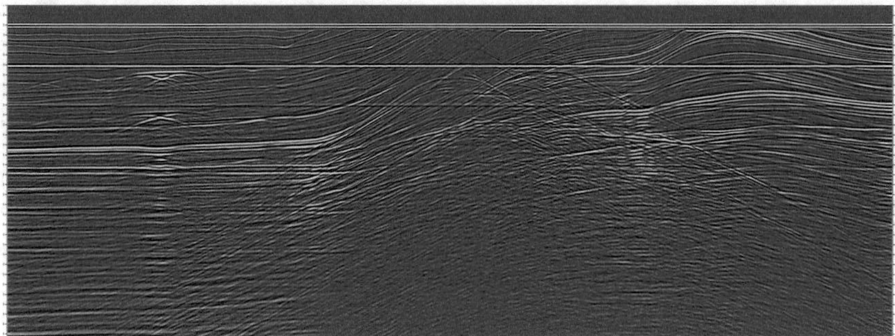

Fig. 7.36 Seismograms constructed from the vertical component of velocity for a flat front, geological model 2D-Marmousi

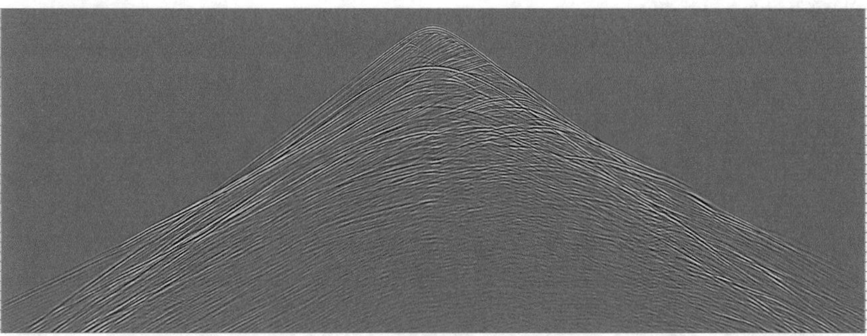

Fig. 7.37 Seismograms constructed from the vertical component of velocity for a point source, geological model 2D-Marmousi

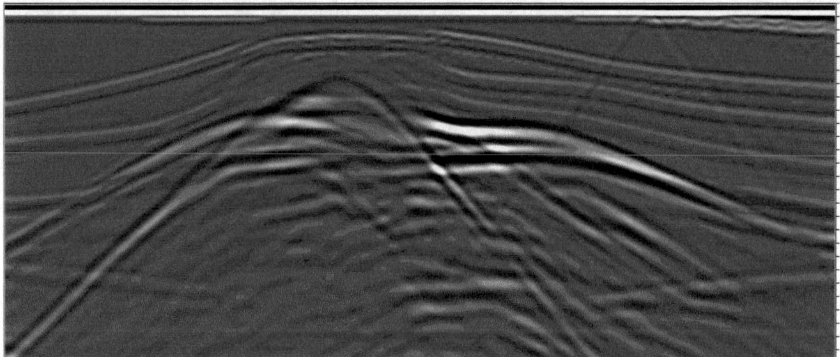

Fig. 7.38 Seismograms constructed from vertical component of the velocity flat front, geological model 3D-SEG

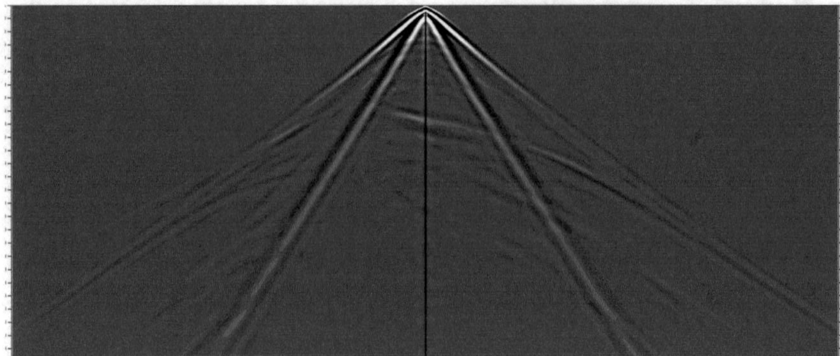

Fig. 7.39 Seismograms constructed from vertical component of the velocity for a point source, geological model 3D-SEG

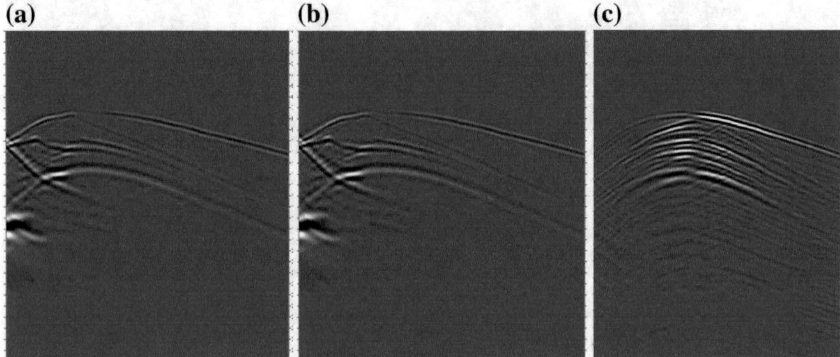

Fig. 7.40 Seismograms constructed from the vertical component of velocity: **a** 3D-SIMPLE model, **b** 3D-FAULTS model, **c** difference seismogram (for a clear visualization in seismic signals the scale 1:100 was used)

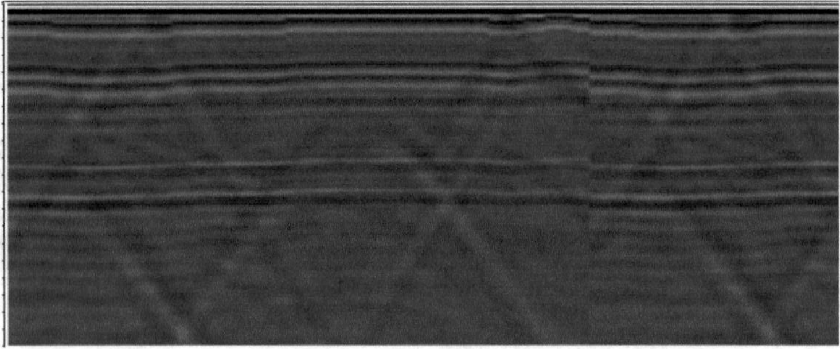

Fig. 7.41 Seismogram constructed by the vertical component of velocity for the case of 3D-DEPTH geological model

Fig. 7.42 Seismogram for 2D seismic survey, 3D-SALT geological model

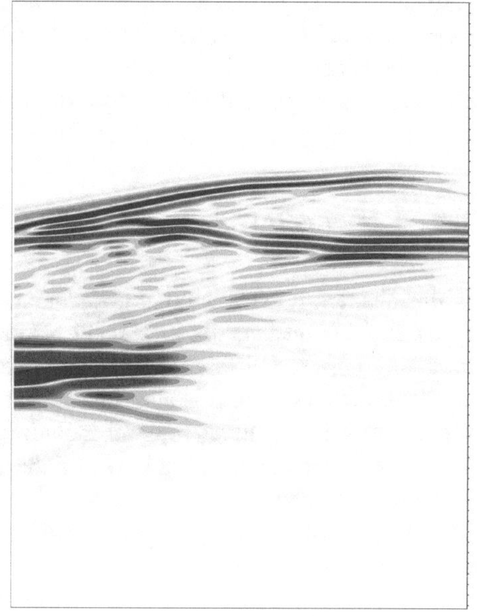

Fig. 7.43 Seismogram for the vertical seismic profiling, 3D-SALT geological model

The calculation of the vertical seismic profiling experiment [30] was performed, in which the receiving points are located in the well every 10 m. Several sources were used on the distance 3500 m with about 100 m. The sources were located on the surface with a Ricker wavelet with a frequency of 30 Hz. The time step is 3 ms, the total number of steps is 1000. Figure 7.43 shows a seismogram for a fixed position of the source of seismic waves.

Table 7.6 Software performance evaluations

Geological model	Nx × Ny × Nz nodes	Nt time steps	Total time (s)	1 core calculation time (h)
2D-Marmousi	3401 × 701	5000	5	0.3
3D-SEG	676 × 676 × 201	4000	4	7
3D-SIMPLE	540 × 560 × 275	1000	3	2

7.3.4 Software Performance Evaluation

The established evaluations of the program time for all the above calculations are presented in Table 7.6.

The conducted evaluations showed that a computational speed is acceptable for software based on the grid-characteristic method. Therefore, the designed software might be used for performing calculations of the modelling of spatial dynamic seismic wave fields and synthetic seismograms for geological models with the complexity characteristic compared with complexity of real oil-bearing deposits.

7.4 Conclusions

In this chapter, it is shown that grid-characteristic method allows to model various types of geological media in 2D and 3D cases taking into account the curvilinear interfaces between different geological rocks, as well as the faults and fractures of the complex topology. It is also shown that it is possible to build the seismograms for various seismic survey systems including 2D seismic survey, 3D seismic survey, and vertical seismic profiling based on the spatial dynamic wave patterns obtained using the grid-characteristic method. Also in this chapter, a comparative analysis of software based on the grid-characteristic method for solving various geophysical problems is carried out.

In this chapter, a study of faults zones of three types of their topology and different type of geological environment within the faults were performed, as well. A detailed analysis of spatial dynamic wave patterns is carried out and predictive conclusions about the nature of the seismograms, which were actually confirmed in the respective seismograms, were developed. The possibility to derive the predictive conclusions about seismograms based on the analysis of spatial wave patterns is demonstrated in this chapter.

Acknowledgements This work has been performed at Moscow Institute of Physics and Technology and was supported by the Russian Science Foundation, grant no. 14-11-00263. This work has been carried out using computing resources of the federal collective usage center Complex for Simulation and Data Processing for Mega-science Facilities at NRC "Kurchatov Institute", http://ckp.nrcki.ru/.

References

1. Larmat C, Montagner JP, Capdeville Y, Banerdt WB, Lognonné P, Vilotte JP (2008) Numerical assessment of the effects of topography and crustal thickness on Martian seismograms using a coupled modal solution–spectral element method. Icarus 196(1):78–89
2. Ceylan S, van Driel M, Euchner F, Khan A, Clinton J, Krischer L, Böse M, Stähler S, Giardini D (2017) From initial models of seismicity, structure and noise to synthetic seismograms for Mars. Space Sci Rev 211(1–4):595–610
3. Golubev VI, Petrov IB, Khokhlov NI (2015) Simulation of seismic processes inside the planet using the hybrid grid-characteristic method. Math Models Comput Simul 7(5):439–445
4. Aki K, Richards PG, Fredman WH (1980) Quantitative seismology theory and methods, vols I and II. San Francisco
5. Korneev V (2007) Slow waves in fractures filled with viscous fluid. Geophysics 73(1):N1–N7
6. Nakagawa S, Korneev VA (2014) Effect of fracture compliance on wave propagation within a fluid-filled fracture. J Acoust Soc Am 135(6):3186–3197
7. Frehner M, Shih PJR, Lupi M (2014) Initiation of Krauklis waves by incident seismic body waves: numerical modeling, laboratory perspectives, and application for fracture-size estimation. SEG Techn Prog Expanded Abs 2014:3422–3427
8. McMechan GA, Mooney WD (1980) Asymptotic ray theory and synthetic seismograms for laterally varying structures: theory and application to the Imperial Valley, California. Bull Seismol Soc Am 70(6):2021–2035
9. Hanyga A, Helle HB (1995) Synthetic seismograms from generalized ray tracing. Geophys Prospect 43(1):51–75
10. Spence GD, Whittall KP, Clowes RM (1984) Practical synthetic seismograms for laterally varying media calculated by asymptotic ray theory. Bull Seismol Soc Am 74(4):1209–1223
11. Kelly KR, Ward RW, Treitel S, Alford RM (1976) Synthetic seismograms: a finite-difference approach. Geophysics 41(1):2–27
12. Ganley DC (1981) A method for calculating synthetic seismograms which include the effects of absorption and dispersion. Geophysics 46(8):1100–1107
13. Bouchon M, Coutant O (1994) Calculation of synthetic seismograms in a laterally varying medium by the boundary element-discrete wavenumber method. Bull Seismol Soc Am 84(6):1869–1881
14. Geller RJ, Ohminato T (1994) Computation of synthetic seismograms and their partial derivatives for heterogeneous media with arbitrary natural boundary conditions using the direct solution method. Geophys J Int 116(2):421–446
15. Geller RJ, Takeuchi N (1995) A new method for computing highly accurate DSM synthetic seismograms. Geophys J Int 123(2):449–470
16. Liu Y, Teng J, Lan H, Si X, Ma X (2014) A comparative study of finite element and spectral element methods in seismic wavefield modeling. Geophysics
17. Biryukov VA, Miryakha VA, Petrov IB, Khokhlov NI (2016) Simulation of elastic wave propagation in geological media: intercomparison of three numerical methods. Comput Math Math Phys 56(6):1086–1095
18. Botter C, Cardozo N, Hardy S, Lecomte I, Escalona A (2014) From mechanical modeling to seismic imaging of faults: a synthetic workflow to study the impact of faults on seismic. Mar Pet Geol 57:187–207
19. Zhu LF, Zheng HE, Xin PAN, Wu XC (2006) An approach to computer modeling of geological faults in 3D and an application. J China Univ Min Technol 16(4):461–465
20. Favorskaya A, Petrov I, Grinevskiy A (2017) Numerical simulation of fracturing in geological medium. Procedia Comput Sci 112:1216–1224
21. Favorskaya AV, Petrov IB, Vasyukov AV, Ermakov AS, Beklemysheva KA, Kazakov AO, Novikov AV (2014) Numerical simulation of wave propagation in anisotropic media. Dokl Math 90(3):778–780
22. Favorskaya A, Petrov I, Golubev V, Khokhlov N (2017) Numerical simulation of earthquakes impact on facilities by grid-characteristic method. Procedia Comput Sci 112:1206–1215

23. Favorskaya AV, Petrov IB (2016) Wave responses from oil reservoirs in the Arctic shelf zone. Dokl Earth Sci 466(2):214–217
24. Favorskaya AV, Petrov IB (2017) Numerical modeling of dynamic wave effects in rock masses. Dokl Math 95(3):287–290
25. Petrov IB, Favorskaya AV, Khokhlov NI, Miryakha VA, Sannikov AV, Golubev VI (2015) Monitoring the state of the moving train by use of high performance systems and modern computation methods. Math Models Comput Simul 7(1):51–61
26. Beklemysheva KA, Favorskaya AV, Petrov IB (2014) Numerical simulation of processes in solid deformable media in the presence of dynamic contacts using the grid-characteristic method. Math Models Comput Simul 6(3):294–304
27. Herwanger J (2014) Seismic geomechanics: how to build and calibrate geomechanical models using 3D and 4D seismic data. In: Education Days Stavanger 2014
28. Brougois A, Bourget M, Lailly P, Poulet M, Ricarte P, Versteeg R (1990) Marmousi, model and data. In: EAEG workshop-practical aspects of seismic data inversion
29. Aminzadeh F, Jean B, Kunz T (1997) 3-D salt and overthrust models. Society of Exploration Geophysicists, USA
30. Hardage BA (1985) Vertical seismic profiling. Lead Edge 4(11):59

Chapter 8
Migration of Elastic Wavefield Using Adjoint Operator and Born Approximation

Oleg Ya. Voynov, Vasiliy I. Golubev, Michael S. Zhdanov and Igor B. Petrov

Abstract This chapter presents a new method of migration of the elastic wavefield. It is based on the Born approximation of the forward modelling operator for the elastic, which is obtained as an extension of the Born approximation for the acoustic field to the case of the elastic wavefield propagation. We present a detail mathematical derivation of the migration operators for the acoustic and elastic cases. The numerical experiments based on these operators are conducted for a set of synthetic multilayered models with curvilinear boundaries between the layers. We also present a comparative study of the migration images produced by the migration of the acoustic and elastic wavefields and examine the sources of false boundaries appeared in some of these images.

Keywords Mathematical modelling · Seismic imaging · Seismic migration Elastic media · Born approximation

O. Y. Voynov (✉) · V. I. Golubev · M. S. Zhdanov · I. B. Petrov
Moscow Institute of Physics and Technology, 9 Institutskiy per, Dolgoprudny, Moscow Region 141701, Russian Federation
e-mail: voinov@phystech.edu

V. I. Golubev · I. B. Petrov
Scientific Research Institute for System Studies of the Russian Academy of Sciences, 36(1), Nahimovskij ave, Moscow 117218, Russian Federation
e-mail: w.golubev@mail.ru

I. B. Petrov
e-mail: petrov@mipt.ru

M. S. Zhdanov
University of Utah, 115 South 1460 East, Rm 383 Salt Lake City, UT 84112, USA
e-mail: michael.zhdanov@utah.edu

A. V. Favorskaya and I. B. Petrov (eds.), *Innovations in Wave Processes Modelling and Decision Making*, Smart Innovation, Systems and Technologies 90,
https://doi.org/10.1007/978-3-319-76201-2_8

8.1 Introduction

The main goal of seismic survey is a creation of a reliable model of the subsurface geological medium. The rigorous approach to the solution of this problem is based on the inversion of the seismic data into the physical properties of the subsurface. However, inversion of seismic data is a very challenging problem, which meets significant difficulties related to the nonuniqueness and instability of mathematical inversion problems. Another approach is based on the migration of the seismic data from the surface of the observation downward within the lower half space. This approach has found wide practical applications due to its relative simplicity and robustness of the produced subsurface seismic images. Untill recently, most of the migration algorithms were based on the solution of the wave equations for the acoustic model of the seismic field. This approach, however, ignores the true physics of the seismic wave propagation within the rock formations. A more accurate approach is based on using the elastic field equations for the solution of this problem. In this chapter, we demonstrate how this approach can be used for developing the algorithms of the elastic field migration in the geological media.

We construct the forward modelling operator and the migration operator (adjoint to the forward one) for an elastic field using the Born approximation. Our derivation follows a straightforward procedure described in [1] for the acoustic case. The validity of the developed method was tested by using the direct simulation of the wave propagation in slightly inhomogeneous media and by analysis of the migration images for synthetic models. A comparison of the results produced for the acoustic and elastic cases is presented as well.

The chapter is organized as follows. The basics of seismic migration are discussed in Sect. 8.2. The derivations of forward modelling and adjoint migration operators for the acoustic and elastic cases are performed in Sects. 8.3 and 8.4, respectively. The examples of migration images produced by the developed algorithms and their comparisons are given in Sect. 8.5. Section 8.6 discusses the synthetic seismograms and the sources of the false boundaries in migration images. Section 8.7 concludes the chapter.

8.2 Basics of Seismic Migration

The principles of geometrical migration of seismic data have been used for interpretation since the 1960s. The theoretical principles of this process based on the use of the wavefront charts and diffraction curves was presented in classical work [2]. Further developments were made in the works of Claerbout [3–5], who first formulated a finite-difference algorithm for migration based on the scalar wave equation. In the works by French [6], Schneider [7], and Zhdanov et al. [8], the Kirchhoff integral method for acoustic wave migration was developed. In the monograph by Zhdanov [9] the Kirchhoff integral method and the corresponding migration algorithms were extended to the elastic wavefields as well. The frequency wavenumber migration algorithm was proposed independently in [10]. In works [11–13], the Born approximation [14] was used to develop the forward modelling operator for a scattering field. The Reverse Time Migration (RTM) was introduced in [15–17]. It relies on the full waveform equation and takes into account the complex phenomena of the wave propagation. The researchers made many efforts to eliminate the drawbacks of the available migration algorithms. For example, in [18] a modification of the Kirchhoff migration operator was proposed to image the steep dips while including high frequencies. A ray-Born method was proposed in [19–21] to take into account a non-uniform background medium.

In a classical marine survey, the seismic source is placed in the water and the pressure fluctuations are recorded by the receivers. Therefore, a purely acoustic-wave equation is a reasonable approximation to describe the P-waves propagation through the water layer that carry major information about the subsurface structure. However, for the sea-bottom receivers and/or land surveys, the S-waves play an important role in the observed seismic wave phenomena. In such a case, applying the migration algorithm based on the elastic wave equation is essential. Most of the publications, cited above, treat the geological formation as an acoustic medium. In this case, only the pressure waves (P-waves) exist in the subsurface volume. However, the process of propagation of seismic waves may be described more accurately using the elastic medium model. One way to consider the different elastic waves is to calculate a migration image using acoustic algorithm for the PP, PS, SP, and SS waves separately [22–26]. However, the use of the full waveform (elastic) migration represents a more rigorous approach. For example, highly dipping subsalt events around the salt keel are found to be imaged by elastic migration much better than by the acoustic one [27].

A significant research has been conducted on the migration of elastic field over the years. Kuo and Dai [28] presented Kirchhoff multicomponent migration for the case of non-coincident source and receiver. This approach was successfully applied to a synthetic Vertical Seismic Profiling (VSP) data [29]. An elastic type of Kirchhoff multicomponent migration was introduced in [9, 30]. Zhe and Greenhalgh proposed the elastic migration by extrapolating the displacement potential [31]. In [32], an algorithm of 3D elastic Kirchhoff prestack depth migration for the VSP data was introduced, which used the complete wavefield recorded by three components as an input. The extension of the Reverse-Time Migration (RTM) to the elastic and anisotropic cases was investigated in [33, 34], respectively. To account for elastic effects, several research groups have developed modified imaging principles using divergence and curl operators to separate compressional and shear waves (e.g., [35, 36]). The Born approximation was applied to the elastic isotropic medium in [37, 38], transversely isotropic medium in [39], and fractured medium in [40]. In these latest works, the migration was treated as a first step of inversion formulated in a least-square sense.

The general mathematical approach to constructing the migration algorithms can be summarized as follows [1]. Let us approximate the forward modelling problem for seismic wavefield by the following linear operator equation:

$$\mathbf{d} = \mathbf{Lm}, \tag{8.1}$$

where **d** are the seismic data, **m** are the model parameters, **L** is the forward modelling linear operator. In this case, the migration field is given by the following formula [41]:

$$\mathbf{m}_{migr} = \mathbf{L}^*\mathbf{d}, \tag{8.2}$$

where the adjoint operator $\mathbf{L}^*$ can be obtained from the equation [1, 42]

$$\left(\mathbf{L}^*\mathbf{d} \cdot \mathbf{m}\right)_m = (\mathbf{d} \cdot \mathbf{Lm})_d. \tag{8.3}$$

Here $(\cdot)$ denotes the dot product in the corresponding Hilbert spaces of the models or data.

The mathematical derivations of the forward and adjoint operators for the acoustic and elastic cases is given in Sects. 8.3 and 8.4, respectively.

8.3 Migration of Acoustic Wavefield

This section presents mathematical derivation of the forward and adjoint operators for the acoustic case. The mathematical model of acoustic medium is given in Sect. 8.3.1. We derive in Sect. 8.3.2, the forward and adjoint operators for quasi-homogeneous space using Born approximation. The formulae for operators are simplified for the case of a point source. In Sect. 8.3.3, a quasi-homogeneous half space is considered.

8.3.1 Acoustic Medium

Consider a simple model of acoustic medium with constant mass density. In this case, the following acoustic wave equation describes a propagation of seismic waves:

$$\Delta P - s^2 \partial_t^2 P = -F^e, \tag{8.4}$$

where P is the pressure field, F^e is the strength of the external source of energy, s is the slowness of the wave propagation, reciprocal to the velocity.

In a certain domain V, slowness of the medium can be represented as a sum of the background and anomalous components. Thus, the total wavefield can be represented as a sum of the incident and scattered fields:

$$s^2 = s_b^2 + \Delta s^2, \quad \Delta s^2\big|_{r \notin V} = 0, \quad P = P^i + P^s,$$

$$\Delta P^i - s_b^2 \partial_t^2 P^i = -F^e, \quad \Delta P^s - s_b^2 \partial_t^2 P^s = \Delta s^2 \partial_t^2 \left(P^i + P^s\right). \tag{8.5}$$

8.3.2 Born Approximation for Homogeneous Space

With the background medium being homogeneous infinite space $\left(s_b = const, V = \mathbb{R}^3\right)$, the fundamental solution of the equations under consideration takes the following form [1]:

$$G\left(\mathbf{r}', t' | \mathbf{r}, t\right) = \frac{\delta\left(t' - t - s_b R'\right)}{4\pi R'}, \tag{8.6}$$

where $R^{\mathrm{i}} = \left|\mathbf{r}^{\mathrm{i}} - \mathbf{r}\right|$.

Therefore, the incident and scattered fields can be written as follows:

$$P^i\left(\mathbf{r}', t\right) = \int_V \frac{F^e\left(\mathbf{r}, t - s_b R'\right)}{4\pi R'} dV,$$

$$P^s\left(\mathbf{r}', t\right) = -\int_V \frac{\partial_t^2\left[P^i\left(\mathbf{r}, t - s_b R'\right) + P^s\left(\mathbf{r}, t - s_b R'\right)\right]}{4\pi R'} \Delta s^2(\mathbf{r})\, dV. \tag{8.7}$$

The forward modelling operator can be obtained from the last expression using Born approximation [1, 43], if the scattered field is negligibly small inside V in comparison with the background field:

$$P^{s,B}\left(\mathbf{r}', t\right) = -\int_V \frac{\partial_t^2 P^i\left(\mathbf{r}, t - s_b R'\right)}{4\pi R'} \Delta s^2(\mathbf{r})\, dV. \tag{8.8}$$

Let D be a Hilbert space of acoustic field data recorded over time interval, T, on the surface of observations, S, with the following metric:

$$(\mathbf{d}_1 \cdot \mathbf{d}_2) = \int_S \int_T d_1\left(\mathbf{r}', t\right) d_2\left(\mathbf{r}', t\right) dt\, dS, \tag{8.9}$$

and M be a Hilbert space of medium models given in the domain V with the metric

$$(\mathbf{m}_1 \cdot \mathbf{m}_2) = \int_V \Delta s_1^2(\mathbf{r})\, \Delta s_2^2(\mathbf{r})\, dV. \tag{8.10}$$

In this case, the migration operator can be represented as follows [1, 42]:

$$\Delta s^2_{\text{migr}}(\mathbf{r}) = -\int_S \int_T \frac{\partial_t^2 P^i\left(\mathbf{r}, t - s_b R'\right)}{4\pi R'} d\left(\mathbf{r}', t\right) dt\, dS\cdot \tag{8.11}$$

We can obtain a simplified expression for the migration operator for the case, where the external energy source is a point source:

$$F^e(\mathbf{r}, t) = \delta(\mathbf{r} - \mathbf{r}_0)\, f(t)\,. \tag{8.12}$$

In this case, an expression for the incident field takes the following form:

$$P^i(\mathbf{r}, t) = \frac{f(t - s_b R_0)}{4\pi R_0}. \tag{8.13}$$

The forward modelling and migration operators are given by the following formulae, respectively:

$$P^{s,B}\left(\mathbf{r}', t\right) = -\int_V \frac{f''\left(t - s_b R' - s_b R_0\right)}{16\pi^2 R' R_0} \Delta s^2(\mathbf{r}) dV \tag{8.14}$$

and

$$\Delta s^2_{\text{migr}}(\mathbf{r}) = -\int_S \int_T \frac{f''\left(t - s_b R' - s_b R_0\right)}{16\pi^2 R' R_0} d\left(\mathbf{r}', t\right) dt\, dS. \tag{8.15}$$

8.3.3 Homogeneous Half Space

With the background medium being a homogeneous infinite half space ($s_b = const, V = \{(x, y, z) : z \geq 0\}$) and the surface $z \geq 0$ being a free surface, the fundamental solution of the equations under consideration takes the following form:

$$G\left(\mathbf{r}', t' | \mathbf{r}, t\right) = \frac{\delta\left(t' - t - s_b R'\right)}{4\pi R'} - \frac{\delta\left(t' - t - s_b \underline{R'}\right)}{4\pi \underline{R'}},$$

$$\underline{\mathbf{r}} = (x, y, -z)^{\mathrm{T}}, \quad \underline{R^{\mathrm{i}}} = \left|\mathbf{r}^{\mathrm{i}} - \underline{\mathbf{r}}\right|, \tag{8.16}$$

where notation $\underline{\mathbf{r}}$ for a "mirrored" position of vector is introduced.

As in the case of homogeneous full space, one can obtain the expressions for forward modelling and migration operators, as follows:

$$P^{s,B}\left(\mathbf{r}',t\right) = -\int\limits_V \left[\begin{array}{l} \frac{f''\left(t-s_b R'-s_b R_0\right)}{R' R_0} - \frac{f''\left(t-s_b R'-s_b \underline{R_0}\right)}{R' \underline{R_0}} \\ + \frac{f''\left(t-s_b \underline{R'}-s_b \underline{R_0}\right)}{\underline{R'}\,\underline{R_0}} - \frac{f''\left(t-s_b \underline{R'}-s_b R_0\right)}{\underline{R'} R_0} \end{array}\right] \frac{\Delta s^2(\mathbf{r}) dV}{16\pi^2}, \tag{8.17}$$

$$\Delta s^2\left(\mathbf{r}\right) = -\int\limits_S \int\limits_T \left[\begin{array}{l} \frac{f''\left(t-s_b R'-s_b R_0\right)}{R' R_0} - \frac{f''\left(t-s_b R'-s_b \underline{R_0}\right)}{R' \underline{R_0}} \\ + \frac{f''\left(t-s_b \underline{R'}-s_b \underline{R_0}\right)}{\underline{R'}\,\underline{R_0}} - \frac{f''\left(t-s_b \underline{R'}-s_b R_0\right)}{\underline{R'} R_0} \end{array}\right] \frac{d\left(\mathbf{r}',t\right) dt\, ds}{16\pi^2}. \tag{8.18}$$

8.4 Migration of Elastic Wavefield

In this section, we present a mathematical derivation of the forward modelling and adjoint operators for the elastic medium. The mathematical model of elastic medium is given in Sect. 8.4.1. In Sect. 8.4.2, the forward and adjoint operators are derived for a quasi-homogeneous medium using the Born approximation. In Sect. 8.4.3, the formulae for the operators are simplified for the case of a polarized point source. In Sect. 8.4.4, a quasi-homogeneous half space is considered.

8.4.1 Elastic Medium

Consider a medium, where a propagation of seismic waves is described by the Lamé equation [1]:

$$\mathbf{\Lambda u} - \frac{\partial^2 \mathbf{u}}{\partial t^2} = -\frac{1}{\rho}\mathbf{f}^{\mathrm{e}}, \quad \mathbf{\Lambda} = c_p^2 \nabla\nabla\cdot - c_s^2 \nabla\times\nabla\times, \tag{8.19}$$

where c_p, c_s are the pressure and shear wave velocities, respectively; ρ is a mass density of the medium; $\mathbf{f}^{\mathrm{e}}$ is a strength of the external force per unit volume applied to the elastic body; and $\mathbf{u}$ is a displacement field.

Let us use for the certain vector fields (potential or solenoidal) and also for the certain velocities (of the pressure or shear waves) the lower Greek letter subscript, e.g. $\mathbf{u}_\alpha, \alpha \in \{p, s\}$, where $\mathbf{u}_p$ is potential field, $\mathbf{u}_s$ is solenoidal field; and $c_\alpha, \alpha \in \{p, s\}$, where c_p and c_s are the velocities of the pressure and shear waves, respectively. As in the acoustic case, the parameters of the medium can be represented in domain V as a sum of the background and anomalous components, and the total wavefield is a sum of the incident and scattered fields:

$$\begin{gathered} c_\alpha^2 = c_{\alpha,b}^2 + \Delta c_\alpha^2,\ \Delta c_\alpha^2\big|_{r\notin V} = 0, \quad \mathbf{\Lambda} = \mathbf{\Lambda}_b + \Delta\mathbf{\Lambda}, \quad \mathbf{u} = \mathbf{u}^i + \mathbf{u}^{\mathrm{s}}, \\ \mathbf{\Lambda}_b \mathbf{u}^i - \frac{\partial^2 \mathbf{u}^i}{\partial t^2} = -\frac{1}{\rho}\mathbf{f}^{\mathrm{e}}, \quad \mathbf{\Lambda}_b \mathbf{u}^s - \frac{\partial^2 \mathbf{u}^s}{\partial t^2} = -\Delta\mathbf{\Lambda}(\mathbf{u}^i + \mathbf{u}^{\mathrm{s}}). \end{gathered} \tag{8.20}$$

The following notations will be used: $s = c^{-1}$, $\mathbf{D}_p^i = \nabla^i \nabla^i \cdot$, $\mathbf{D}_s^i = -\nabla^i \times \nabla^i \times$, $\nabla^i = \left(\partial_{x^i} \partial_{y^i} \partial_{z^i}\right)^{\mathrm{T}}$, $R^{\mathrm{i}} = \left|\mathbf{r}^{\mathrm{i}} - \mathbf{r}\right|$, $R_{\mathrm{i}} = |\mathbf{r}_{\mathrm{i}} - \mathbf{r}|$.

8.4.2 Born Approximation for Homogeneous Space

With the background medium being homogeneous infinite space $\left(c_{\alpha,b} = const,\, V = \mathbb{R}^3\right)$, the fundamental solution of the equations under consideration takes the following form [1]:

$$\mathbf{G}_\alpha^L = \mathbf{D}_\alpha \boldsymbol{\Gamma}_\alpha = \mathbf{D}_\alpha' \boldsymbol{\Gamma}_\alpha, \quad \boldsymbol{\Gamma}_\alpha = \left\{\chi\left(t' - t - s_{\alpha,b} R'\right) - \chi\left(t' - t\right)\right\} \frac{\mathbf{I}}{4\pi R'}, \tag{8.21}$$

where $\mathbf{I}$ is an identity tensor, and $\chi(t) = \max(0, t)$.

Hence, the incident and scattered fields are given by the following formulae:

$$\mathbf{u}_\alpha^i\left(\mathbf{r}', t'\right) = \int\limits_V \int\limits_{-\infty}^{+\infty} \frac{1}{\rho(\mathbf{r})} \mathbf{f}^{\mathrm{e}}(\mathbf{r}, t) \cdot \mathbf{G}_\alpha^L\left(\mathbf{r}', t' | \mathbf{r}, t\right) dt\, dV, \tag{8.22}$$

$$\mathbf{u}_\alpha^s\left(\mathbf{r}', t'\right) = \int\limits_V \int\limits_{-\infty}^{+\infty} \left\{\Delta\boldsymbol{\Lambda}(\mathbf{r})\left[\mathbf{u}^i(\mathbf{r}, t) + \mathbf{u}^s(\mathbf{r}, t)\right]\right\} \cdot \mathbf{G}_\alpha^L\left(\mathbf{r}', t' | \mathbf{r}, t\right) dt\, dV. \tag{8.23}$$

If the scattered field is negligibly small inside V in comparison with the background field, the forward modelling operator can be obtained from the last expression using the Born approximation. Considering the properties of the incident field $\mathbf{D}_\alpha \mathbf{u}_\alpha^i = \nabla^2 \mathbf{u}_\alpha^i$, we obtain

$$\mathbf{u}_\alpha^{s,B}\left(\mathbf{r}', t'\right) = \sum_\beta \int\limits_V \int\limits_{-\infty}^{+\infty} \Delta c_\beta^2(\mathbf{r})\, \nabla^2 \mathbf{u}_\beta^i(\mathbf{r}, t) \cdot \mathbf{G}_\alpha^L\left(\mathbf{r}', t' | \mathbf{r}, t\right) dt\, dV \cdot \tag{8.24}$$

Let D be a Hilbert space of the wavefield data recorded at times T on the surface of observations, S, with the following metric:

$$(\mathbf{d}_1 \cdot \mathbf{d}_2) = \int\limits_S \int\limits_T \mathbf{d}_1\left(\mathbf{r}', t\right) \cdot \mathbf{d}_2\left(\mathbf{r}', t\right) dt\, dS, \tag{8.25}$$

and M be a Hilbert space of the models given in domain V with the metric

$$(\mathbf{m}_1 \cdot \mathbf{m}_2) = \int\limits_V \left[\Delta c_{p,1}^2(\mathbf{r})\, \Delta c_{p,2}^2(\mathbf{r}) + \Delta c_{s,1}^2(\mathbf{r})\, \Delta c_{s,2}^2(\mathbf{r})\right] dV. \tag{8.26}$$

Thus, the migration operator is given by the following formula:

$$\Delta c^2_{\alpha,\mathrm{migr}}(\mathbf{r}) = \int\limits_S \int\limits_T \int\limits_{-\infty}^{+\infty} \left\{\nabla^2 \mathbf{u}^i_\alpha(\mathbf{r},t)\right\} \cdot \mathbf{G}^L\left(\mathbf{r}',t'|\mathbf{r},t\right) \cdot \mathbf{d}\left(\mathbf{r}',t'\right) dt\, dt'\, ds \cdot \quad (8.27)$$

8.4.3 A Polarized Point Source

Let us confine ourselves to consideration of the case, where the external energy source is a polarized point source:

$$\mathbf{f}^e(\mathbf{r},t) = \delta(\mathbf{r}-\mathbf{r}_0)\, F(t)\,\mathbf{f} = \delta(\mathbf{r}-\mathbf{r}_0)\, f''(t)\,\mathbf{f}. \quad (8.28)$$

Let us restrict our consideration to wavelets, $f(t)$, with the property

$$\left[f\left(t'-t\right) - f'\left(t'-t\right)t\right]\Big|^0_{t=+\infty} = f\left(t'\right). \quad (8.29)$$

Considering the vector identity for an arbitrary function $g(\mathbf{r})$ and a constant vector $\mathbf{f}$,

$$\mathbf{f}\cdot\left[\mathbf{D}_\alpha\left(g(\mathbf{r})\,\mathbf{I}\right)\right] = \mathbf{D}_\alpha\left(g(\mathbf{r})\,\mathbf{f}\right), \quad (8.30)$$

and the identity,

$$\int\limits_{-\infty}^{+\infty} \chi\left(t'-t\right) f''(t)\, dt = \left[f\left(t'-t\right) - f'\left(t'-t\right)t\right]\Big|^0_{t=+\infty} = f\left(t'\right), \quad (8.31)$$

expression (Eq. 8.22) for the incident field can be cast in the following form:

$$\mathbf{u}^i_\alpha(\mathbf{r},t) = \mathbf{D}_\alpha \boldsymbol{\phi}_\alpha(\mathbf{r},t),\; \boldsymbol{\phi}_\alpha(\mathbf{r},t) = \frac{f\left(t - s_{\alpha,b}R_0\right) - f(t)}{4\pi\rho(\mathbf{r}_0)\,R_0}\mathbf{f}. \quad (8.32)$$

Considering the following identity,

$$\nabla^2 \frac{f\left(t - s_{\alpha,b}R_0\right) - f(t)}{R_0} = \frac{F\left(t - s_{\alpha,b}R_0\right)}{c^2_{\alpha,b}R_0}, \quad (8.33)$$

one can simplify expressions $\nabla^2 \mathbf{u}^i_\alpha$ as follows:

$$\nabla^2 \mathbf{u}^i_\alpha(\mathbf{r},t) = \mathbf{D}_\alpha \frac{F\left(t - s_{\alpha,b}R_0\right)\mathbf{f}}{4\pi c^2_{\alpha,b}\rho(\mathbf{r}_0)\,R_0} = \frac{1}{\rho(\mathbf{r}_0)}\mathbf{D}^0_\alpha \frac{F\left(t - s_{\alpha,b}R_0\right)\mathbf{f}}{4\pi c^2_{\alpha,b}R_0}. \quad (8.34)$$

Therefore, we have:

$$\int\limits_{-\infty}^{+\infty} \nabla^2 \mathbf{u}^i_\beta (\mathbf{r}, t) \cdot \mathbf{G}^L_\alpha (\mathbf{r}', t' | \mathbf{r}, t) dt = \frac{1}{\rho (\mathbf{r}_0)} \mathbf{D}'_\alpha \mathbf{D}^0_\beta \frac{f\left(t' - s_{\beta,b} R_0 - s_{\alpha,b} R'\right)}{16\pi^2 c^2_{\beta,b} R_0 R'} \mathbf{f}. \tag{8.35}$$

Thus, formula for forward operator takes the following form:

$$\mathbf{u}^{s,B}_\alpha \left(\mathbf{r}', t'\right) = \sum_\beta \frac{\mathbf{D}'_\alpha \mathbf{D}^0_\beta}{\rho (\mathbf{r}_0)} \int\limits_V \Delta c^2_\beta (\mathbf{r}) \frac{f\left(t' - s_{\beta,b} R_0 - s_{\alpha,b} R'\right)}{16\pi^2 c^2_{\beta,b} R_0 R'} \mathbf{f} dV, \tag{8.36}$$

and formula for migration operator can be written as follows:

$$\Delta c^2_{\beta,\mathrm{migr}} (\mathbf{r}) = \sum_\alpha \int\limits_S \int\limits_T \frac{\mathbf{d}\left(\mathbf{r}', t'\right)}{\rho (\mathbf{r}_0)} \cdot \mathbf{D}'_\alpha \mathbf{D}^0_\beta \frac{f\left(t' - s_{\beta,b} R_0 - s_{\alpha,b} R'\right)}{16\pi^2 c^2_{\beta,b} R_0 R'} \mathbf{f} dt' dS \cdot \tag{8.37}$$

8.4.4 Homogeneous Half Space

With the background medium being a homogeneous infinite half space $\left(c_{\alpha,b} = const, V = \{(x, y, z) : z \geq 0\}\right)$ and the surface $z \geq 0$ being a free surface, the fundamental solution of equations under consideration takes the following form:

$$\mathbf{G}^{L,H}_\alpha \left(\mathbf{r}', t' | \mathbf{r}, t\right) = \mathbf{D}'_\alpha \left[\mathbf{\Gamma}_\alpha - \underline{\mathbf{\Gamma}_\alpha}\right], \quad \underline{\mathbf{\Gamma}_\alpha} = \left\{\chi\left(t' - t - s_{\alpha,b} \underline{R'}\right) - \chi\left(t' - t\right)\right\} \frac{\mathbf{I}}{4\pi \underline{R'}}$$

$$\underline{\mathbf{r}} = (x, y - z)^{\mathrm{T}}, \quad \underline{R}^{\mathrm{i}} = \left|\mathbf{r}^{\mathrm{i}} - \underline{\mathbf{r}}\right| \cdot \tag{8.38}$$

As in the case of homogeneous full space, one can obtain the expressions for the incident wavefield:

$$\mathbf{u}^i_\alpha (\mathbf{r}, t) = \mathbf{D}^0_\alpha \frac{f\left(t - s_{\alpha,b} R_0\right) - f(t)}{4\pi \rho (\mathbf{r}_0) R_0} \mathbf{f} - \mathbf{D}^0_{\underline{\alpha}} \frac{f\left(t - s_{\alpha,b} R_0\right) - f(t)}{4\pi \rho (\mathbf{r}_0) \underline{R_0}} \mathbf{f}. \tag{8.39}$$

Finally, the forward modelling and migration operators can be summarized as follows:

$$\mathbf{u}^{s,B}_\alpha \left(\mathbf{r}', t'\right) = \sum_\beta \frac{\mathbf{D}'_\alpha}{\rho (\mathbf{r}_0)} \int\limits_V \Delta c^2_\beta (\mathbf{r})$$

$$\left\{\begin{array}{l}\mathbf{D}_{\beta}^{0}\left[\dfrac{f\left(t'-s_{\beta,b}R_0-s_{\alpha,b}R'\right)}{16\pi^2c_{\beta,b}^2R_0R'}-\dfrac{f\left(t'-s_{\beta,b}R_0-s_{\alpha,b}\underline{R'}\right)}{16\pi^2c_{\beta,b}^2R_0\underline{R'}}\right]\\ \quad+\mathbf{D}_{\beta}^{\underline{0}}\left[\dfrac{f\left(t'-s_{\beta,b}\underline{R_0}-s_{\alpha,b}\underline{R'}\right)}{16\pi^2c_{\beta,b}^2\underline{R_0}\underline{R'}}-\dfrac{f\left(t'-s_{\beta,b}\underline{R_0}-s_{\alpha,b}R'\right)}{16\pi^2c_{\beta,b}^2\underline{R_0}R'}\right]\end{array}\right\}\mathbf{f}dV, \quad (8.40)$$

$$\Delta c_{\beta,\mathrm{migr}}^2(\mathbf{r})=\sum_{\alpha}\int_S\int_T\frac{\mathbf{d}\left(\mathbf{r}',t'\right)}{\rho\left(\mathbf{r}_0\right)}.\mathbf{D}'_{\alpha}$$

$$\left\{\begin{array}{l}\mathbf{D}_{\beta}^{0}\left[\dfrac{f\left(t'-s_{\beta,b}R_0-s_{\alpha,b}R'\right)}{16\pi^2c_{\beta,b}^2R_0R'}-\dfrac{f\left(t'-s_{\beta,b}R_0-s_{\alpha,b}\underline{R'}\right)}{16\pi^2c_{\beta,b}^2R_0\underline{R'}}\right]\\ \quad+\mathbf{D}_{\beta}^{\underline{0}}\left[\dfrac{f\left(t'-s_{\beta,b}\underline{R_0}-s_{\alpha,b}\underline{R'}\right)}{16\pi^2c_{\beta,b}^2\underline{R_0}\underline{R'}}-\dfrac{f\left(t'-s_{\beta,b}\underline{R_0}-s_{\alpha,b}R'\right)}{16\pi^2c_{\beta,b}^2\underline{R_0}R'}\right]\end{array}\right\}\mathbf{f}dt'dS, \quad (8.41)$$

where $\mathbf{D}_{\beta}^{\underline{0}}$ is obtained from $\mathbf{D}_{\beta}^{0}$ by substitution $\partial_{z_0}\rightarrow-\partial_{z_0}$.

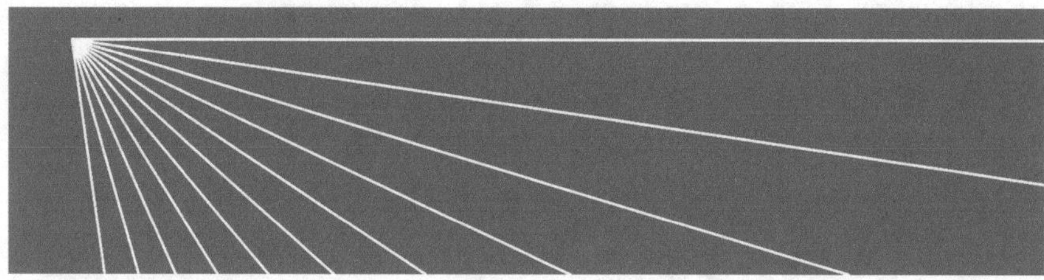

Fig. 8.1 Model "Dipping Reflectors"

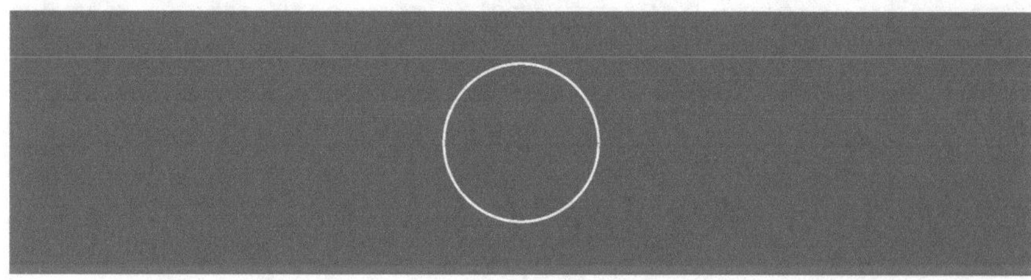

Fig. 8.2 Model "Circle"

8.5 Comaprison of Migration Images for Acoustic and Elastic Models

In this section, we present a comparison of the acoustic and elastic Born migration algorithms. For this purpose, we run a number of numerical experiments of for-

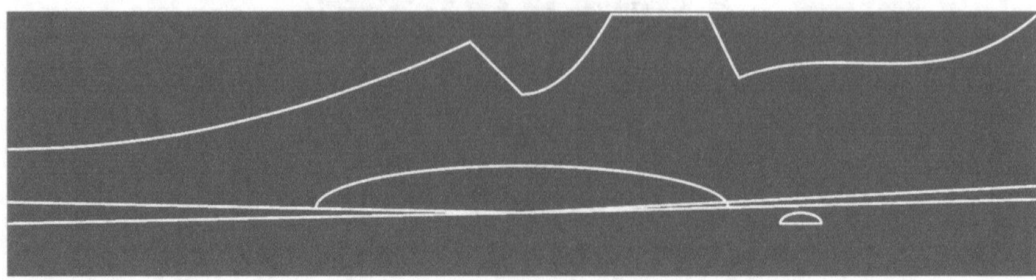

Fig. 8.3 Model "Marmousi"

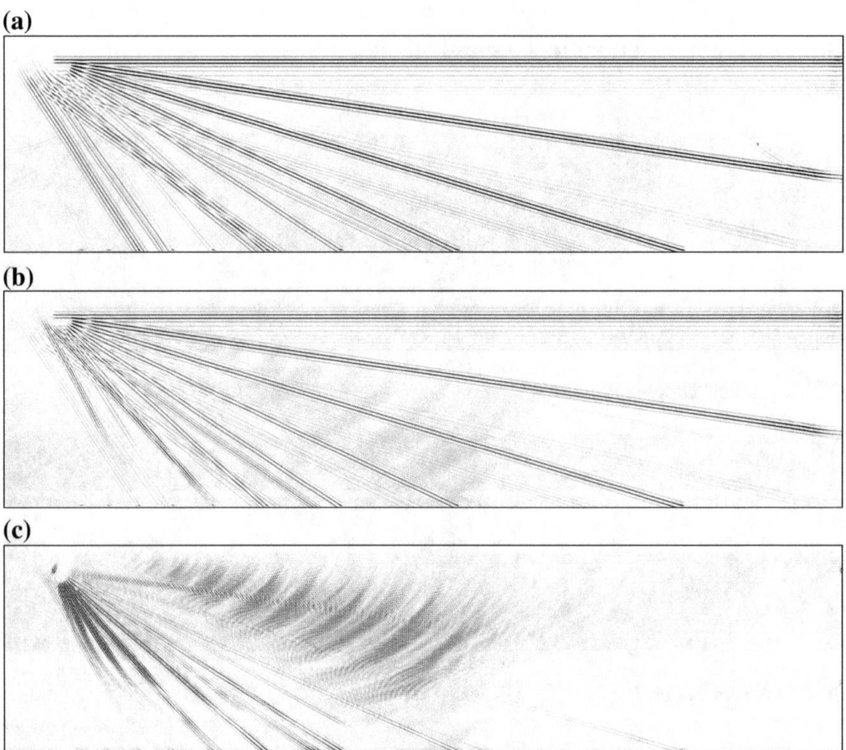

Fig. 8.4 Migration images for "Dipping Reflectors": **a** acoustic migration image, **b** p-component, **c** s-component

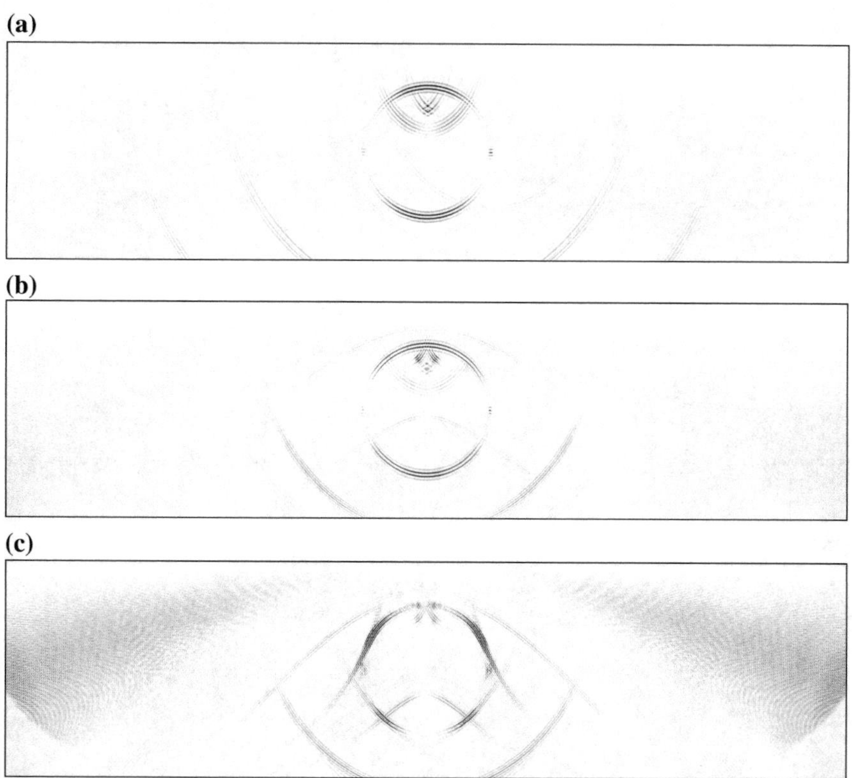

Fig. 8.5 Migration images for "Circle": **a** acoustic migration image, **b** p-component, **c** s-component

ward modelling and migration imaging for weakly inhomogeneous media using the computer codes based on Eqs. 8.40, 8.41 and 8.18.

The models of the media used in these numerical tests are shown in Figs. 8.1, 8.2 and 8.3. In these figures, the white color corresponds to the anomalous features of the medium, and the grey color corresponds to the background medium. All computations were carried out for 2D models (with one of the horizontal coordinates fixed, i.e. $y = 0$). Each model is 10 km wide, 2.5 km deep and has the background values of pressure and shear wave velocities equal to 2.5 km/s and 1.25 km/s, respectively. The inhomogeneities have the value of contrast $\Delta c_{\alpha}^2/c_{\alpha,b}^2$ equal to 1%. The mass density of media at the level of geophones equals to 2500 kg/m^3.

For these models, synthetic zero-offset data were obtained based on Eq. 8.40 with the following parameters: the vertically polarized $\mathbf{f} = (0, 0, 1)^T$ geophones were located at a level of $z = 15$ m every 10 m across the horizontal axis; the source pulse was the Ricker wavelet $F(t) = \left(1 - 2\pi^2 f_M^2 t^2\right) e^{-\pi^2 f_M^2 t^2}$, $f_M = 25$ Hz; the data were recorded each 2 ms during a period of 4 s. A detail discussion of the seismic sections is given in Sect. 8.6. It should be noted that for 2D models, the y-component of the data is equal to zero.

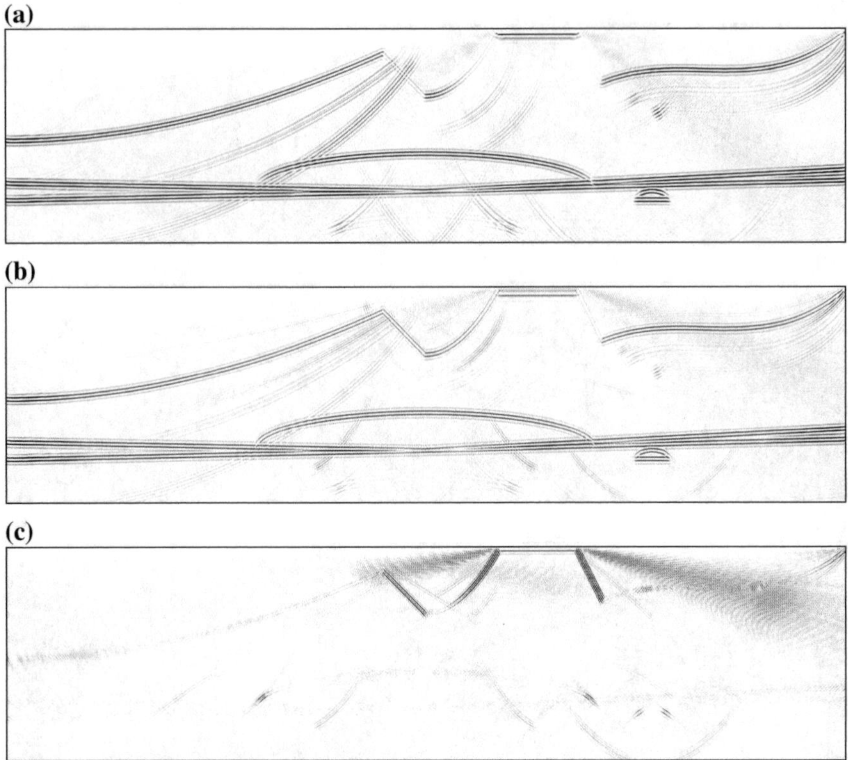

Fig. 8.6 Migration images for "Marmousi": **a** acoustic migration image, **b** p-component, **c** s-component

The migration images for the z-component of the scattered field (the x and y components are considered being equal to zero) are shown in Figs. 8.4, 8.5 and 8.6. Each figure consists of three parts: the acoustic migration images (Figs. 8.4a, 8.5 and 8.6a) obtained based on Eq. 8.18 and the elastic migration images obtained based on Eq. 8.41 for p-component (Figs. 8.4b, 8.5 and 8.6b) and for s-component (Figs. 8.4c, 8.5 and 8.6c). The step of the discretization grid was 10 m.

The data shown in Figs. 8.4, 8.5 and 8.6 were initially processed. To eliminate the time fade-out of the data and the depth fade-out of the images, they were weighted according to the following formulae:

$$d^{\text{adjusted}}\left(\mathbf{r}', t'\right) = d\left(\mathbf{r}', t'\right) / \sqrt{\mathbf{L}\left(\mathbf{r}', t'|\mathbf{r}\right)\mathbf{L}^*\left(\mathbf{r}|\mathbf{r}', t'\right)},$$

$$m^{\text{adjusted}}\left(\mathbf{r}\right) = m\left(\mathbf{r}\right) / \sqrt{\mathbf{L}^*\left(\mathbf{r}|\mathbf{r}', t'\right)\mathbf{L}\left(\mathbf{r}', t'|\mathbf{r}\right)}. \tag{8.42}$$

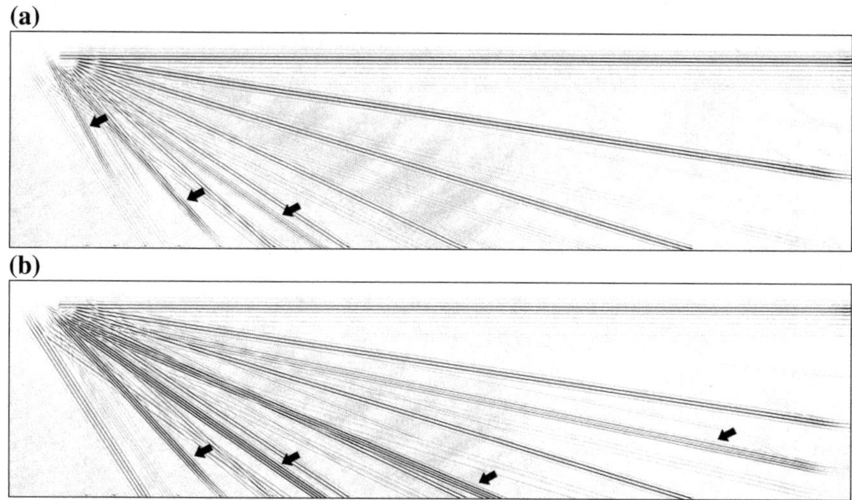

Fig. 8.7 The p-components of elastic migration images for model "Dipping Reflectors" calculated for: **a** vertical component, **b** vertical and horizontal components (false boundaries are marked with arrows)

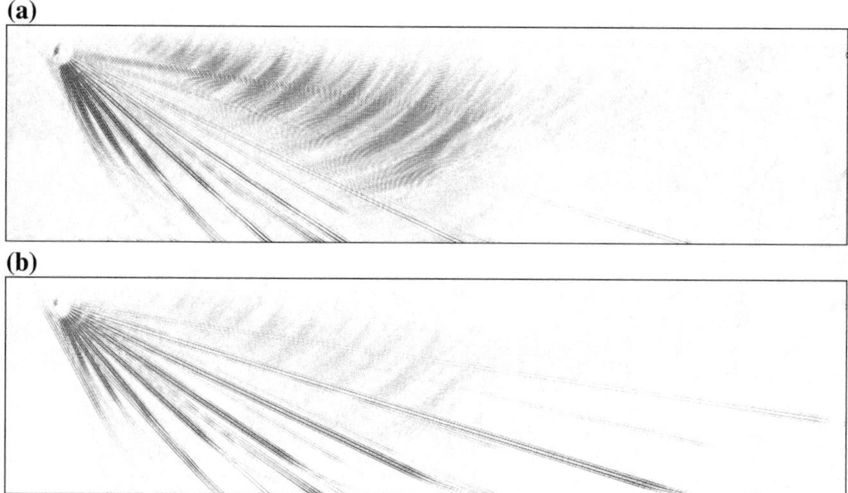

Fig. 8.8 The s-components of elastic migration images for model "Dipping Reflectors" calculated for: **a** vertical component, **b** vertical and horizontal components

For Model 3, the maximum values of a, p, and s images were calculated, then the minimal of them was denoted as K_n, finally, all three images were divided by a factor of $0.5K_n$ to make a better comparison

One can see that both methods locate the interfaces, in general. The acoustic and p-component images have a poor resolution of the steep interfaces (though, p-

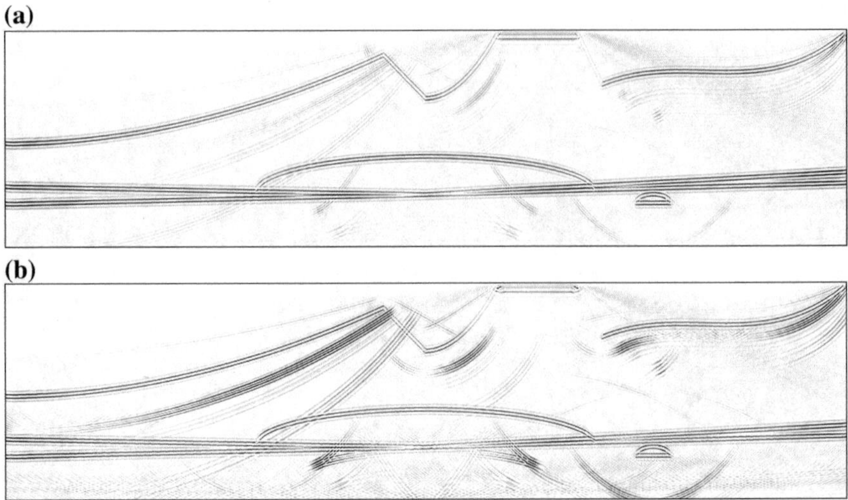

Fig. 8.9 The p-components of elastic migration images for model "Marmousi" calculated for: **a** vertical component, **b** vertical and horizontal components

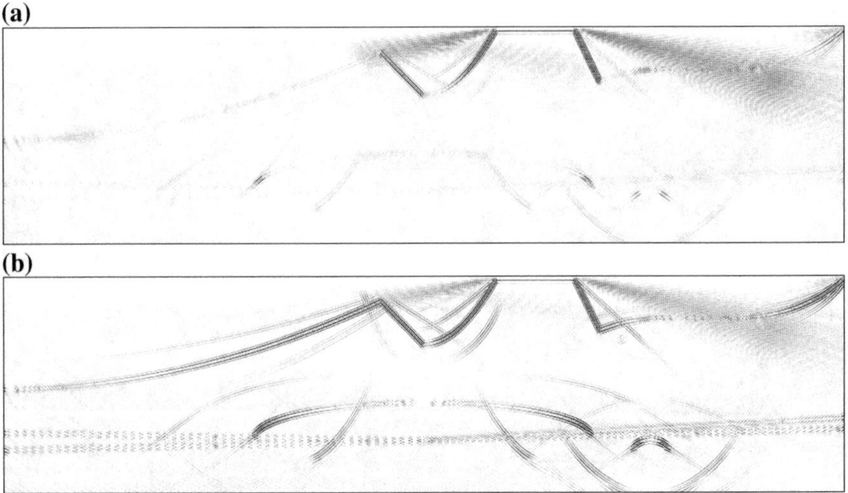

Fig. 8.10 The s-components of elastic migration images for model "Marmousi" calculated for: **a** vertical component, **b** vertical and horizontal components

component has slightly better resolution) and a good resolution of sloping, while the s-component image has the opposite properties. Both methods show false boundaries, which are more strongly pronounced in the acoustic images than in the elastic ones. The sources of the false boundaries are discussed in Sect. 8.6.

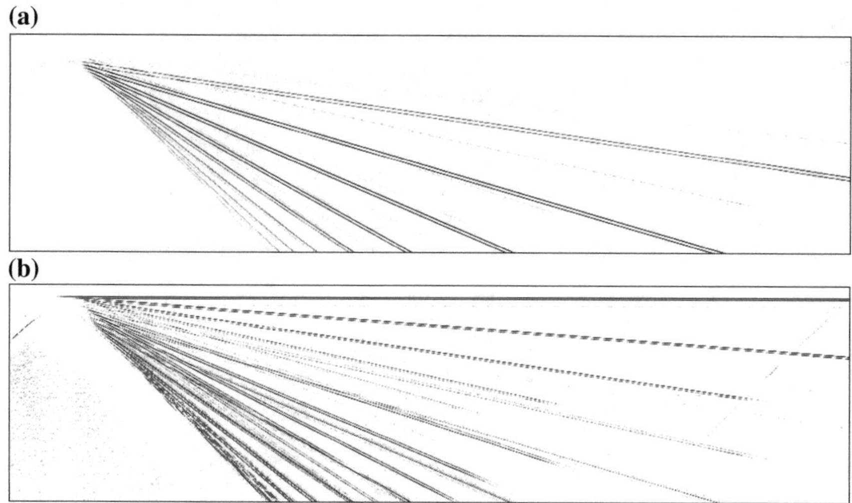

Fig. 8.11 Zero-offset data for model "Dipping Reflectors": **a** x-component, **b** z-component

To use additional information about the medium contained in the horizontal components of the scattered field, one can apply the elastic migration method to all three components. The migration images calculated in this way are present in Figs. 8.7, 8.8, 8.9 and 8.10. Each figure consists of an image calculated for the vertical component only (Figs. 8.7a, 8.8, 8.9 and 8.10a) and an image calculated for both the vertical and horizontal components (Figs. 8.7b, 8.8, 8.9 and 8.10b).

These figures show that such an approach increases the magnitude of the false boundaries in the p-component, but improves the image produced by the s-component. Thus, we can suggest a two-pass migration algorithm, which uses the p-component of the z-component migration only and the s-component for the full wavefield migration only.

8.6 Synthetic Seismograms and the Sources of False Boundaries

Figures 8.11, 8.12 and 8.13 show the seismograms of the x- and z-components for the models discussed in Sect. 8.5. All images in these figures were processed as follows: (a) initially, the maximum value of an image, K_n, was calculated; (b) then the x-component images were divided by a factor of $0.5K_n$ and the z-component images were divided by a factor of $0.2K_n$.

To understand the sources of the false boundaries in migration images, let us discuss in detail a simple model shown in Fig. 8.14 and the corresponding zero-offset data shown in Fig. 8.15. This model contains one flat inclined boundary only,

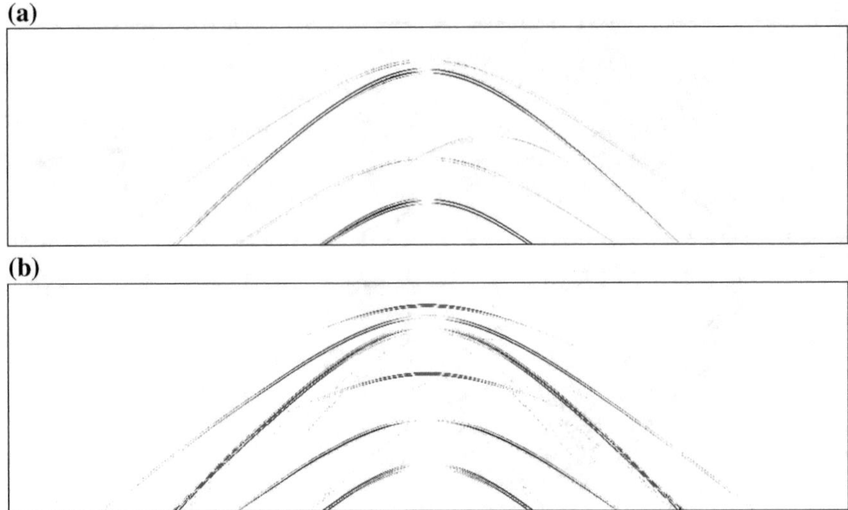

Fig. 8.12 Zero-offset data for model "Circle": **a** x-component, **b** z-component

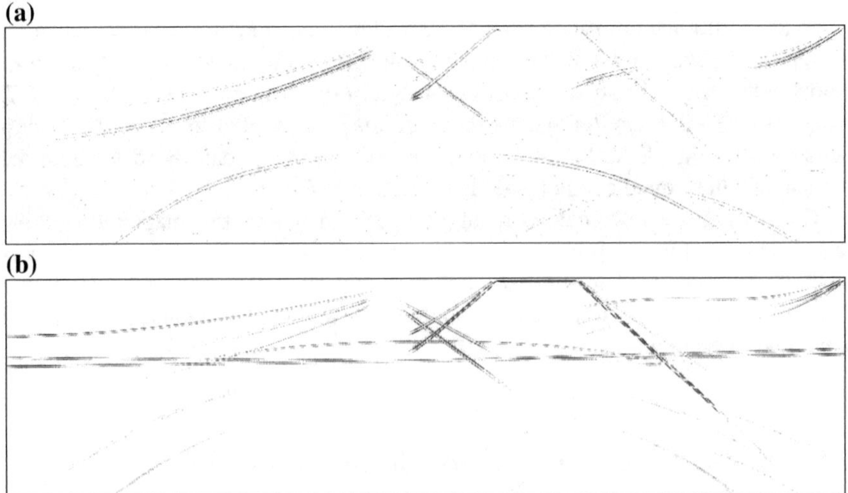

Fig. 8.13 Zero-offset data for model "Marmousi": **a** x-component, **b** z-component

but each component of the scattered field has three reflections, which, from the top to the bottom, are caused by the p-waves, $p \leftrightarrow s$ transitional waves, and by the s-waves, respectively. Each reflection consists of two parts: the earliest one is caused by the direct waves and the latest one is caused by the waves reflected from the surface of observation.

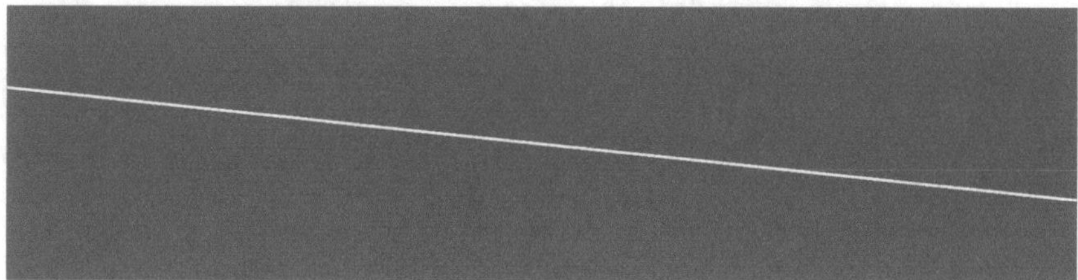

Fig. 8.14 Model with one dipping reflector

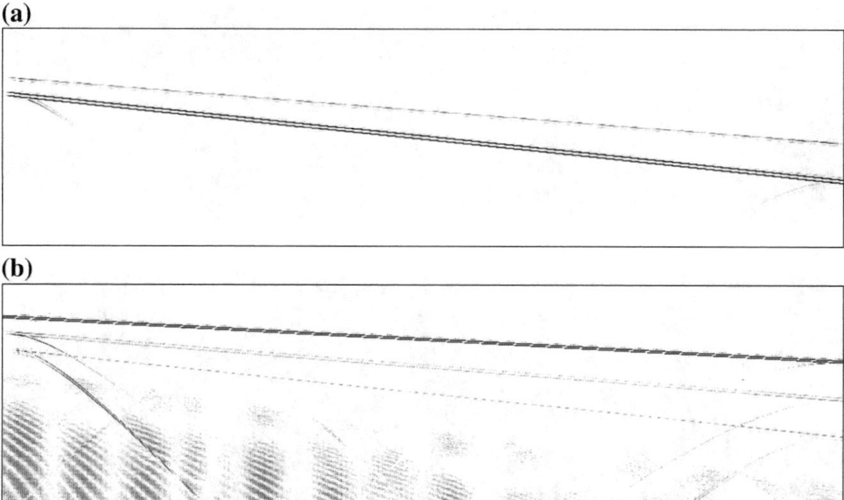

Fig. 8.15 Zero-offset data for model with one dipping reflector: **a** p-component, **b** s-component

Let's check where each of these reflections moves after the migration. In Fig. 8.16, the color of a boundary corresponds to its source, parts of images are interleaved to prevent overlapping, and false boundaries are marked by numbers in the images produced by the p- and s-componens.

Since the acoustic migration algorithm uses just one wave velocity, the delay of the reflection defines uniquely the depth of the image of the corresponding boundary.

The same applies to boundaries 4, 5 and 6, 7 in the migration image of the p-component of the elastic wavefield. Boundaries 1 and 2 are caused by the p-waves, and boundary 3 is caused by the s-waves, back-propagated with the velocity of transitional waves.

In the elastic migration image produced by the s-component, the p-waves become s-waves, which causes boundaries 1 and 3; p-waves become transitional, which causes boundary 2; the transitional waves become s-waves, which causes boundary 4, and the s-waves become transitional, which causes boundaries 5 and 6.

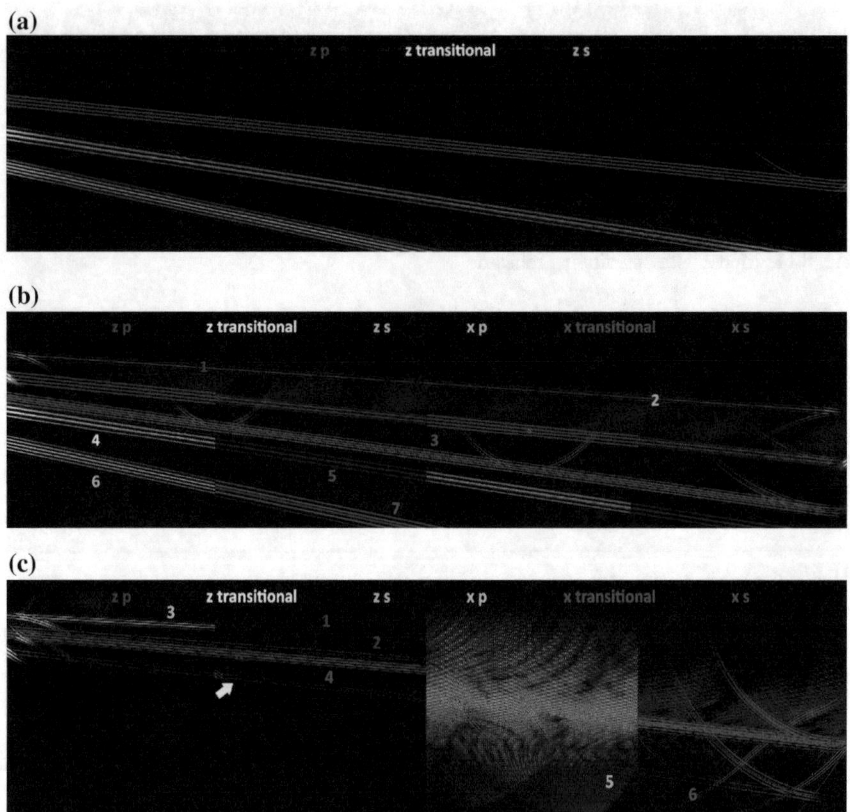

Fig. 8.16 Different boundaries in migration image: **a** acoustic, **b** elastic p, **c** elastic s

8.7 Conclusions

In this chapter, we have derived the expressions for forward seismic modelling and migration operators for quasi-homogeneous acoustic and elastic media based on the Born approximation and considering a permanently polarized point source. For a set of low-contrast geological models, a comparison between acoustic and elastic methods has shown that the straightforward expression of migration operator for the acoustic case can be easily extended to the elastic isotropic case as well. The migration images in the elastic case show the correct location of the reflection boundaries and produce less pronounced false boundaries as compared to the images in the acoustic case.

Acknowledgements This work has been performed at Moscow Institute of Physics and Technology and was supported by the Russian Science Foundation, grant no. 14-11-00263. This work has been carried out using computing resources of the federal collective usage center Complex for Simulation and Data Processing for Mega-science Facilities at NRC "Kurchatov Institute", http://ckp.nrcki.ru/.

References

1. Zhdanov MS (2002) Geophysical inverse theory and regularization problems. Methods in geochemistry and geophysics, vol 36. Elsevier, Netherlands
2. Hagedoorn JG (1954) A process of seismic reflection interpretation. Geophys Prosp 2(2):85–127
3. Claerbout JF (1971) Toward a unified theory of reflector mapping. Geophysics 36(3):467–481
4. Claerbout JF, Doherty SM (1972) Downward continuation of moveout corrected seismograms. Geophysics 37(5):741–768
5. Claerbout JF (1970) Coarse grid calculations of wave in inhomogeneous media with application to delineation of complicated seismic structure. Geophysics 35(3):407–418
6. French WS (1975) Computer migration of oblique seismic reflection profiles. Geophysics 40(6):961–980
7. Schneider WA (1978) Integral formulation for migration in two and three dimensions. Geophysics 43(1):49–76
8. Zhdanov MS, Matusevich VY, Frenkel MA (1988) Seismic and electromagnetic migration, Nauka, Moscow
9. Zhdanov MS (1988) Integral transforms in geophysics. Springer Verlag, Heidelberg
10. Stolt RH (1978) Migration by Fourier transform. Geophysics 43(1):23–48
11. Clayton RW, Stolt RH (1981) A Born-WKBJ inversion method for acoustic reflection data. Geophysics 46(11):1559–1567
12. Cohen JK, Bleistein N (1979) Velocity inversion procedure for acoustic waves. Geophysics 44(6):1077–1087
13. Beydoun W, Tarantola A (1988) First Born and Rytov approximations: modeling and inversion conditions in a canonical example. J Acoust Soc Am 83(3):1045–1055
14. Morse PM, Feshbach H (1953) Methods of theoretical physics. McGraw shill Book Co. Inc, New York
15. Mechan G (1982) Determination of source parameters by wavefield extrapolation. Geophys J Int 71(3):613–628
16. Mechan G (1983) Migration by extrapolation of time-dependent boundary values. Geophys Prospect 31(3):413–420
17. Baysal E, Kosloff D, Sherwood J (1983) Reverse time migration. Geophysics 48(11):1514–1524
18. Gray SH (1992) Frequency-selective design of the Kirchhoff migration operator. Geophys Prospect 40(5):565–571
19. Moser TJ (2012) Review of ray-Born forward modeling for migration and diffraction analysis Stud Geophys Geod 56(2):411–432
20. Cervený V, Klimeš L, Pšenčík I (2007) Seismic ray method: recent developments. Adv Geophys 48:1–126
21. Thierry P, Operto S, Lambaré G (1999) Fast 2-D ray—Born migration/inversion in complex media. Geophysics 64(1):162–181
22. Sun R, McMechan G, Hsiao HH, Chow J (2004) Separating P- and S-waves in prestack elastic seismograms using divergence and curl. Geophysics 69:286–297
23. Amundsen L, Reitan A (1995) Decomposition of multicomponent sea-floor data into upgoing and downgoing P- and S-waves. Geophysics 60(2):563–572
24. Amano H (1995) An analytical solution to separate P-waves and S-waves in VSP wavefields. Geophysics 60(4):955–967
25. Sun R, McMechan GA (2001) Scalar reverse-time depth migration of prestack elastic seismic data. Geophysics 66(5):1519–1527
26. Whitmore ND, Marfurt KJ (1988) Method for depth imaging multicomponent seismic data. U.S. patent 4, pp 766–574
27. Jiao K, Huang W, Vigh D, Kapoor J, Coates R, Starr EW, Cheng X (2012) Elastic migration for improving salt and subsalt imaging and inversion. Steeples D (ed) SEG Technical program expanded abstracts. Society of exploration geophysicists, U.S., pp 1–5

28. Kuo JT, Dai T (1984) Kirchhoff elastic wave migration for the case of noncoincident source and receiver. Geophysics 49(8):1223–1238
29. Keho TH, Wu R-S (1987) Elastic Kirchhoff migration for vertical seismic profiles. 67th Ann Int Mtg, Soc. Expl. Geophys., Expanded Abstracts, pp 774–776
30. Hokstad K (2000) Multicomponent Kirchhoff migration. Geophysics 65(3):861–873
31. Zhe J, Greenhalgh SA (1997) Prestack multicomponent migration. Geophysics 62(2):598–613
32. Gherasim M (2005) 3-D VSP elastic kirchhoff pre-stack depth migration. Monk D (ed) SEG Technical program expanded abstracts. Society of Exploration Geophysicists, U.S., pp 2649–2652
33. Luo Y, Tromp J, Denel B, Calandra H (2013) 3D coupled acoustic-elastic migration with topography and bathymetry based on spectral-element and adjoint methods. Geophysics 78(4):193–202
34. Lu R, Yan J, Traynin P, Anderson JE, Dickens T (2010) Elastic RTM: anisotropic wave-mode separation and converted-wave polarization correction. Levin S (ed) SEG Technical program expanded abstracts. Society of Exploration Geophysicists, U.S., pp 3171–3175
35. Xie X, Wu R (2005) Multicomponent prestack depth migration using the elastic screen method. Geophysics 70(1):30–37
36. Yan J, Sava P (2008) Isotropic angle-domain elastic reverse-time migration. Geophysics 73(6):229–239
37. Beydoun WB, Mendes M (1989) Elastic ray-Born L2-migration/inversion. Geophys J Int 97(1):151–160
38. Beylkin G, Burridge R (1990) Linearized inverse scattering problems in acoustics and elasticity. Wave Motion 12(1):15–52
39. Eaton DWS, Stewart RR (1990) 2-1/2 D elastic ray-Born migration/inversion theory for transversely isotropic media. CREWES Proj Res Rep 2:361–377
40. Bansal R, Mrinal K (2010) Sen Ray-Born inversion for fracture parameters. Geophys J Int 180(3):1274–1288
41. Clærbout JF (1992) Earth soundings analysis: processing versus inversion, vol 6. Blackwell Scientific Publications, London
42. Zhdanov MS (2015) Inverse theory and applications in geophysics, Elsevier, Netherlands, 41
43. Hudson JA, Heritage JR (1981) The use of the Born approximation in seismic scattering problems. Geophys J Int 66(1):221–240

Chapter 9
Migration of Elastic Fields Based on Kirchhoff and Rayleigh Integrals

Alena V. Favorskaya and Michael S. Zhdanov

Abstract Seismic method is one of the main methods of geophysical exploration. Interpretation of seismic data requires their transformation into the images of the subsurface geological formation. These images can be produced by migration of the observed seismic data from the surface of observation in the lower half space. The traditional migration algorithms are based on the solution of the wave equation for the seismic field in the reverse time. In this chapter, we develop an algorithm of migration based on the elastic field equations, which describes more accurately the seismic field propagation than the simple acoustic wave equation. Our method uses the Kirchhoff and Rayleigh integral formulas for elastic wavefields. Considering that the elastic model approximates the propagation of the seismic fields in the geological media better than the acoustic model, the developed approach can significantly improve the quality of interpretation of seismic data in complex geological formations.

Keywords Numerical simulation · Migration · Kirchhoff integrals
Rayleigh integrals · Grid-characteristic method · Elastic field · Green tensor
Converging wave · Diverging wave · Adjoint Green tensor

A. V. Favorskaya (✉) · M. S. Zhdanov
Moscow Institute of Physics and Technology, 9 Intitutsky per., Dolgoprudny 141700, Russian Federation
e-mail: aleanera@yandex.ru

A. V. Favorskaya
Scientific Research Institute for System Studies of the Russian Academy of Sciences, 36(1) Nahimovskij av., Moscow 117218, Russian Federation

M. S. Zhdanov
University of Utah, 115 South 1460 East, Rm 383, Salt Lake City, UT 84112, USA
e-mail: michael.zhdanov@utah.edu

A. V. Favorskaya and I. B. Petrov (eds.), *Innovations in Wave Processes Modelling and Decision Making*, Smart Innovation, Systems and Technologies 90,
https://doi.org/10.1007/978-3-319-76201-2_9

9.1 Introduction

Migration transformation is widely used in seismic exploration for imaging geological cross sections. One of the methods of obtaining the migration field is based on the Rayleigh integral formula, which appears from the Kirchhoff integral formula in a case of the plane observation surface. One can find a detailed review of the migration process in Chap. 8.

For example, in a case of acoustic wavefield equation, the classical Kirchhoff formula can be used to obtain the reverse-time migration algorithm [1]. One can find several methods of seismic migration using Kirchhoff formulas for the acoustic field in [2], where a case of the zero offset between the receivers and transmitters was considered. The migration algorithms for different offsets between the sources and receivers are discussed in a number of publications [3–6]. In [7], a case of migration in anisotropic medium is discussed. Papers [8–10] consider the case of the vector field migration instead of a scalar field. The integral formulas for scattering of elastic waves are discussed in [11–13].

Monograph [1] introduces the Kirchhoff formulas in the case of a general elastic field equation. The integral Kirchhoff formula requires knowing both the elastic field and its normal derivative on the surface of integration, S. However, it can be demonstrated that the boundary values of the wavefield and its normal derivative on the surface S are not independent [12]. In fact, they are related by the corresponding integral equation. Using this relationship, in a case of the plane observation surface S, one can transform the integral Kirchhoff formula into the Rayleigh formula, which operates with the components of the field themselves.

A comparison of the solutions of the wavefield equations obtained using this Rayleigh formula and using the grid-characteristic method [14–19] can be found in Sect. 9.6. Note that, the Rayleigh formula can be used for obtaining the migration field as well [1].

Section 9.2 presents the fundamental elastic field equations. The analytical expressions for Rayleigh formula are developed in Sect. 9.4. The discretization of these expressions is discussed in Sect. 9.5. Then the analytical algorithm is developed in Sect. 9.7 by substituting in these analytical expressions the Adjoint Green tensor discussed in Sect. 9.3 and by conducting the corresponding analytical calculations. The main stages of these calculations are discussed in Sect. 9.8. The results of migration imaging using the developed algorithm are presented in Sect. 9.9. Section 9.10 provides a conclusion to the Chapter.

9.2 Equations of the Elastic Wavefields

The system of equations describing the elastic wave propagation can be written as follows [1, 12]:

$$\mathbf{L}\vec{U} - \frac{\partial^2 \vec{U}}{\partial t^2} = 0, \tag{9.1}$$

where $\vec{U}$ is the displacement vector; $\mathbf{L}$ is the Lamé's differential operator:

$$\mathbf{L} = c_p^2 \nabla\nabla \cdot - c_s^2 \nabla \times \nabla \times \tag{9.2}$$

and c_p, c_s are the velocities of the compressional and shear waves (P-waves and S-waves), respectively. These velocities are related to the Lamé's coefficients (λ and μ) by the following formulae:

$$c_p = \sqrt{\frac{\lambda + 2\mu}{\rho}} \tag{9.3}$$

$$c_s = \sqrt{\frac{\mu}{\rho}} \tag{9.4}$$

where ρ is the density of the media.

There is a one-to-one relationship between the displacement vector $\vec{U}$, the velocity vector $\vec{v}$, and the stress tensor $\boldsymbol{\sigma}$, which can be expressed as shown in Eqs. 9.5–9.11.

$$\sigma_{11} = \lambda\left(\frac{\partial U_1}{\partial x} + \frac{\partial U_2}{\partial y} + \frac{\partial U_3}{\partial z}\right) + 2\mu\frac{\partial U_1}{\partial x} \tag{9.5}$$

$$\sigma_{22} = \lambda\left(\frac{\partial U_1}{\partial x} + \frac{\partial U_2}{\partial y} + \frac{\partial U_3}{\partial z}\right) + 2\mu\frac{\partial U_2}{\partial y} \tag{9.6}$$

$$\sigma_{33} = \lambda\left(\frac{\partial U_1}{\partial x} + \frac{\partial U_2}{\partial y} + \frac{\partial U_3}{\partial z}\right) + 2\mu\frac{\partial U_3}{\partial z} \tag{9.7}$$

$$\sigma_{12} = \mu\left(\frac{\partial U_1}{\partial y} + \frac{\partial U_2}{\partial x}\right) \tag{9.8}$$

$$\sigma_{13} = \mu\left(\frac{\partial U_1}{\partial z} + \frac{\partial U_3}{\partial x}\right) \tag{9.9}$$

$$\sigma_{23} = \mu\left(\frac{\partial U_2}{\partial z} + \frac{\partial U_3}{\partial y}\right) \tag{9.10}$$

$$\vec{v} = \frac{\partial \vec{U}}{\partial t} \tag{9.11}$$

9.3 Green's Tensors

The Green's tensor for the elastic field equation (Eq. 9.1) describes the propagation of elastic waves generated by a point pulse force. It plays a very important role both in the solution of the elastic wave equations and in the development of the migration methods.

This Green's tensor can be written in the following form:

$$\mathbf{G}^{L}\left(\vec{r},t\left|\vec{r}',t'\right.\right)=\mathbf{G}^{L(p)}\left(\vec{r},t\left|\vec{r}',t'\right.\right)+\mathbf{G}^{L(s)}\left(\vec{r},t\left|\vec{r}',t'\right.\right)=\begin{bmatrix} g_{11} & g_{12} & g_{13} \\ g_{21} & g_{22} & g_{23} \\ g_{31} & g_{32} & g_{33} \end{bmatrix}, \tag{9.12}$$

where we use the following notations:

$$g_{11}=-\frac{\partial^2 g_p}{\partial x^2}-\frac{\partial^2 g_s}{\partial y^2}-\frac{\partial^2 g_s}{\partial z^2} \tag{9.13}$$

$$g_{22}=-\frac{\partial^2 g_p}{\partial y^2}-\frac{\partial^2 g_s}{\partial x^2}-\frac{\partial^2 g_s}{\partial z^2} \tag{9.14}$$

$$g_{33}=-\frac{\partial^2 g_p}{\partial z^2}-\frac{\partial^2 g_s}{\partial x^2}-\frac{\partial^2 g_s}{\partial y^2} \tag{9.15}$$

$$g_{12}=-\frac{\partial^2 g_p}{\partial x\partial y}+\frac{\partial^2 g_s}{\partial x\partial y} \tag{9.16}$$

$$g_{13}=-\frac{\partial^2 g_p}{\partial x\partial z}+\frac{\partial^2 g_s}{\partial x\partial z} \tag{9.17}$$

$$g_{23}=-\frac{\partial^2 g_p}{\partial y\partial z}+\frac{\partial^2 g_s}{\partial y\partial z} \tag{9.18}$$

In the above formulae, symbols g_p and g_s denote the following functions. In a case of diverging waves, g_p and g_s are given by the following formulae:

$$g_p\left(\vec{r},t\left|\vec{r}',t'\right.\right)=\frac{1}{4\pi\left|\vec{r}-\vec{r}'\right|}\left(\chi\left(t-t'\right)-\chi\left(t-t'-\frac{\left|\vec{r}-\vec{r}'\right|}{c_p}\right)\right) \tag{9.19}$$

$$g_s\left(\vec{r},t\left|\vec{r}',t'\right.\right)=\frac{1}{4\pi\left|\vec{r}-\vec{r}'\right|}\left(\chi\left(t-t'\right)-\chi\left(t-t'-\frac{\left|\vec{r}-\vec{r}'\right|}{c_s}\right)\right). \tag{9.20}$$

In a case of converging waves, g_p and g_s are given by the following expressions:

$$g_p\left(\vec{r},t\left|\vec{r}',t'\right.\right)=\frac{1}{4\pi\left|\vec{r}-\vec{r}'\right|}\left(\chi\left(t-t'\right)-\chi\left(t-t'+\frac{\left|\vec{r}-\vec{r}'\right|}{c_p}\right)\right) \tag{9.21}$$

$$g_s\left(\vec{r},t\left|\vec{r}',t'\right.\right)=\frac{1}{4\pi\left|\vec{r}-\vec{r}'\right|}\left(\chi\left(t-t'\right)-\chi\left(t-t'+\frac{\left|\vec{r}-\vec{r}'\right|}{c_s}\right)\right). \tag{9.22}$$

In a case of diverging waves and in a case of adjoint Green's tensor, g_p and g_s can be written as follows:

$$g_p\left(\vec{r},t\left|\vec{r}',t'\right.\right)=\frac{1}{4\pi\left|\vec{r}-\vec{r}'\right|}\left(\chi\left(-t+t'\right)-\chi\left(-t+t'-\frac{\left|\vec{r}-\vec{r}'\right|}{c_p}\right)\right) \tag{9.23}$$

$$g_s\left(\vec{r},t\left|\vec{r}',t'\right.\right)=\frac{1}{4\pi\left|\vec{r}-\vec{r}'\right|}\left(\chi\left(-t+t'\right)-\chi\left(-t+t'-\frac{\left|\vec{r}-\vec{r}'\right|}{c_s}\right)\right). \tag{9.24}$$

Finally, in a case of converging waves and for an adjoint Green's tensor, functions g_p and g_s have the following form:

$$g_p\left(\vec{r},t\left|\vec{r}',t'\right.\right)=\frac{1}{4\pi\left|\vec{r}-\vec{r}'\right|}\left(\chi\left(-t+t'\right)-\chi\left(-t+t'+\frac{\left|\vec{r}-\vec{r}'\right|}{c_p}\right)\right) \tag{9.25}$$

$$g_s\left(\vec{r},t\left|\vec{r}',t'\right.\right)=\frac{1}{4\pi\left|\vec{r}-\vec{r}'\right|}\left(\chi\left(-t+t'\right)-\chi\left(-t+t'+\frac{\left|\vec{r}-\vec{r}'\right|}{c_s}\right)\right). \tag{9.26}$$

9.4 Rayleigh Integral Formula

The analytical expressions for Rayleigh formula for elastic waves can be obtained based on the corresponding Kirchhoff integral formula, which provides a solution for a boundary value problem for the elastic field equations [1, 12]:

$$\begin{aligned}&-\iint\limits_{dD}\int\limits_{-\infty}^{\infty}\left(c_p^2\vec{n}\cdot\left(\vec{U}\left(\nabla\cdot\mathbf{G}^L\right)-\mathbf{G}^L\left(\nabla\cdot\vec{U}\right)\right)\right.\\&\left.+c_s^2\vec{n}\cdot\left(\vec{U}\times\left(\nabla\times\mathbf{G}^L\right)+\left(\nabla\times\vec{U}\right)\times\mathbf{G}^L\right)\right)dtds=\begin{cases}\vec{U}\left(\vec{r}',t'\right),\vec{r}'\in D\cup dD\\0,\vec{r}'\notin D\cup dD\end{cases}\end{aligned} \tag{9.27}$$

The Kirchhoff integral formula, Eq. 9.27, shows that the elastic displacement field can be reconstructed everywhere inside the domain D from the known values of these fields and their normal derivatives at the domain boundary dD. It can be demonstrated that the boundary values of the wavefield and its normal derivative on the boundary are not independent, but in fact satisfy some integral equations derived from the Kirchhoff integral formula, Eq. 9.27 [12, 13]. The solution of these integral equations proves to be a difficult problem in the case of an arbitrary surface dD. However, if the surface dD is a horizontal plane, these integral formulae can be simplified, which results in the so-called Rayleigh integral formula [1, 12, 13]. We present a short summary of the derivation of the Rayleigh integral formula below.

Let us assume in Eq. 9.27 that dD is the plane surface, $z = 0$. Under this assumption, after long but straightforward algebra, we can obtain the following analytical expressions for the scalar components of the displacement field:

$$U_1\left(\vec{r}',t'\right) = -\frac{c_s^2}{2\pi}\frac{\partial^3}{\partial x'^2\partial z'}I_{1p} - \frac{c_s^2}{2\pi}\frac{\partial^3}{\partial y'^2\partial z'}I_{1s} - \frac{c_s^2}{2\pi}\frac{\partial^3}{\partial z'^3}I_{1s} - \frac{c_s^2}{2\pi}\frac{\partial^3}{\partial x'\partial y'\partial z'}I_{2p} + \frac{c_s^2}{2\pi}\frac{\partial^3}{\partial x'\partial y'\partial z'}I_{2s} - \frac{c_p^2}{2\pi}\frac{\partial^3}{\partial x'\partial z'^2}I_{3p} + \frac{c_p^2}{2\pi}\frac{\partial^3}{\partial x'\partial z'^2}I_{3s} \tag{9.28}$$

$$U_2\left(\vec{r}',t'\right) = -\frac{c_s^2}{2\pi}\frac{\partial^3}{\partial x'\partial y'\partial z'}I_{1p} + \frac{c_s^2}{2\pi}\frac{\partial^3}{\partial x'\partial y'\partial z'}I_{1s} - \frac{c_s^2}{2\pi}\frac{\partial^3}{\partial y'^2\partial z'}I_{2p} - \frac{c_s^2}{2\pi}\frac{\partial^3}{\partial x'^2\partial z'}I_{2s} - \frac{c_s^2}{2\pi}\frac{\partial^3}{\partial z'^3}I_{2s} - \frac{c_p^2}{2\pi}\frac{\partial^3}{\partial y'\partial z'^2}I_{3p} + \frac{c_p^2}{2\pi}\frac{\partial^3}{\partial y'\partial z'^2}I_{3s} \tag{9.29}$$

$$U_3\left(\vec{r}',t'\right) = -\frac{c_s^2}{2\pi}\frac{\partial^3}{\partial x'\partial z'^2}I_{1p} + \frac{c_s^2}{2\pi}\frac{\partial^3}{\partial x'\partial z'^2}I_{1s} - \frac{c_s^2}{2\pi}\frac{\partial^3}{\partial y'\partial z'^2}I_{2p} + \frac{c_s^2}{2\pi}\frac{\partial^3}{\partial y'\partial z'^2}I_{2s} - \frac{c_p^2}{2\pi}\frac{\partial^3}{\partial z'^3}I_{3p} - \frac{c_p^2}{2\pi}\frac{\partial^3}{\partial x'^2\partial z'}I_{3s} - \frac{c_p^2}{2\pi}\frac{\partial^3}{\partial y'^2\partial z'}I_{3s} \tag{9.30}$$

In Eqs. 9.28–9.30, I_{ip} denotes the following integrals given by Eqs. 9.31 and 9.32.

$$I_{ip} = \iint\limits_S \frac{1}{\left|\vec{r}-\vec{r}'\right|}\int\limits_0^\infty U_i\left(\vec{r},t\right) g_p\left(\vec{r},t\left|\vec{r}',t'\right.\right) dtds \tag{9.31}$$

$$I_{is} = \iint\limits_S \frac{1}{\left|\vec{r}-\vec{r}'\right|}\int\limits_0^\infty U_i\left(\vec{r},t\right) g_s\left(\vec{r},t\left|\vec{r}',t'\right.\right) dtds \tag{9.32}$$

Substituting the expressions for the Green's tensor introduced in Sect. 9.3 into Eqs. 9.31 and 9.32, one can find the following expressions for the above integrals:

$$I_{ip} = \iint\limits_S \frac{1}{\left|\vec{r}-\vec{r}'\right|}\int\limits_0^\infty \left(U_i\left(\vec{r},t+t'\right) - U_i\left(\vec{r},t+t'-\frac{\left|\vec{r}-\vec{r}'\right|}{c_p}\right)\right) tdtds \tag{9.33}$$

$$I_{is} = \iint\limits_S \frac{1}{\left|\vec{r}-\vec{r}'\right|}\int\limits_0^\infty \left(U_i\left(\vec{r},t+t'\right) - U_i\left(\vec{r},t+t'-\frac{\left|\vec{r}-\vec{r}'\right|}{c_s}\right)\right) tdtds \tag{9.34}$$

The following notations will also be used to develop the numerical method based on Rayleigh integrals:

$$I_p = \frac{1}{b_k} \iint\limits_S \frac{1}{\left|\vec{r} - \vec{r}'\right|} \int\limits_0^\infty \left(U_k\left(x, y, 0, t + t'\right) - U_k\left(x, y, 0, t + t' - \frac{\left|\vec{r} - \vec{r}'\right|}{c_p}\right)\right) tdtds \tag{9.35}$$

$$I_s = \frac{1}{b_k} \iint\limits_S \frac{1}{\left|\vec{r} - \vec{r}'\right|} \int\limits_0^\infty \left(U_k\left(x, y, 0, t + t'\right) - U_k\left(x, y, 0, t + t' - \frac{\left|\vec{r} - \vec{r}'\right|}{c_s}\right)\right) tdtds \tag{9.36}$$

$$I_{ps} = I_p - I_s \tag{9.37}$$

Note that in a case of using the Ricker wavelet for the source

$$f(t) = \left(1 - \frac{180}{\alpha^2}\left(t - \frac{\alpha}{2}\right)^2\right) \exp\left(-\frac{90}{\alpha^2}\left(t - \frac{\alpha}{2}\right)^2\right) \tag{9.38}$$

integral $\int_0^\infty f\left(t + t'\right) tdt$ can be solved analytically as shown below:

$$\int\limits_0^\infty f\left(t + t'\right) tdt = -\frac{\alpha^2}{180} \exp\left(-\frac{90}{\alpha^2}\left(b - \frac{\alpha}{2}\right)^2\right). \tag{9.39}$$

Thus, we arrive at the following equation:

$$\int\limits_0^\infty \left(f\left(t + t'\right) - f\left(t + t' - \frac{\left|\vec{r} - \vec{r}'\right|}{c_s}\right)\right) tdt$$
$$= \frac{\alpha^2}{180} \exp\left(-\frac{90}{\alpha^2}\left(t' - \frac{\left|\vec{r} - \vec{r}'\right|}{c_s} - \frac{\alpha}{2}\right)^2\right) - \frac{\alpha^2}{180} \exp\left(-\frac{90}{\alpha^2}\left(t' - \frac{\alpha}{2}\right)^2\right). \tag{9.40}$$

Note that in a special case of the plane wave, one can use the following equations to calculate the integrals, introduced above:

$$I_p = \int\limits_0^X \int\limits_0^Y \frac{1}{r}\left(\frac{\alpha^2}{180} \exp\left(-\frac{90}{\alpha^2}\left(t' - \frac{r}{c_p} - \frac{\alpha}{2}\right)^2\right) - \frac{\alpha^2}{180} \exp\left(-\frac{90}{\alpha^2}\left(t' - \frac{\alpha}{2}\right)^2\right)\right) dxdy \tag{9.41}$$

$$I_s = \int\limits_0^X \int\limits_0^Y \frac{1}{r}\left(\frac{\alpha^2}{180} \exp\left(-\frac{90}{\alpha^2}\left(t' - \frac{r}{c_s} - \frac{\alpha}{2}\right)^2\right) - \frac{\alpha^2}{180} \exp\left(-\frac{90}{\alpha^2}\left(t' - \frac{\alpha}{2}\right)^2\right)\right) dxdy \tag{9.42}$$

In Eqs. 9.41–9.42, r is given by Eq. 9.43.

$$r = \sqrt{(x' - x)^2 + (y' - y)^2 + (z')^2} \tag{9.43}$$

9.5 Discretization of the Derivatives

One of the main problems in developing a numerical representation of the formulae discussed in Sect. 9.4 is related to the discretization of the spatial derivatives appearing in Eqs. 9.28–9.30. One can calculate these derivatives by the following finite-difference approximations:

$$\begin{aligned}(U_1)^L_{IJK} = &\left(c_s^2 b_1 \left(\mathbf{D}_{xxz}\left(I_p\right)^L_{IJK} + \mathbf{D}_{yyz}\left(I_s\right)^L_{IJK} + \mathbf{D}_{zzz}\left(I_s\right)^L_{IJK}\right)\right.\\ &\left.+ c_s^2 b_2 \mathbf{D}_{xyz}\left(I_{ps}\right)^L_{IJK} + c_p^2 b_3 \mathbf{D}_{xzz}\left(I_{ps}\right)^L_{IJK}\right)\frac{1}{2\pi h^3}\end{aligned} \tag{9.44}$$

$$\begin{aligned}(U_2)^L_{IJK} = &\left(c_s^2 b_2 \left(\mathbf{D}_{xxz}\left(I_s\right)^L_{IJK} + \mathbf{D}_{yyz}\left(I_p\right)^L_{IJK} + \mathbf{D}_{zzz}\left(I_s\right)^L_{IJK}\right)\right.\\ &\left.+ c_s^2 b_1 \mathbf{D}_{xyz}\left(I_{ps}\right)^L_{IJK} + c_p^2 b_3 \mathbf{D}_{yzz}\left(I_{ps}\right)^L_{IJK}\right)\frac{1}{2\pi h^3}\end{aligned} \tag{9.45}$$

$$\begin{aligned}(U_3)^L_{IJK} = &\left(c_p^2 b_3 \left(\mathbf{D}_{xxz}\left(I_s\right)^L_{IJK} + \mathbf{D}_{yyz}\left(I_s\right)^L_{IJK} + \mathbf{D}_{zzz}\left(I_p\right)^L_{IJK}\right)\right.\\ &\left.+ c_s^2 b_1 \mathbf{D}_{xzz}\left(I_{ps}\right)^L_{IJK} + c_s^2 b_2 \mathbf{D}_{yzz}\left(I_{ps}\right)^L_{IJK}\right)\frac{1}{2\pi h^3},\end{aligned} \tag{9.46}$$

where we use the following finite-difference operators:

$$\begin{aligned}\mathbf{D}_{xxz} W^L_{IJK} = &\left(W^L_{I+2,J,K+1} - 2W^L_{I+1,J,K+1} + W^L_{I,J,K+1}\right)\\ &- \left(W^L_{I+2,J,K} - 2W^L_{I+1,J,K} + W^L_{I,J,K}\right)\end{aligned} \tag{9.47}$$

$$\begin{aligned}\mathbf{D}_{yyz} W^L_{IJK} = &\left(W^L_{I,J+2,K+1} - 2W^L_{I,J+1,K+1} + W^L_{I,J,K+1}\right)\\ &- \left(W^L_{I,J+2,K} - 2W^L_{I,J+1,K} + W^L_{I,J,K}\right)\end{aligned} \tag{9.48}$$

$$\mathbf{D}_{zzz} W^L_{IJK} = W^L_{I,J,K+3} - 3W^L_{I,J,K+2} + 3W^L_{I,J,K+1} - W^L_{I,J,K} \tag{9.49}$$

$$\begin{aligned}\mathbf{D}_{xyz} W^L_{IJK} = &W^L_{I+1,J+1,K+1} - W^L_{I+1,J+1,K} - W^L_{I,J+1,K+1}\\ &- W^L_{I+1,J,K+1} + W^L_{I+1,J,K} + W^L_{I,J+1,K} + W^L_{I,J,K+1} - W^L_{I,J,K}\end{aligned} \tag{9.50}$$

$$\begin{aligned}\mathbf{D}_{xzz} W^L_{IJK} = &\left(W^L_{I+1,J,K+2} - 2W^L_{I+1,J,K+1} + W^L_{I+1,J,K}\right)\\ &- \left(W^L_{I,J,K+2} - 2W^L_{I,J,K+1} + W^L_{I,J,K}\right)\end{aligned} \tag{9.51}$$

$$\begin{aligned}\mathbf{D}_{yzz} W^L_{IJK} = &\left(W^L_{I,J+1,K+2} - 2W^L_{I,J+1,K+1} + W^L_{I,J+1,K}\right)\\ &- \left(W^L_{I,J,K+2} - 2W^L_{I,J,K+1} + W^L_{I,J,K}\right).\end{aligned} \tag{9.52}$$

In Eqs. 9.44–9.46, h denotes the spatial step. In Eqs. 9.47–9.52, W represents one of the integrals, I_p, I_s, I_{ps}. In Eqs. 9.44–9.52, index L corresponds to time t, while indices I, J, and K correspond to the spatial coordinates, X, Y, and Z, respectively.

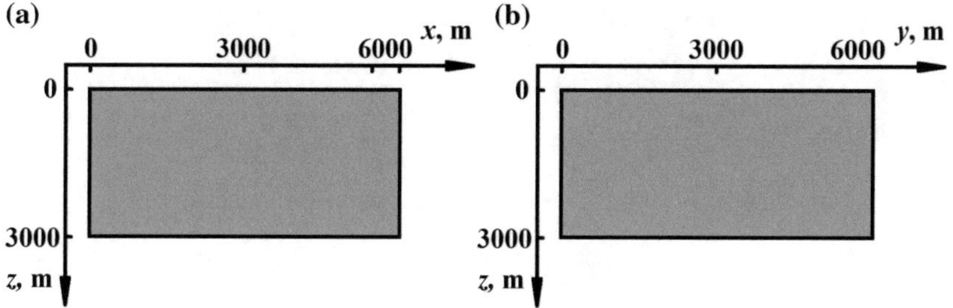

Fig. 9.1 The modelling domain for a case of the plane wave propagation along: **a** OXZ plane, **b** OYZ plane

9.6 Comparison of the Solutions Obtained Using the Rayleigh Formula and the Grid-Characteristic Method

In this section we consider two simple scenarios of elastic wave propagation. One scenario represents the plane wave propagation; another scenario is a case of the wave generated by the Ricker wavelet in the circle.

In a case of the source represented by the Ricker wavelet in the circle **C** located in the plane $z = 0$, with the center (x_0, y_0) and a radius of 100 m, the following equation holds:

$$U_k(x, y, 0, t) = \begin{cases} b_k f(t), (x, y) \in \mathbf{C} \\ 0, (x, y) \notin \mathbf{C} \end{cases} \tag{9.53}$$

In a case of the plane wave, the given solution at the boundary takes the following form for $x \in [0, X]$, $y \in [0, Y]$:

$$U_k(x, y, 0, t) = b_k f(t) \tag{9.54}$$

In Eqs. 9.53 and 9.54, $f(t)$ denotes the Ricker wavelet.

We assume that the velocity of P-waves equals to 5000 m/s, the velocity of S-waves equals to 3100 m/s, and the destiny is of 2500 kg/m^3 for the elastic medium under consideration. The grid of 121 × 121 × 121 nodes with the spatial step of 50 m and the time step of 0.01 s was used for obtaining the solution by the grid-characteristic method using 201 time steps. We also used the grid of 61 × 61 × 31 nodes with the spatial step of 100 m, and the time step of 0.02 s, for obtaining the solution by Rayleigh formula using 81 time steps. One can see this modelling domain in Fig. 9.1.

In a case of the plane wave, b_1, b_2, b_3 satisfy Eq. 9.55.

$$(b_1, b_2, b_3) = (0, 0, 1) \tag{9.55}$$

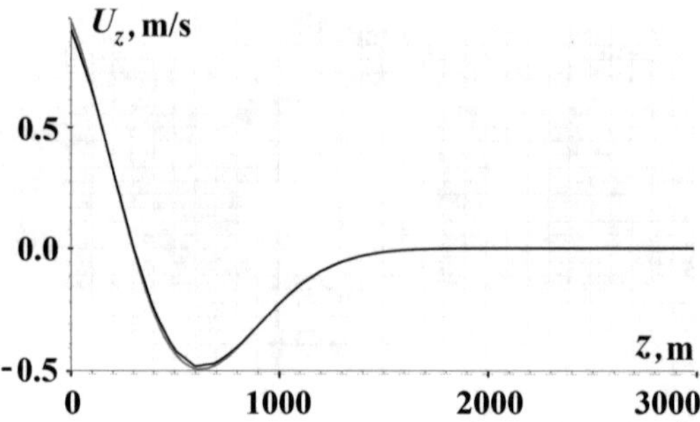

Fig. 9.2 Plane wave model. The plots of the vertical component of the displacement field vector along the Z-axis for the time moment of 0.58 s. The black solid line shows the solution obtained using the Rayleigh formula. The grey solid line presents the solution produced by the grid-characteristic method

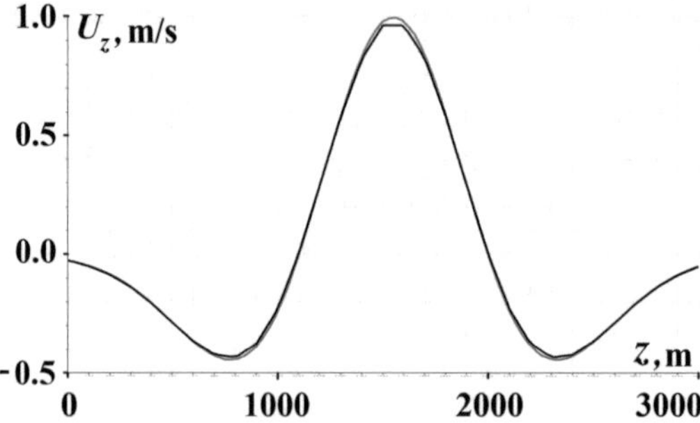

Fig. 9.3 Plane wave model. The plots of the vertical component of the displacement field vector along the Z-axis for the time moment of 0.92 s. The black solid line shows the solution obtained using the Rayleigh formula. The grey solid line presents the solution produced by the grid-characteristic method

Figures 9.2, 9.3, and 9.4 show the plots of the Z-component of the displacement field vector along the Z-coordinate in a case of the plane wave. The black solid line shows the solution obtained using the Rayleigh formula. The grey solid line presents the solution produced by the grid-characteristic method. These curves were plotted along the line parallel to the Z-axis and passing through the center of the modelling domain. Figures 9.2, 9.3, and 9.4 present the curves for the time moments of 0.58, 0.92, and 1.08 s, respectively.

In a case of Ricker wavelet in a circle, the parameters b_1, b_2, b_3 satisfy one of the following equations:

$$(b_1, b_2, b_3) = (0, 0, 1) \tag{9.56}$$

$$(b_1, b_2, b_3) = (1, 2, 3) \tag{9.57}$$

In this case, we have used a grid with $401 \times 401 \times 201$ nodes with the spatial step of 50 m and the time step of 0.01 s for obtaining the solution by grid-characteristic

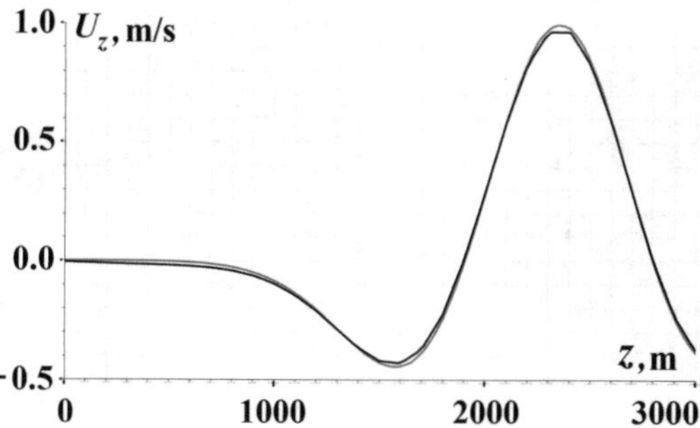

Fig. 9.4 Plane wave model. The plots of the vertical component of the displacement field vector along the Z-axis for the time moment of 1.08 s. The black solid line shows the solution obtained using the Rayleigh formula. The grey solid line presents the solution produced by the grid-characteristic method

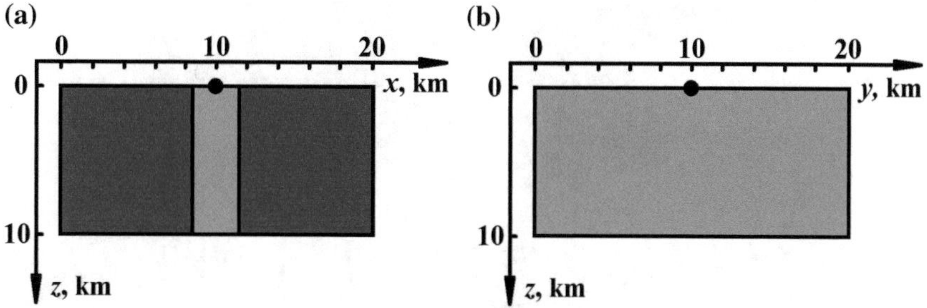

Fig. 9.5 The modelling domain in a case of the Ricker wavelet in the circle (the position of the source is shown by a bold black dot); the areas colored by grey and dark grey colors show the region of integration for the case of using grid-characteristic method, the area colored by the grey color only corresponds to the case of using Rayleigh formula: **a** OXZ plane, **b** OYZ plane

method using 263 time steps. The grid of 31 × 201 × 101 nodes with the spatial step of 100 m and the time step of 0.02 s was used for obtaining the solution by Rayleigh formula using 131 time steps (see Fig. 9.5).

Figures 9.6 and 9.7 show the wave patterns of the displacement vector components U_y and U_z, respectively, for the case of coefficients given by Eq. 9.56 at time moment of 2 s Sections of 1.5 km × 8 km × 10 km modelling domain are shown in these figures. The black lines show the boundaries of the entire modelling domain. Figures 9.6a and 9.7a present the solutions obtained by Rayleigh formula, while Figs. 9.6b and 9.7b show the solutions produced by the grid-characteristic method.

Note that due to the use of the Green's tensor discussed in Appendix A in the solutions by Rayleigh formula, there is no head wave presents [20]. In the solutions by the grid-characteristic method one can find this wave in Figs. 9.6b and 9.7b.

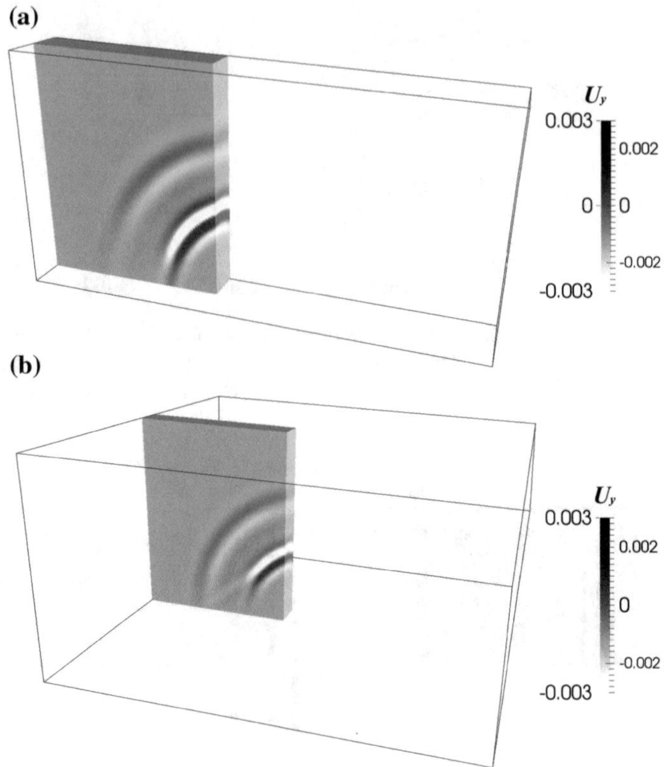

Fig. 9.6 Wave patterns of the Y-component of the displacement field for time moment of 2 s: **a** Rayleigh integral method, **b** grid-characteristic method

9.7 Migration Based on Rayleigh Integral Formula

The process of migration includes two elements: (1) backward extrapolation of the scattered wavefields (i.e. continuation of the waves in the direction opposite to that of their actual propagation and in the reverse time) and (2) synthesis of the medium image as a snapshot (at time $t = 0$) of the spatial structure of the wavefield produced by backward extrapolation. These elements form the foundation of the majority of algorithms of time section migration (e.g. Berkhout [21, 22], Claerbout [23], and Zhdanov [1]).

In this section, we consider the algorithm of the backward extrapolation of the scattered elastic wavefields in the reverse time based on the Rayleigh integral formula, which was introduced in Sect. 9.4 above.

To simplify the analytical expressions and the subsequent development of the software, we use the following notations in the algorithm of the elastic field migration:

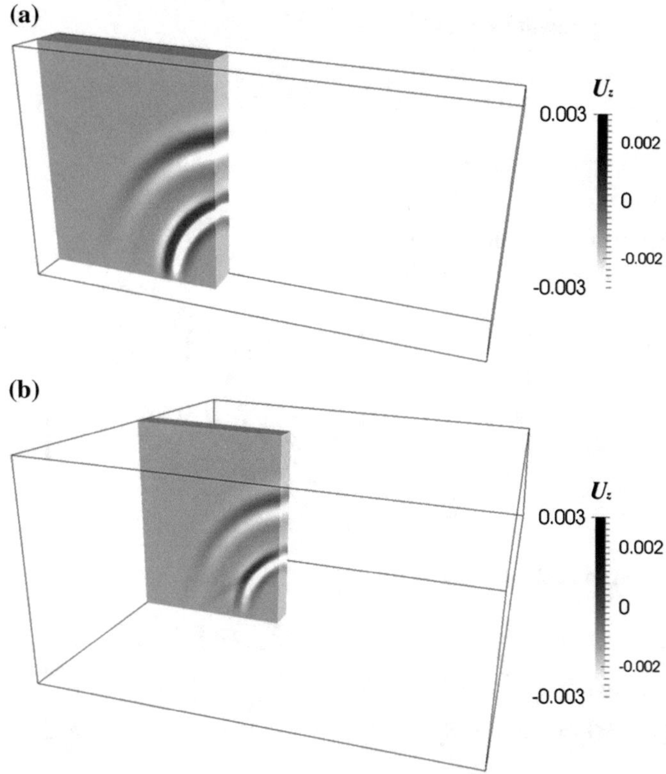

Fig. 9.7 Wave patterns of the Z-component of the displacement field at time moment of 2 s: **a** Rayleigh integral method, **b** grid-characteristic method

$$Q_{An} = \left(\frac{c_B}{r}\right)^2 \int\limits_0^{\frac{r}{c_A}} U_n\left(\vec{r}', \tau\right) \tau d\tau \tag{9.58}$$

$$S_{An} = \left(\frac{c_B}{c_A}\right)^2 U_n\left(\vec{r}', \frac{r}{c_A}\right) \tag{9.59}$$

where index n takes the following values:

$$n = 1, 2, 3 \tag{9.60}$$

and index A stands for P or S:

$$A = P, \mathrm{S} \tag{9.61}$$

If $n = 3$, index $B = \mathrm{P}$; if $n \in \{1, 2\}$, index $B = \mathrm{S}$.

The following notations are also used:

$$X = x - x' \tag{9.62}$$

$$Y = y - y' \tag{9.63}$$

$$Z = z \tag{9.64}$$

Indices a, b take the following values: X, Y, Z.
Thus, the following notations can be used:

$$R_{z3}^{nA} = 3\frac{1}{r^2}\frac{Z}{r}\left\{\left(3 - 5\left(\frac{Z}{r}\right)^2\right)Q_{An} + \left(2\left(\frac{Z}{r}\right)^2 - 1\right)S_{An}\right\} \tag{9.65}$$

$$R_{aab}^{nA} = \frac{1}{r^2}\frac{b}{r}\left\{\left(3 - 7\left(\frac{a}{r}\right)^2\right)Q_{An} + \left(2\left(\frac{a}{r}\right)^2 - 1\right)S_{An}\right\} \tag{9.66}$$

$$R_{xyz}^{nA} = -3\frac{1}{r^2}\frac{X}{r}\frac{Y}{r}\frac{Z}{r}Q_{An} \tag{9.67}$$

Finally, one can calculate the following expressions:

$$J_1(x, y, z) = -R_{xxz}^{1P} - R_{yyz}^{1S} - R_{z3}^{1S} - R_{xyz}^{2P} + R_{xyz}^{2S} - R_{zzx}^{3P} + R_{zzx}^{3S} \tag{9.68}$$

$$J_2(x, y, z) = -R_{yyz}^{2P} - R_{xxz}^{2S} - R_{z3}^{2S} - R_{xyz}^{1P} + R_{xyz}^{1S} - R_{zzy}^{3P} + R_{zzy}^{3S} \tag{9.69}$$

$$J_3(x, y, z) = -R_{z3}^{3P} - R_{xxz}^{3S} - R_{yyz}^{3S} - R_{zzx}^{1P} + R_{zzx}^{1S} - R_{zzy}^{2P} + R_{zzy}^{2S} \tag{9.70}$$

Based on Eqs. 9.68–9.70, one can find the result of migration of the elastic wavefield using the following formulae:

$$U_n^{\mathrm{MIGR}}(x, y, z) = \frac{1}{2\pi}\iint\limits_S J_n ds \tag{9.71}$$

The absolute value of the migrated elastic field can be found as follows:

$$U_{\mathrm{MIGR}}(x, y, z) = \sqrt{\left(U_1^{\mathrm{MIGR}}(x, y, z)\right)^2 + \left(U_2^{\mathrm{MIGR}}(x, y, z)\right)^2 + \left(U_3^{\mathrm{MIGR}}(x, y, z)\right)^2} \tag{9.72}$$

9.8 Migration Imaging Conditions

In this section, we discuss the second element of the migration process—synthesis of the medium image based on the elastic field migration (imaging conditions). The analytical expressions for a snapshot (at time $t = 0$) of the spatial structure of the migration field can be found based on the Rayleigh formula as well.

Let the seismograms be known for a time interval from 0 to T. Then the integrals for the conjugate field at the point $t = 0$ are written by Eqs. 9.73–9.78:

$$I_{1p} = \iint_S \frac{1}{\left|\vec{r} - \vec{r}'\right|} \int_0^{\frac{\left|\vec{r} - \vec{r}'\right|}{c_p}} U_1\left(\vec{r}, \tau\right)\left(\tau - \frac{\left|\vec{r} - \vec{r}'\right|}{c_p}\right) d\tau ds \tag{9.73}$$

$$I_{2p} = \iint_S \frac{1}{\left|\vec{r} - \vec{r}'\right|} \int_0^{\frac{\left|\vec{r} - \vec{r}'\right|}{c_p}} U_2\left(\vec{r}, \tau\right)\left(\tau - \frac{\left|\vec{r} - \vec{r}'\right|}{c_p}\right) d\tau ds \tag{9.74}$$

$$I_{3p} = \iint_S \frac{1}{\left|\vec{r} - \vec{r}'\right|} \int_0^{\frac{\left|\vec{r} - \vec{r}'\right|}{c_p}} U_3\left(\vec{r}, \tau\right)\left(\tau - \frac{\left|\vec{r} - \vec{r}'\right|}{c_p}\right) d\tau ds \tag{9.75}$$

$$I_{1s} = \iint_S \frac{1}{\left|\vec{r} - \vec{r}'\right|} \int_0^{\frac{\left|\vec{r} - \vec{r}'\right|}{c_s}} U_1\left(\vec{r}, \tau\right)\left(\tau - \frac{\left|\vec{r} - \vec{r}'\right|}{c_s}\right) d\tau ds \tag{9.76}$$

$$I_{2s} = \iint_S \frac{1}{\left|\vec{r} - \vec{r}'\right|} \int_0^{\frac{\left|\vec{r} - \vec{r}'\right|}{c_s}} U_2\left(\vec{r}, \tau\right)\left(\tau - \frac{\left|\vec{r} - \vec{r}'\right|}{c_s}\right) d\tau ds \tag{9.77}$$

$$I_{3s} = \iint_S \frac{1}{\left|\vec{r} - \vec{r}'\right|} \int_0^{\frac{\left|\vec{r} - \vec{r}'\right|}{c_s}} U_3\left(\vec{r}, \tau\right)\left(\tau - \frac{\left|\vec{r} - \vec{r}'\right|}{c_s}\right) dtds \tag{9.78}$$

Equations 9.73–9.78 should be differentiated in accordance to Eqs. 9.28–9.30. Note that Eqs. 9.79–9.80 are true:

$$I = \int_0^{g(a)} f\left(a, \tau\right) d\tau \tag{9.79}$$

$$\frac{\partial I}{\partial a} = f\left(a, g\left(a\right)\right) \frac{\partial g\left(a\right)}{\partial a} + \int_0^{g(a)} \frac{\partial f\left(a, \tau\right)}{\partial a} d\tau. \tag{9.80}$$

We will also use the following notation:

$$\Phi_{nA} = \frac{(c_B)^2}{r} \int\limits_0^{\frac{r}{c_A}} U_n\left(\vec{r}', \tau\right)\left(\tau - \frac{r}{c_A}\right) d\tau \tag{9.81}$$

In Eq. 9.81, $n = 1, 2, 3$, A = P, S, B = P for $n = 3$, B = S for $n \in \{1, 2\}$ in accordance to notations used in Sect. 9.7. Integral Φ_{nA} should be differentiated by r. Thus, one can obtain Eqs. 9.82–9.84 using Eq. 9.80:

$$\frac{\partial \Phi_{nA}}{\partial r} = -\left(\frac{c_B}{r}\right)^2 \int\limits_0^{\frac{r}{c_A}} U_n\left(\vec{r}', \tau\right) \tau d\tau \tag{9.82}$$

$$\frac{\partial^2 \Phi_{nA}}{\partial r^2} = \frac{2}{r^2}\Phi_{nA} + \frac{2}{c_A}\frac{(c_B)^2}{r^2} \int\limits_0^{\frac{r}{c_A}} U_n\left(\vec{r}', \tau\right) d\tau - \frac{1}{r}\left(\frac{c_B}{c_A}\right)^2 U_n\left(\vec{r}', \frac{r}{c_A}\right) \tag{9.83}$$

$$\frac{\partial^3 \Phi_{nA}}{\partial r^3} = -\frac{3}{r}\frac{\partial^2 \Phi_{nA}}{\partial r^2}. \tag{9.84}$$

One can calculate the following Eqs. 9.85–9.90:

$$\frac{\partial^3 \Phi_{nA}}{\partial z^3} = 3\frac{z}{r}\left\{\left(\frac{z}{r}\right)^2 \frac{1}{r^2}\frac{\partial \Phi_{nA}}{\partial r} - \frac{1}{r^2}\frac{\partial \Phi_{nA}}{\partial r} + \frac{1}{r}\frac{\partial^2 \Phi_{nA}}{\partial r^2} - 2\left(\frac{z}{r}\right)^2 \frac{1}{r}\frac{\partial^2 \Phi_{nA}}{\partial r^2}\right\} \tag{9.85}$$

$$\frac{\partial^3 \Phi_{nA}}{\partial x^2 \partial z} = \frac{Z}{r}\left\{3\left(\frac{X}{r}\right)^2 \frac{1}{r^2}\frac{\partial \Phi_{nA}}{\partial r} - \frac{1}{r^2}\frac{\partial \Phi_{nA}}{\partial r} + \frac{1}{r}\frac{\partial^2 \Phi_{nA}}{\partial r^2} - 2\left(\frac{X}{r}\right)^2 \frac{1}{r}\frac{\partial^2 \Phi_{nA}}{\partial r^2}\right\} \tag{9.86}$$

$$\frac{\partial^3 \Phi_{nA}}{\partial y^2 \partial z} = \frac{Z}{r}\left\{3\left(\frac{Y}{r}\right)^2 \frac{1}{r^2}\frac{\partial \Phi_{nA}}{\partial r} - \frac{1}{r^2}\frac{\partial \Phi_{nA}}{\partial r} + \frac{1}{r}\frac{\partial^2 \Phi_{nA}}{\partial r^2} - 2\left(\frac{Y}{r}\right)^2 \frac{1}{r}\frac{\partial^2 \Phi_{nA}}{\partial r^2}\right\} \tag{9.87}$$

$$\frac{\partial^3 \Phi_{nA}}{\partial x \partial z^2} = \frac{X}{r}\left\{3\left(\frac{Z}{r}\right)^2 \frac{1}{r^2}\frac{\partial \Phi_{nA}}{\partial r} - \frac{1}{r^2}\frac{\partial \Phi_{nA}}{\partial r} + \frac{1}{r}\frac{\partial^2 \Phi_{nA}}{\partial r^2} - 2\left(\frac{Z}{r}\right)^2 \frac{1}{r}\frac{\partial^2 \Phi_{nA}}{\partial r^2}\right\} \tag{9.88}$$

$$\frac{\partial^3 \Phi_{nA}}{\partial y \partial z^2} = \frac{Y}{r}\left\{3\left(\frac{Z}{r}\right)^2 \frac{1}{r^2}\frac{\partial \Phi_{nA}}{\partial r} - \frac{1}{r^2}\frac{\partial \Phi_{nA}}{\partial r} + \frac{1}{r}\frac{\partial^2 \Phi_{nA}}{\partial r^2} - 2\left(\frac{Z}{r}\right)^2 \frac{1}{r}\frac{\partial^2 \Phi_{nA}}{\partial r^2}\right\} \tag{9.89}$$

$$\frac{\partial^3 \Phi_{nA}}{\partial x \partial y \partial z} = 3\frac{X}{r}\frac{Y}{r}\frac{Z}{r}\frac{1}{r^2}\frac{\partial \Phi_{nA}}{\partial r}. \tag{9.90}$$

In Eqs. 9.85–9.90, X, Y, Z are given by Eqs. 9.62–9.64. Thus, one can substitute Eqs. 9.82–9.84 into Eqs. 9.85–9.90 and obtain the analytical algorithm discussed in Sect. 9.7.

9.9 Examples of Migration Imaging

In this Section, we present examples of migration imaging based on the Rayleigh integral formulas discussed in Sects. 9.7–9.8. We use zero-offset seismograms calculated by Born approximation introduced in Chap. 8. The modelling domain (region of integration) was selected with the size of 10 km × 2 m × 2.5 km. We have considered several models shown in the top panels of Figs. 9.6a, 9.7a, 9.8a, and 9.9a. The background velocities of P- and S-waves are equal to 2500 and 1250 m/s respectively. Note that, the velocities used in the migration formulas Eqs. 9.71–9.72 should be twice smaller than the real ones. The inclusions shown by white color in these figures have the squared velocities exceeded those of the background values by 1%. The density equals to 2500 kg/m^3. The sources and receivers are located 15 m under the earth surface every 10 m horizontally. The Ricker wavelet with the frequency of 25 Hz and the polarization (0, 0, 1) was used. The receivers recorded the data every 2 ms during 4 s time interval.

Four different models were considered shown in Figs. 9.8a, 9.9a, 9.10 a, and 9.11a, respectively. Figures 9.8b, 9.9b, 9.10b, and 9.11b present the results of the migration of the three-component data U_x, U_y, and U_z based on Rayleigh integral formula. Figures 9.8c, 9.9c, 9.10c, 9.11c show the results of migrations of one-component data U_z based on Rayleigh integral formula.

Note that instead of one true boundary in the migration field, one can see five boundaries. However, the positions of these false boundaries are theoretically predictable. To predict these false boundaries, let us consider the reflector located at a depth h.

The velocities of P- and S-waves are related by the following equation:

$$c_p = 2c_s \tag{9.91}$$

There are four types of waves scattered from the reflector.

1. PP-waves, observed at the following time moment:

$$t_{PP} = \frac{h}{c_p} + \frac{h}{c_p} = \frac{2h}{c_p} = \frac{h}{c_s} \tag{9.92}$$

2. PS-waves, observed at the following time moment:

$$t_{PS} = \frac{h}{c_p} + \frac{h}{c_s} = \frac{h}{c_p} + \frac{2h}{c_p} = \frac{3h}{c_p} = \frac{3}{2}\frac{h}{c_s} \tag{9.93}$$

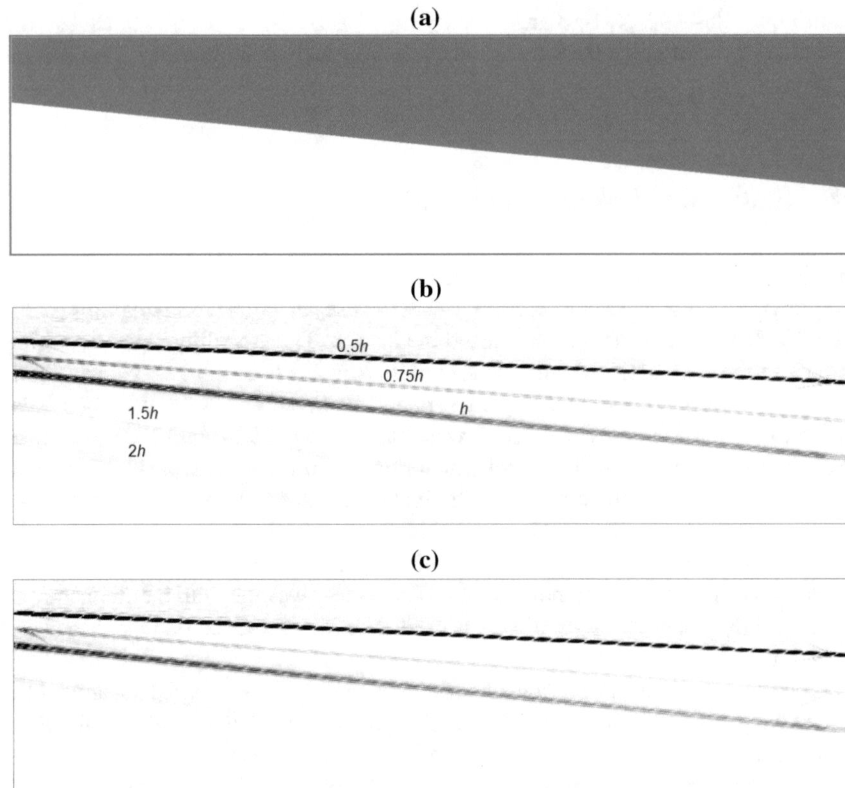

Fig. 9.8 The first model and the results of the migration based on the Rayleigh integral formula: **a** the model, **b** migration image based on three-component seismograms, **c** migration image based on one-component seismogram

3. SP-waves, observed at the following time moment:

$$t_{SP} = \frac{h}{c_s} + \frac{h}{c_p} = \frac{2h}{c_p} + \frac{h}{c_p} = \frac{3h}{c_p} = \frac{3}{2}\frac{h}{c_s} \tag{9.94}$$

4. SS-waves observed at the following time moment:

$$t_{SS} = \frac{h}{c_s} + \frac{h}{c_s} = 2\frac{h}{c_s} \tag{9.95}$$

The migration algorithm restores the boundaries based on P- and S-waves. Thus, there are 8 types of restored boundaries:

1. Boundary with the depth restored based on PP waves using P-waves. The depth is given by the following formula:

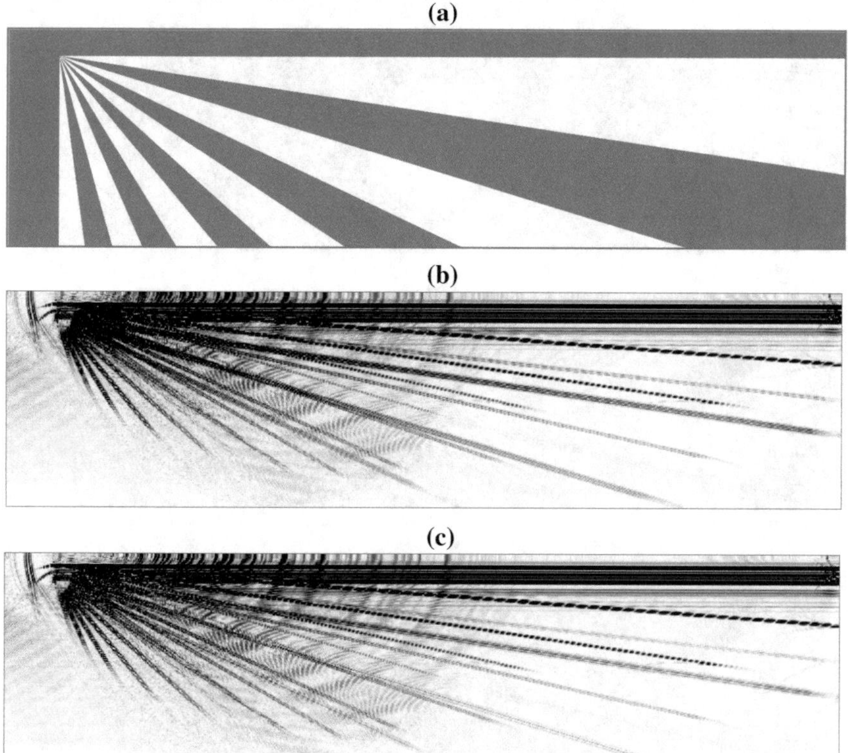

Fig. 9.9 The second model and the migration based on the Rayleigh integral formula: **a** the model, **b** migration image based on three-component seismograms, **c** migration image based on one-component seismogram

$$h_{PP}^{P} = \frac{c_p}{2} t_{PP} = \frac{c_p}{2} \frac{h}{c_s} = h \quad (9.96)$$

2. Boundary with the depth restored based on PP waves using S-waves. The depth is given by the following formula:

$$h_{PP}^{S} = \frac{c_s}{2} t_{PP} = \frac{c_s}{2} \frac{h}{c_s} = \frac{h}{2} = 0.5h \quad (9.97)$$

3. Boundary with the depth restored based on PS waves using P-waves. The depth is given by the following formula:

$$h_{PS}^{P} = \frac{c_p}{2} t_{PS} = \frac{c_p}{2} \frac{3}{2} \frac{h}{c_s} = \frac{3}{2} h = 1.5h \quad (9.98)$$

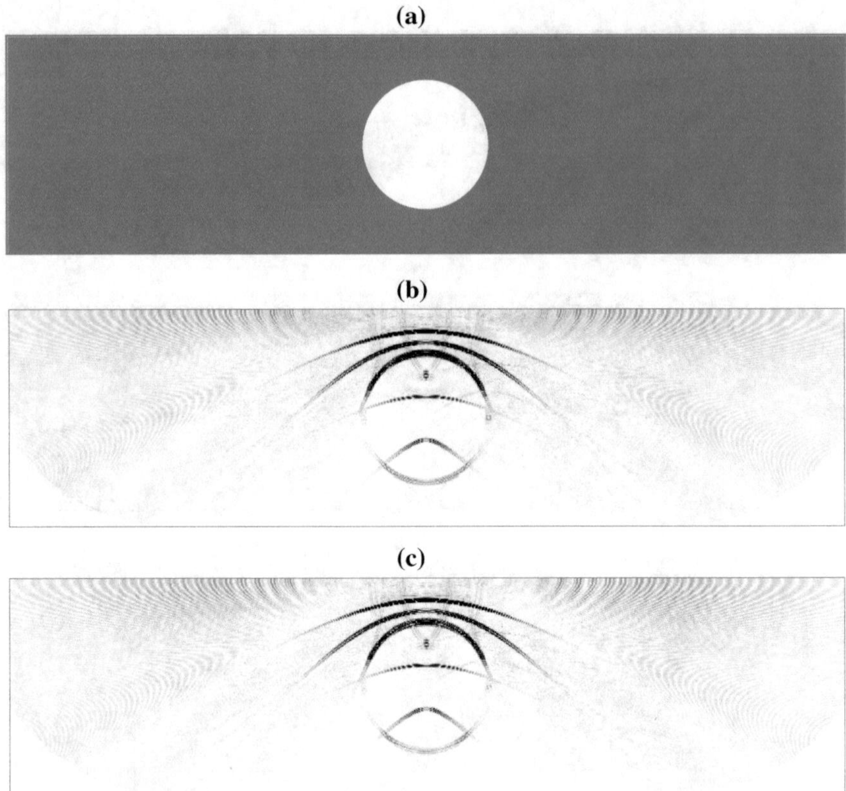

Fig. 9.10 The third model and the migration based on the Rayleigh integral formula: **a** the model, **b** migration image based on three-component seismograms, **c** migration image based on one-component seismogram

4. Boundary with the depth restored based on PS waves using S-waves. The depth is given by the following formula:

$$h_{PS}^{S} = \frac{c_s}{2} t_{PS} = \frac{c_s}{2} \frac{3}{2} \frac{h}{c_s} = \frac{3}{4} h = 0.75h \tag{9.99}$$

5. Boundary with the depth restored based on SP waves using P-waves. The depth is given by the following formula:

$$h_{SP}^{P} = \frac{c_p}{2} t_{SP} = \frac{c_p}{2} \frac{3}{2} \frac{h}{c_s} = \frac{3}{2} h = 1.5h \tag{9.100}$$

6. Boundary with the depth restored based on SP waves using S-waves. The depth is given by the following formula:

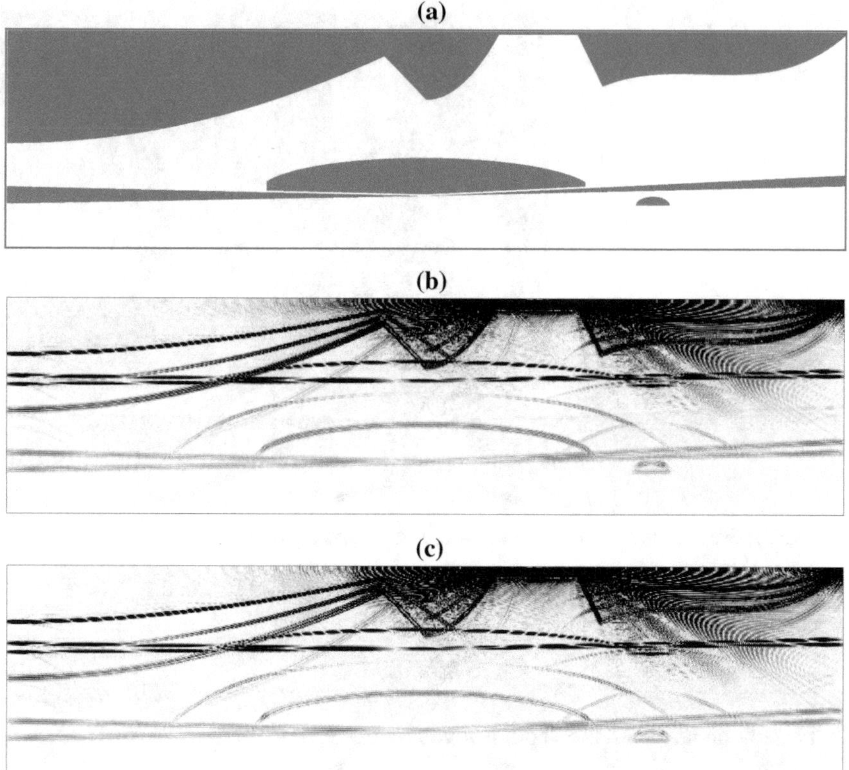

Fig. 9.11 The forth model and the migration based on the Rayleigh integral formula: **a** the model, **b** migration image based on three-component seismograms, **c** migration image based on one-component seismogram

$$h_{SP}^{S} = \frac{c_s}{2} t_{SP} = \frac{c_s}{2} \frac{3}{2} \frac{h}{c_s} = \frac{3}{4} h = 0.75h \tag{9.101}$$

7. Boundary with the depth restored based on SS waves using P-waves. The depth is given by the following formula:

$$h_{SS}^{P} = \frac{c_P}{2} t_{SS} = \frac{c_P}{2} \frac{2h}{c_s} = 2h \tag{9.102}$$

8. Boundary with the depth restored based on SS waves using S-waves. The depth is given by the following formula:

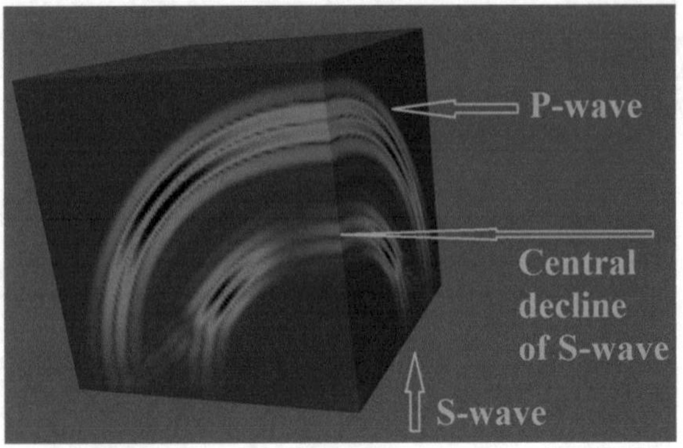

Fig. 9.12 The central decline of the amplitude of S-wave

$$h_{SS}^{S} = \frac{c_s}{2} t_{SS} = \frac{c_s}{2} \frac{2h}{c_s} = h \tag{9.103}$$

Thus, there are five different values of the boundary depth.

1. True boundary with the depth is given by the following formula:

$$h = h_{PP}^{P} = h_{SS}^{S} \tag{9.104}$$

2. False boundary with the depth is given by Eq. 9.105.

$$0.5h = h_{PP}^{S} \tag{9.105}$$

3. False boundary with the depth is given by Eq. 9.106.

$$0.75h = h_{PS}^{S} = h_{SP}^{S} \tag{9.106}$$

4. False boundary with the depth is given by Eq. 9.107.

$$1.5h = h_{PS}^{P} = h_{SP}^{P} \tag{9.107}$$

5. False boundary with the depth is given by Eq. 9.108.

$$2h = h_{SS}^{P} \tag{9.108}$$

The impact of S-waves in the zero-offset seismograms is known to be diminished due to the central decline of the amplitude of S-wave as shown in Fig. 9.12.

In summary, we have demonstrated that there could be three types of the boundary brightness in the migration image:

1. Two bright boundaries due to the strong impact of PP-wave.

   1.1 True boundary with the depth given by Eq. 9.104 mostly due to PP-waves.
   1.2 False boundary with the depth given by Eq. 9.105.

2. Two boundaries with the middle brightness. PS-waves are S-waves with the central decline but formed as a reflection of P-waves without the central decline. SP-waves are P-waves without the central decline but formed as a reflection of S-waves with the central decline. Otherwise, SS-waves are S-waves with central decline formed as a reflection of S-waves with the central decline as well.

   2.1 False boundary with the depth given by Eq. 9.106.
   2.2 False boundary with the depth given by Eq. 9.107.

3. There is one boundary with a low brightness—a false boundary with the depth given by Eq. 9.108.

Practically, there is the following structure of the boundaries brightness, as one can see in Fig. 9.6b.

1. There are two bright boundaries.

   1.1. True boundary with the depth is given by Eq. 9.104.
   1.2. False boundary with the depth is given by Eq. 9.105.

2. There is one boundary with the middle brightness. False boundary with the depth is given by Eq. 9.106.
3. There are two boundaries with a low brightness.

   3.1. False boundary with the depth is given by Eq. 9.107.
   3.2. False boundary with the depth is given by Eq. 9.108.

Some observed differences between the theoretical and practical structures of the boundary brightness are caused by diminishing the brightness of the lower boundaries.

9.10 Conclusions

In this Chapter, we have developed the analytical expressions of Rayleigh integral formula for the elastic wavefield. We have also studied the method of modelling the elastic wave propagation based on the Rayleigh formula. A comparison between the solutions obtained by this method and by grid-characteristic method was done, which demonstrated a good qualitative and quantitative agreement between the results produced by both methods.

The corresponding analytical algorithm for migration of the elastic wavefield based on the Rayleigh integral formula was developed as well. This approach to the elastic field migration was numerically tested on several zero-offset seismograms computed using the Born approximation. We have demonstrated that there are one true and four types of false boundaries in the resulting migration images with the theoretically predicted locations. Future work will be aimed at developing the methods of eliminating these false boundaries. Note that the developed method can be used for migration of both one-component and three-component seismic data.

Acknowledgements The research was supported by the grant of the President of the Russian 458 Federation No. MK1831.2017.9. This work has been carried out using computing resources of the federal collective usage center Complex for Simulation and Data Processing for Mega-science Facilities at NRC "Kurchatov Institute", http://ckp.nrcki.ru/.

References

1. Zhdanov MS (2002) Geophysical inverse theory and regularization problems. Elsevier, Netherlands
2. Trorey AW (1970) A simple theory for seismic diffractions. Geophysics 35(5):762–784
3. Trorey AW (1977) Diffraction for arbitrary source/receiver locations. Geophysics 42(6):1177–1182
4. Duzhirin A (2003) Decoupled elastic prestack depth migration. J Appl Geophys 5(4):369–389
5. Xue A, McMechan GA (2000) Prestack elastic Kirchhoff migration for multicomponent seismic data in variable velocity media. SEG Technical Program Expanded Abstracts, pp 449–452
6. Kuo JT, Dai T (1984) Kirchhoff elastic wave migration for the case of noncoincident source and receiver. Geophysics 49(8):1223–1238
7. Zhang LY, Liu Y (2008) Anisotropic converted wave amplitude-preserving prestack time migration by the pseudooffset method. Appl Geophys 5(3):204–211
8. Wang DP (2004) Vector 3C3D VSP Kirchhoff migration. In: 74th Annual International Meeting, Society of Exploration Geophysicists, Expanded Abstracts, pp 2458–2461
9. Du Q, Hou B (2008) Elastic Kirchhoff migration of vectorial wave-fields. Appl Geophys 5(4):284–293
10. Keho TH, Wu RS (1987) Elastic Kirchhoff migration for vertical seismic profiles. In: 57th Annual International Meeting, Society of Exploration Geophysicists, Expanded Abstracts pp 774–776
11. Pao YH, Varatharajulu V (1976) Huygens' principle, radiation conditions and integral formulas for the scattering of elastic waves. J Acoust Soc Am 59:1361–1371
12. Zhdanov MS (1988) Integral transforms in geophysics. Springer-Verlag, New York
13. Zhdanov MS, Matusevich VYu, Frenkel MA (1988) Seismic and electromagnetic migration. Nauka, Moscow (in Russian)
14. Favorskaya AV, Petrov IB (2016) A study of high-order grid-characteristic methods on unstructured grids. Numer Anal Appl 9(2):171–178
15. Khokhlov NI, Petrov IB (2016) On one class of high-order compact grid-characteristic schemes for linear advection. Russ J Numer Anal Math Modell 31(6):355–368
16. Favorskaya A, Petrov I, Golubev V, Khokhlov N (2017) Numerical simulation of earthquakes impact on facilities by grid-characteristic method. Procedia Comput Sci 112:1206–1215
17. Golubev VI, Petrov IB, Khokhlov NI (2016) Compact grid-characteristic schemes of higher orders of accuracy for a 3D linear transport equation. Math Models Comput Simul 8(5):577–584
18. Muratov MV, Petrov IB, Sannikov AV, Favorskaya AV (2014) Grid-characteristic method on unstructured tetrahedral meshes. Comput Math Math Phys 54(5):837–847
19. Petrov IB, Favorskaya AV, Muratov MV, Biryukov VA, Sannikov AV (2014) Grid-characteristic method on unstructured tetrahedral grids. Dokl Math 90(3):781–783

20. Kennett B (2009) Seismic wave propagation in stratified media. ANU E Press, Australia
21. Berkhout AJ (1980) Seismic migration-imaging of acoustic energy by wave field extrapolation. Elsevier, Amsterdam
22. Berkhout AJ (1984) Seismic migration-imaging of acoustic energy by wave field extrapolation. B: Practical Aspects. Elsevier, Amsterdam
23. Claerbout JF (1985) Imaging the Earth's interior. Blackwell Scientific Publications, Oxford

Subject Index

A. V. Favorskaya and I. B. Petrov (eds.), *Innovations in Wave Processes Modelling and Decision Making*, Smart Innovation, Systems and Technologies 90,
https://doi.org/10.1007/978-3-319-76201-2

Editors

Dr. Alena V. Favorskaya works as a researcher and Associated Professor at Moscow Institute of Physics and Technology, Russian Federation. Dr. Favorskaya is an advisor of more than 10 students.

Her main research interests include computational mathematics, numerical methods, wave processes modelling, computer modelling in geophysics, seismic prospecting, railway industry, high-order interpolation, and wave processes analysis for a large area of practical applications. She is the author and co-author of more than 100 publications in these fields. Dr. Alena Favorskaya developed techniques in computational mathematics including intelligent high-order interpolation on unstructured tetrahedral grids (2009), grid-characteristic method on unstructured tetrahedral grids (2009), combined GCM-SPH method (2012), intelligent boundary condition of dynamical friction (2011), intelligent interface corrector between elastic and acoustic media (2013), numerical method for finding seismic migration field based on Kirchhoff and Rayleigh formulae for elastic media (2015), the method of wave processes analysis for a large area of practical applications called Wave Logica (2017).

Dr. Favorskaya was rewarded by Award of the President of the Russian Federation for Young Scientists in 2012, 2013, 2014, by Award of the Government of the Russian Federation for Young Scientists in 2013, 2014, by IBM Ph.D. Fellowship Award in 2015, by the grant of the President of the Russian Federation in 2017.

A. V. Favorskaya and I. B. Petrov (eds.), *Innovations in Wave Processes Modelling and Decision Making*, Smart Innovation, Systems and Technologies 90,
https://doi.org/10.1007/978-3-319-76201-2

Dr. Igor B. Petrov is a Head of Department of Computer Science and Computational Mathematics and Full Professor at Moscow Institute of Physics and Technology, Russian Federation. He supervised more than 60 Ph.D. students.

His main research interests are computational mathematics, numerical methods, wave processes modelling, computer modelling in geophysics, medicine, railway industry, seismic prospecting. He is the author or the co-author of more than 200 publications (book chapters, articles, papers, and certificates of software tools) in these fields.

Prof. Petrov developed grid-characteristic numerical method for solving hyperbolic system of equations and for full wave modelling of dynamical three-dimensional processes. Also he developed a method of smooth particles for solving meteorite and asteroid protection problems. He obtained numerical solutions for problems of high-speed interaction of deformable bodies, the impact of charged particle beams and X-ray on metal and composite shells, the diffraction of elastic-plastic waves in cavities in groundwater environments, the deformation of shells under the influence of the aerodynamic flow of gases, of biomechanics with medical applications in areas such as ophthalmology, cardiology, traumatology.

Prof. Petrov is a member of New York Academy of Sciences and a member of Advisory Board of Russian Foundation for Basic Research. He was awarded by Order for Merit of Second Degree Medal of Russian Federations in 1999. Prof. Petrov became a Corresponding Member of Russian Academy of Science in 2011.

Printed in the United States
By Bookmasters